Strand

Symons Hall

The Econometric Analysis
of Time Series

The Econometric Analysis of Time Series

A C HARVEY

The London School of Economics

A HALSTED PRESS BOOK

John Wiley & Sons
New York

First published 1981 by
PHILIP ALLAN PUBLISHERS LIMITED
MARKET PLACE
DEDDINGTON
OXFORD OX5 4SE

Published in the USA by
HALSTED PRESS, A DIVISION OF
JOHN WILEY & SONS INC.,
NEW YORK

Library of Congress Cataloging in Publication Data
Harvey, Andrew C.
   The econometric analysis of time series.
   1. Econometrics   2. Time-series analysis
I. Harvey, Andrew C.   II. Title
HB139   330'.01'5195   80-84067
ISBN 0-470-27104-3

Contents

v

9. Simultaneous Equation Models   309

Preface

As the title suggests, this book concentrates primarily on the econometric theory relating to time series data. However, because the initial chapters deal with general principles, there are few areas of the subject which are not touched on, either directly or indirectly. In fact, the topics covered correspond roughly to those which are included in the 'Econometric Theory' course given to final year undergraduates at the London School of Economics.

The book may be regarded as a treatise on the statistical aspects of economic model building. There are three major themes. The first is concerned with the way in which recent advances in time series analysis have affected the development of a theory of dynamic econometrics. The second is concerned with setting out an integrated approach to the problems of estimation and testing based on the method of maximum likelihood. Because of the tremendous advances in computer technology and accompanying development of algorithms for numerical optimisation, such an approach is now feasible whereas fifteen years ago it was not. The final theme is the attempt to present a reasonably coherent strategy for model selection. Finding a suitable specification is the most difficult part of any exercise in econometric model building. Unfortunately this aspect of the subject is still in its infancy, and the material presented here represents only the beginnings of a solution.

Questions relating to the estimation, testing and specification of dynamic models which are not based on any behavioural theory are discussed in much greater detail in the companion volume, 'Time Series Models'. On occasion reference will be made to this work as 'TSM'. Although this present book is designed to be self-contained, a student wishing to study econometric theory in any depth would be well advised to become familiar with the material in TSM.

The emphasis in the text is on creating an understanding of the main ideas and concepts in econometrics, rather than in presenting a series of rigorous proofs. As a result, mathematical and statistical prerequisites are not unduly heavy. It is assumed that the reader has attended an introductory course in matrix algebra, but heavy matrix manipulations are avoided wherever possible. Certain key topics, such as matrix differentiation, are dealt with in the Introduction and the Appendix. On the statistical side, it is assumed that the reader is familiar with the basic ideas of statistical inference and with regression.

Equations are numbered according to the section. The chapter number is omitted except when referring to an equation in another chapter. Examples are numbered within each section and are referenced in the same way as equations. Tables and figures are numbered consecutively throughout each chapter and are independent of the section in which they appear. The 'Notes' at the end of each chapter are primarily for references not given in the text, and for further reading.

Certain sections are starred (*). These sections contain material which is more difficult or more esoteric, or both. They can be omitted without any loss of continuity, although for a graduate course most of them would be included.

The exercises at the end of each chapter are specifically designed to test the understanding of various points in the text. They should not be regarded as examination-type questions. The book by Phillips and Wickens 'Exercises in Econometrics' provides an excellent source for further questions on the material presented here, as well as allowing the reader the opportunity to extend his knowledge on certain topics.

Bold face type-setting is not used in the text. As a general rule lower case letters denote scalars or vectors, while, with a few obvious exceptions, upper case letters denote matrices.

My understanding of econometrics has benefited considerably from discussions with colleagues at the LSE. In particular, the ideas of Denis Sargan and David Hendry have had a profound influence on my attitude to the subject, although they themselves are in no way responsible for the way in which I have expressed my views in this book. Similarly, all the colleagues and friends who were kind enough to comment on various draft chapters are exempt from blame for any blunders I may have committed. In this respect special thanks must go to Dick Baillie, James Davidson, Katarina Juselius and Hugh Wills.

The bulk of the manuscript was typed at LSE by Maggie Robertson and Sue Kirkbride, although Sue Pratt, Hazel Rice and Anne Usher also shared in the workload. Some of the earlier drafts were typed in

Vancouver by Maryse Ellis. I'm grateful to them all for their efficiency in typing a difficult manuscript. Last, but by no means least, I would like to thank Lavinia Harvey for her help in typing, referencing and indexing.

London
March 1980

1
Introduction

1. Estimation, Testing and Model Selection

Econometrics is concerned with the estimation of relationships suggested by economic theory. This serves two purposes. Firstly, it allows economic hypotheses to be tested empirically. Secondly, it provides a framework for making rational and consistent predictions. These two objectives are, of course, closely related, since the ability to make successful predictions provides the crucial test of any theory.

Econometric models may be fitted to both time series and cross-section observations. A household budget survey provides an example of cross-sectional data and a typical economic hypothesis might concern the expenditure of households on a commodity such as food. Suppose that economic theory specifies a behavioural relationship in which expenditure on food depends on household income and the size of the family. If the relationship between these variables is taken to be linear, the model becomes

$$y = \delta + \beta_1 x_1 + \beta_2 x_2 \tag{1.1}$$

where y is household expenditure on food, x_1 is household income and x_2 is household size. Given estimates of the parameters, δ, β_1 and β_2, predictions of food expenditure for households not in the sample can be made on the basis of information about their income and size.

Estimates of the parameters in (1.1) can be obtained by the method of least squares. This may be justified on purely pragmatic grounds: an attempt is being made to fit the model as closely as possible to the observations. In carrying out this exercise, it is implicitly recognised that the relationship in (1.1) will not hold exactly for every household in the sample. However, the deterministic model makes no explicit allowance for these discrepancies.

1

Thus, while the estimates of δ, β_1 and β_2 will clearly be different if they are computed from another sample, no attempt is made to assess the variability which can be expected. In a similar way no attempt is made to assess the margin of error associated with predictions. It is therefore impossible to test economic hypotheses in any meaningful way using this model.

The solution to the above problem is to construct a statistical model. If a 'j' subscript indicates an observation on the jth household in a sample of size N, the statistical model suggested by (1.1) is

$$y_j = \delta + \beta_1 x_{1j} + \beta_2 x_{2j} + u_j, \qquad j = 1, \ldots, N \qquad (1.2)$$

where u_j is a *disturbance* term. This represents the influence of all the variables excluded from the model, including unobservables such as individual tastes. The introduction of such a term allows y_j to be treated as a random variable. Conditional on the values of x_{1j} and x_{2j}, certain assumptions may be made concerning the distribution of y_j and it is these assumptions which complete the specification of the statistical, or stochastic, model.

Given the probability structure of the model, it becomes possible to deduce the properties of estimators and test statistics and to assess their value. The test statistics may be used to confront specific economic hypotheses with the empirical evidence presented by the data. More generally, they provide a guide to choosing between the different specifications suggested by economic theory. It is this interaction between economic theory and statistical inference which provides the key to successful work in applied economics. The purpose of the first five chapters of this book is to lay the statistical foundations for such work by examining the principles for the construction of efficient estimators and by considering the ways in which test statistics may be constructed and applied in model specification.

Estimation

For a model like (1.2), the method of ordinary least squares has a strong intuitive appeal. Indeed it can be justified on statistical grounds under very mild conditions regarding the distribution of the dependent variable. However, for many of the models encountered in econometrics, least squares is not an appropriate estimation technique. Econometric models typically consist of sets of equations which incorporate feedback effects from one variable to another. These are known as simultaneous equation systems. Treating the estimation of a single equation from such a system as an exercise in multiple regression will, in general, lead to estimators with poor statistical properties. Furthermore, even when a model can be conceived as a single equation, it will often be of a form which does not fit naturally into a regression framework. Some of the dynamic models which are applied to time series data fall into this category.

A more general approach to estimating the parameters of a statistical model is to apply the method of *maximum likelihood*. The essence of this approach is to write down the joint density function of the observations and then to treat it as a function of the unknown parameters. Interpreted in this way, it becomes a likelihood function which indicates the plausibility of different parameter values, given the particular set of observations obtained. The most plausible values are taken to be those which maximise the likelihood function. These are the maximum likelihood estimators and under suitable conditions they are statistically efficient. However, as a general rule, this property can be demonstrated only in large samples. Hence the discussion of some of the basic aspects of asymptotic theory to be found in Section 1.4.

A key feature of maximum likelihood estimation concerns the form of the distribution of the observations. Unless there is prior information to suggest otherwise, this is generally taken to be normal. In a regression model, the interpretation of the disturbance term as a composite of omitted variables means that the assumption of normality can, to some extent, be justified by the central limit theorem. A more pragmatic reason for its adoption is that the likelihood function is usually relatively easy to handle mathematically. However, if the normality assumption is inappropriate, the maximum likelihood estimator can have very poor properties, even though the remaining features of the model are correctly specified. This point is examined in Section 3.7, and a number of estimators which are *robust* to the form of the distribution are introduced. Although robust procedures are not widely used in econometrics at present, there is clearly a case for their adoption. Indeed the whole area of robustness deserves more attention. As a general concept, robustness concerns the sensitivity of particular estimation techniques to all the assumptions of the model, not just those concerned with the shape of the distribution.

The maximum likelihood estimators are generally obtained by differentiating the likelihood function with respect to the unknown parameters. Setting each of the derivatives equal to zero yields the likelihood equations. If these equations are nonlinear, which is usually the case, they must be solved by an iterative procedure. At one time this was considered a serious drawback to maximum likelihood estimation. However, with the rapid advances in computer technology and the accompanying development of numerical optimisation procedures, these arguments have lost much of their force. This is not to say that maximising a likelihood function is now a straightforward affair. A number of pitfalls await the unwary and it is as well to have some understanding of what is involved in the optimisation procedures which are used. Thus, having described the statistical theory of maximum likelihood estimation in Chapter 3, various methods of computing the estimators are reviewed in Chapter 4. However, although Chapter 4 is devoted primarily to numerical methods, an important theoretical result emerges in the last section. It is shown there that numerical procedures specifically geared towards maximum likelihood estimation can be used as the basis for constructing estimators which, although not iterative, are fully efficient. These are known as *two-step estimators* and such estimators are often proposed in econometrics without their relationship to maximum likelihood being made explicit. Indeed, two-step estimators constructed on a relatively *ad hoc* basis often turn out not to have the desirable properties associated with maximum likelihood.

Although the stress in this book is on maximum likelihood, the method of least squares is dealt with at some length in Chapter 2. The material here covers the classical linear regression model, the generalised linear regression model and (non-simultaneous) systems of regression equations. The properties of ordinary least squares and generalised least squares estimators are established, and the method of *instrumental variables* introduced. As with least squares estimators, instrumental variable estimators use only information contained in the first two moments of the data. Nevertheless they can be extremely useful, particularly in large scale econometric models. Furthermore, instrumental variables estimates can often be used as starting values for maximum likelihood iterative procedures. From the theoretical point of view, the method of instrumental variables provides the framework for a unified treatment of certain estimation procedures in simultaneous equations systems. Such systems are discussed in Chapter 9. There, both two-stage and three-stage least squares are treated as instrumental variable estimators, and this interpretation then leads to the construction of similar estimators in more complex models.

Testing

In a classical linear regression model, tests of hypotheses may be based on the t- and F-statistics. These are described in Chapter 2, but in Chapter 5 it is shown that they may be developed within the more general framework of maximum likelihood. The basic test

procedure is based on the likelihood ratio statistic, the null
hypothesis being that certain restrictions on the model are valid. As
with the maximum likelihood estimator, the likelihood ratio test has
certain desirable statistical properties. However, it suffers from the
disadvantage that the model must be estimated both with and
without the restrictions implied by the null hypothesis. Fortunately,
two alternative test procedures with similar statistical properties are
available. These are the Wald and Lagrange multiplier tests. The Wald
test requires only that the unrestricted model be estimated, whereas
the Lagrange multiplier test is based on the restricted model.

The likelihood ratio, Wald and Lagrange multiplier tests form the
basis of classical hypothesis testing. However, they cannot always be
applied. When comparing two models, neither of which is a special
case of the other, non-nested procedures must be employed. These
procedures are also described in Chapter 5, although it is worth
noting that they too are developed in a likelihood framework.

Methods for determining how well a model fits the data are also
considered in Chapter 5. These are known as diagnostic checks, and
include graphical procedures as well as formal test statistics. They
are discussed primarily within the context of linear regression,
although their application is much wider.

Model Selection

A strategy for model selection is developed in Chapter 5 and
elaborated further in Chapter 8. The main points at issue concern the
ways in which the formal test procedures and diagnostic checks can
be used to arrive at the most appropriate specification. However,
before considering the statistical basis for model selection, the
fundamental question of what constitutes a 'good' model must first
be answered. Five criteria will be distinguished. These may be
characterised as parsimony, identifiability, goodness of fit,
theoretical consistency, and predictive power.

Parsimony   The mechanisms by which economic variables are
generated are invariably complex. When a model is constructed, it is
not intended to be an accurate description of the real world. On the
contrary, the aim is to simplify the underlying processes in such a
way that only the essential features are brought out. In the words of
Milton Friedman (1953, p. 14): 'A hypothesis is important if it
'explains' much by little . . .'. Similar sentiments are expressed by
Popper (1959, p. 142) when he says that: 'Simple statements . . . are
to be prized more highly than less simple ones because they tell us

more; because their empirical content is greater; and because they
are better testable'.

From the statistical point of view, the key feature of a simple
model is that it contains a small number of parameters. This is often
known as the principle of parsimony, and it is well illustrated by
(1.1). In explaining household expenditure on food, we are well
aware of influences other than income and household size.
Occupation, education, social class and age are all relevant factors,
but in formulating (1.1) it is hypothesised that income and household
size are the dominating influences. Very little explanatory power is
lost by relegating the other factors to the disturbance term, but a
good deal is gained by focusing attention on the variables which
really matter.

Identifiability   A model is not identifiable if more than one set of
parameter values is consistent with the data. This concept is fundamental to any statistical model and it is defined more precisely in
Section 3.6. The main point to note for the moment is that if a
model is not identifiable, the estimates cannot, in general, be
interpreted in any meaningful way. In econometrics, identifiability is
particularly important in connection with simultaneous equation
models and a good deal of the discussion in Chapter 9 is concerned
with rules for determining whether a given equation is identifiable. If
it is not, there is no point in attempting to estimate it, even with a
large number of observations.

Although identifiability is a precisely defined statistical concept, it
is related to parsimony. As a general rule, the more parsimonious a
model, the less likely it is to suffer from problems of identifiability.

Goodness of Fit   The object of building a model like (1.1) is to
explain the movements in the dependent variable. If the model is at
all successful, the prediction based on x_{1t} and x_{2t} should be
reasonably close to the observed value, y_t, for all points in the
sample. The goodness of fit of a regression model may be characterised by a measure such as the coefficient of multiple correlation, R^2,
and similar criteria may be constructed for other classes of models.
However, other considerations arise in determining whether the
movements in a variable, or variables, are adequately captured by the
model. In particular, the *residuals* from the fitted model should be
approximately random. If they are not, there is some systematic
aspect of behaviour which is not being picked up. The purpose of
'diagnostic checks' is basically to assess the adequacy of a model in
this respect.

Theoretical Consistency   A fourth requirement of a good model is that it should be consistent with what is known *a priori*, irrespective of whether that knowledge stems from economic theory or common logic. In the case of (1.1), for example, an equation in which the coefficient of either x_1 or x_2 was negative would have to be viewed with suspicion.

Predictive Power   The final criterion for judging a model is the accuracy of its predictions. To quote again from Friedman (1953, p. 7): '... the only relevant test of the validity of a hypothesis is comparison of its predictions with experience'. It is important, therefore, to be clear about what is meant by predictive power in this context; in particular, it must be distinguished from the earlier criterion of 'goodness of fit'.

In the cross-sectional example, (1.2), a model is specified and estimated on the basis of a sample of N observations. The accuracy of predictions within the sample is captured by a goodness of fit measure such as R^2. The important question, however, is whether the estimated model can predict food expenditure for households *outside* the sample, given the relevant information on income and household size. Success in this respect is much more convincing than the attainment of a good fit within the sample, since the preferred specification is being validated against a set of completely fresh observations. Similar considerations apply in modelling time series. The usual strategy is to examine the accuracy of predictions for observations arising after the sample, i.e. in the *post-sample* period.

A number of criteria can be formulated for assessing predictive power outside the sample period. One obvious possibility is the sum of squares of the prediction errors, and this is the measure proposed at the end of Chapter 5. Once such a measure has been adopted, some operational content can be given to the idea that a model should be retained until it is confronted with an alternative which yields more accurate predictions.

2. Time Series Observations

The basic techniques of estimation, testing and specification are developed in Chapters 2 to 5. These techniques are applicable to both cross-section and time series problems. However, the main interest is in time series, and the last four chapters are primarily concerned with applying the methodology in this direction. The variable of interest, y_t, is observed at equally spaced intervals over the period $t = 1, \ldots, T$, together with a set of possible explanatory variables. Econometric theory is concerned with the issues surrounding the estimation and specification of a suitable model for explaining and predicting the behaviour of y_t. More generally y_t is a vector of observations and the model constructed takes the form of a system of equations.

The element of time introduces a new dimension into modelling. It raises important questions concerning the economic interpretation of a model, particularly with respect to equilibrium and steady-state growth, and it brings in a whole range of statistical considerations concerned with modelling variables which do not adjust instantaneously to changes in other variables. The models proposed in time series analysis form the statistical basis for the models used in dynamic econometrics.

Statics and Dynamics

The model

$$y_t = \beta x_t + u_t \qquad (2.1)$$

is static. If x changes, y immediately responds and no further change takes place in y if x then remains constant. The system is therefore always observed in an equilibrium position.

A dynamic element may be injected into (2.1) by introducing lagged values of the explanatory variable. A very simple modification is

$$y_t = \beta_0 x_t + \beta_1 x_{t-1} + u_t \qquad (2.2)$$

If x increases by one unit, the expected value of y increases immediately by β_0 units, but the full change of $\beta_0 + \beta_1$ units is only felt after one whole time period has elapsed.

An alternative way of introducing dynamic effects into a model is by means of a lagged dependent variable. Consider the specification

$$y_t = \alpha y_{t-1} + \beta x_t + u_t \qquad (2.3)$$

where α and β are parameters. Suppressing the disturbance term and substituting repeatedly for lagged values of y gives

$$y_t = \beta x_t + \beta \alpha x_{t-1} + \beta \alpha^2 x_{t-2} + \cdots \qquad (2.4)$$

Provided $|\alpha| < 1$, the influence of values of x in the remote past is negligible. In these circumstances the model has an equilibrium solution, since if x_t is constant at $x_t = \bar{x}$ the right hand side of (2.4) is an infinite geometric progression. On summing these terms the equilibrium value of y is given by

$$\bar{y} = \beta \bar{x}/(1 - \alpha) \qquad (2.5)$$

If x changes, y moves gradually towards its new equilibrium level. The closer $|\alpha|$ is to one the more slowly this adjustment takes place.

Systems of Equations

A simple supply and demand model consists of three equations. The first is the supply curve, which determines the way in which the quantity supplied, q^s, responds to a change in price, p. The second is the demand curve, which relates the quantity demanded, q^D, to price, while the third is the market clearing condition. This is simply an identity which sets demand equal to supply.

Under the assumption that the demand and supply curves are both linear, a possible deterministic specification is:

$$q_t^S = \gamma_{11} p_t + \beta_{11} + \beta_{12} x_t \qquad (2.6a)$$

$$q_t^D = \gamma_{21} p_t + \beta_{21} \qquad (2.6b)$$

$$q_t^S = q_t^D \qquad (2.6c)$$

It will be observed that supply is assumed to depend on another variable other than price. The variable x is unlike p and q in that it is not generated within the system. It is *exogenous* rather than *endogenous* and the significance of this distinction will become apparent in due course. The immediate point to note is that the three equations in (2.6) can be reduced to two. Setting $q_t^S = q_t^D = q_t$, yields

$$S: \quad q_t = \gamma_{11} p_t + \beta_{11} + \beta_{12} x_t \qquad (2.7a)$$

$$D: \quad q_t = \gamma_{21} p_t + \beta_{21} \qquad (2.7b)$$

If x_t is constant, the position of the supply curve is fixed, and it may be traced out from (2.6a). The demand curve may also be traced out and figure 1.1 shows the familiar supply and demand diagram of economic theory. The equilibrium solution is at the intersection of the demand and supply curves, and so long as x_t remains constant, this remains unchanged. Both q and p are endogenous, and once x_t is fixed, their values are jointly determined by the market clearing condition. In other words they are determined *simultaneously*. Thus although p_t appears on the right hand side of both equations, its status is exactly the same as q_t.

If x_t moves to a new level, the supply curve shifts to S' and a new equilibrium solution is obtained. Because the system in (2.6) is a static one, the movement to this new position takes place immediately, and no further movement takes place if the new level of x_t is maintained.

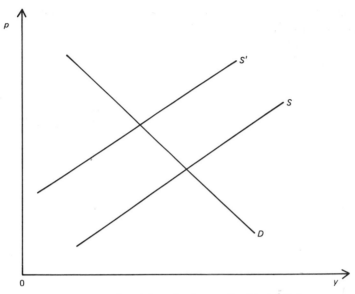

Figure 1.1   *Demand and Supply Curves for Model 2.6*

Each time x_t changes, a new equilibrium position is reached and values of p_t and q_t trace out the demand curve. However, the form of the supply curve cannot be determined on the basis of these observations. Even though the model is deterministic, there is a problem of identifiability.

If the exogenous variable were attached to the demand equation, rather than the supply equation, the supply curve would be traced out. However, the demand curve would not now be obtainable. Interestingly enough, this position can change in a dynamic model. Consider the following modification:

$$q_t^S = \gamma_{11} p_{t-1} + \beta_{11} \tag{2.8a}$$

$$q_t^D = \gamma_{21} p_t + \beta_{21} + \beta_{22} x_t \tag{2.8b}$$

$$q_t^S = q_t^D \tag{2.8c}$$

The demand equation is still static, but the supply equation is dynamic insofar as the quantity supplied does not adjust instantaneously to a change in price. The consequences of this are shown in figure 1.2. Suppose that the market is initially in equilibrium at (p_0, q_0), but that at $t = 1$, the level of x changes to $\bar{x}$, causing the demand curve to shift outwards. The effect of this change is to increase price, but because the quantity supplied responds only to the price in the *previous* time period, q remains at q_0. Hence the immediate effect is to push the price up from p_0 to p_1, as shown on the diagram. In the next time period the quantity supplied is determined by the price, p_1. However, putting this quantity, q_2, on the market causes the price to fall and at time $t = 2$ the observed price and quantity are p_2 and q_2 respectively where

$$q_2 = \gamma_{11} p_1 + \beta_{11}$$

and

$$p_2 = q_2 - \beta_{21} - \beta_{22} \bar{x}$$

The above process of adjustment continues with p_t and q_t gradually moving towards the new equilibrium position, (p^*, q^*). Figure 1.2 is often known as a 'cobweb' diagram for obvious reasons. The observed values of q_t and p_t are marked with crosses and it will be observed that from $t = 1$ onwards, all of these lie on the new demand curve. The equilibrium values (p_0, q_0) and (p^*, q^*) both lie on the supply curve. This suggests that introducing a dynamic element into the model in this way solves the identifiability problem. Information on both the demand and supply curves is available, and this can be used to derive the values of the parameters in each equation.

The question of whether an equilibrium solution actually exists is no longer a trivial one. If figure 1.2 is drawn with the demand curve steeper, or as steep, as the supply curve, it will be found that p_t and q_t no longer converge to the new equilibrium solution. Translated

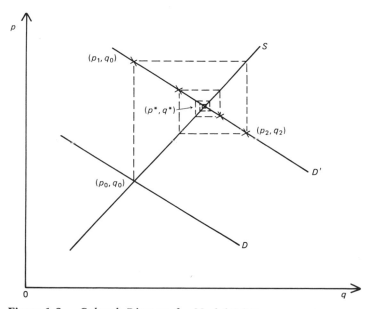

Figure 1.2   *Cobweb Diagram for Model 1.2.8*

into the terms of (2.8), this means that the condition

$$|\gamma_{21}| > \gamma_{11} \qquad (2.9)$$

must hold if the model is to be stable. (It is implicitly assumed that $\gamma_{11} > 0$ and $\gamma_{21} < 0$.) If model (2.8) were estimated and (2.9) did not hold, the appropriateness of the specification would have to be seriously questioned.

The final point to note about this model concerns the availability of data. Suppose that (2.8) is a reasonable reflection of the way the market works when the time period is one month, but that the variables are only recorded every six months. If changes in x_t take place relatively infrequently, all that will tend to be observed is a series of equilibrium positions. Thus, in terms of the observation period, the model is effectively a simultaneous one, and the demand curve cannot be identified.

Time Series Models

A univariate time series model attempts to explain the behaviour of a variable, y_t, in terms of its own past. The most fruitful approach is to regard y_t as a stochastic process, in which the observations evolve over time according to certain probabilistic laws. The first order autoregressive model,

$$y_t = \phi y_{t-1} + \epsilon_t \qquad (2.10)$$

is a simple example. The term ϵ_t denotes a sequence of uncorrelated disturbances with mean zero and constant variance.

If ϕ is less than one in absolute value, the observations generated by (2.10) fluctuate around a mean of zero. If, on the other hand, $|\phi|$ is greater than one, the model is explosive. A model which satisfies the condition $|\phi| < 1$ is said to be stationary. Stationarity is an important concept in time series modelling, although very few time series display the characteristics which it implies. Nevertheless, the theory surrounding stationary time series can often be applied to nonstationary series by the simple expedient of taking first or second differences. Thus suppose

$$\Delta y_t = \phi \Delta y_{t-1} + \epsilon_t \qquad (2.11)$$

where $\Delta y_t = y_t - y_{t-1}$. If $|\phi| < 1$, Δy_t is stationary even though y_t is not.

The dynamic regression model, (2.3), may be regarded as a generalisation of (2.10). Any knowledge concerning estimation or testing in (2.10) is therefore directly relevant to (2.3). Furthermore, any other model constructed in the same spirit as (2.10) will obviously have important implications for the development of a dynamic econometric methodology.

The most straightforward way of extending (2.10) is to let y_t depend on past values up to a lag length of, say, p. This leads to the pth order autoregressive process

$$y_t = \phi_1 y_{t-1} + \phi_2 y_{t-2} + \cdots + \phi_p y_{t-p} + \epsilon_t \qquad (2.12)$$

where $\phi_1, \ldots, \phi_p$ are parameters. An alternative way of capturing the dependence of y_t on its past is by a moving average model. The first order *moving average* process is defined by

$$y_t = \epsilon_t + \theta \epsilon_{t-1} \qquad (2.13)$$

Although this is formulated in terms of two unobservable random disturbances, ϵ_t and ϵ_{t-1}, repeated substitution gives

$$y_t = - \sum_{j=1}^{\infty} (-\theta)^j y_{t-j} + \epsilon_t \qquad (2.14)$$

provided that $|\theta| < 1$. Model (2.13) is therefore an autoregressive process with an infinite number of lags, but the coefficients, $\phi_1, \phi_2, \ldots$, are constrained so as to depend on only one parameter.

The different characteristics of autoregressive and moving average processes may be combined in a mixed model,

$$y_t = \phi_1 y_{t-1} + \cdots + \phi_p y_{t-p} + \epsilon_t + \theta_1 \epsilon_{t-1} + \cdots + \theta_q \epsilon_{t-q} \qquad (2.15)$$

This is known as an *autoregressive-moving average* (ARMA) process, and it represents the most flexible class of models for univariate time series. Provided $\theta_1, \ldots, \theta_q$ satisfy certain constraints, (2.15) can be expressed as an autoregressive process in the same way as (2.13). By selecting a suitable model from the class of ARMA processes, y_t may be modelled in terms of its past using a minimum number of parameters.

The above considerations extend to multivariate time series. If y_t is a vector of observations, it may be modelled efficiently in terms of its own past by a *multivariate* ARMA process,

$$y_t = \Phi_1 y_{t-1} + \cdots + \Phi_p y_{t-p} + \epsilon_t + \Theta_1 \epsilon_{t-1} + \cdots + \Theta_q \epsilon_{t-q} \qquad (2.16)$$

The parameters are now contained within matrices, $\Phi_1, \ldots, \Phi_p$, $\Theta_1, \ldots, \Theta_q$, while ϵ_t is a vector of random disturbances.

Data Generation

The cross-sectional model, (1.2), applies to a sample of N households selected at random. The basic assumption underlying statistical inference in such a model is that the sample obtained is simply one of an infinite number of possible samples. The whole exercise of drawing a sample and estimating parameters on the basis of it could be repeated as many times as we wished.

Similar considerations are adopted in modelling a time series. The observations are regarded as a single *realisation* from some underlying data generation process. Unlike the cross-sectional case, however, only one set of observations can ever be obtained in practice. Nevertheless it is possible to regard the underlying data generation process as being capable, in principle, of producing an infinite number of realisations over that same time period. It is on this basis that the properties of dynamic models are derived and statistical inferences made.

In a pure time series model, such as (2.10), this is all that needs to be said. The model generates a set of observations $y_1, \ldots, y_T$, the values of which depend on the random disturbances, $\epsilon_1, \ldots, \epsilon_T$, and on the initial value, y_0. For a regression model like (2.3) there is the additional complication of the explanatory variable. This can be regarded as fixed or stochastic. If it is fixed, the observed values would be reproduced if another set of observations on y were generated. If it is stochastic its values would change in repeated realisations, thereby introducing an additional source of variation into y_t. In a cross-sectional model, such as (1.2), it is fairly easy to visualise both fixed and stochastic explanatory variables: the survey could be repeated with the same households or with a completely different sample of households. For a dynamic model the distinction is less clear cut, but as a rule it is more reasonable to regard the explanatory variables as being stochastic. However, from the statistical point of view the distinction between stochastic and non-stochastic variables is often unimportant. If no information is lost by conditioning on an explanatory variable, i.e. by treating it *as if* it were fixed, that variable is said to be *exogenous*. Unless otherwise stated, an exogenous variable will be denoted by the symbol 'x', as will a variable which truly can be considered fixed in repeated samples.

Prediction and Forecasting

Model (2.3) embodies a *behavioural* hypothesis in that it attempts to explain the movements in y_t in terms of the movements in x_t. A test of the validity of the theory linking y to x is made by comparing post-sample observations on y with the predictions made by equation (2.3) on the basis of post-sample observations on x. This is sometimes known as *ex post* prediction. It should be carefully distinguished from *ex ante* predictions, in which unconditional predictions of future values of y are made. An exercise of this kind requires predictions for future values of x.

Unconditional predictions are called *forecasts*. The use of the term prediction without any qualification implies that future values of the exogenous variables are known. Of course for pure time series models, such as (2.15), there is no distinction between conditional and unconditional predictions, and the terms forecast and prediction may be used interchangeably.

3. Mathematical and Statistical Preliminaries

It is assumed that the reader has a grounding in probability and statistical inference, and is familiar with calculus and matrix algebra. This section is merely intended to review certain key concepts in these areas and to introduce some notational conventions.

A familiarity with asymptotic theory is not taken for granted and Section 4 is devoted to a relatively heuristic introduction to the subject.

Matrix Algebra

A thorough understanding of the basic operations in matrix algebra, such as addition and inversion, is necessary. In addition, the reader should be aware of the concept of the trace of a matrix, i.e. the sum of the elements on the leading diagonal, and the definitions of a positive definite (p.d.) and positive semi-definite (p.s.d.) matrix. A symmetric matrix A is said to be p.d. if $x'Ax > 0$, where x is any vector such that $x \neq 0$; it is p.s.d. if $x'Ax \geq 0$. Kronecker (tensor) products are reviewed in the Appendix.

The factorisation of a p.d. matrix appears in the text in a number of places. If A denotes such a matrix it can always be factored as $A = L'L$, where L is non-singular. One way of carrying out this factorisation is by means of the *Cholesky decomposition*. In this case L is lower triangular, with all the elements on the leading diagonal positive; see, for example, Wolfe (1978, pp. 275–277). An alternative formulation is $A = \bar{L}'D\bar{L}$, where $\bar{L}$ is a lower triangular matrix with ones in the leading diagonal and D is a diagonal matrix.

Vector and matrix differentiation play an important role in estimation and the following definitions will be used. If $f(x)$ is a function of an $n \times 1$ vector of variables, $x = (x_1, x_2, \ldots, x_n)'$, the $n \times 1$ vector of partial first derivatives is

$$\frac{\partial f}{\partial x} = \left(\frac{\partial f}{\partial x_1}, \frac{\partial f}{\partial x_2}, \ldots, \frac{\partial f}{\partial x_n} \right) \tag{3.1}$$

The notation $\partial f / \partial x'$ indicates the row vector $(\partial f / \partial x)'$. The $n \times n$ matrix of second order derivatives is

$$\frac{\partial^2 f}{\partial x \partial x'} = \begin{bmatrix} \dfrac{\partial f}{\partial x_1^2} & \dfrac{\partial f}{\partial x_1 \partial x_2} & \cdots & \dfrac{\partial f}{\partial x_1 \partial x_n} \\[2ex] \dfrac{\partial f}{\partial x_2 \partial x_1} & \dfrac{\partial f}{\partial x_2^2} & \cdots & \dfrac{\partial f}{\partial x_2 \partial x_n} \\[2ex] \vdots & \vdots & & \vdots \\[2ex] \dfrac{\partial f}{\partial x_n x_1} & \dfrac{\partial f}{\partial x_n x_2} & \cdots & \dfrac{\partial f}{\partial x_n^2} \end{bmatrix} \tag{3.2}$$

This matrix, which is known as the *Hessian*, is symmetric if the second order derivatives are continuous functions of the x's.

Some specific results on vector and matrix differentiation are given in the Appendix.

Random Variables and the Expectation Operator

The probability density function (p.d.f.) of a random variable, y, will be denoted by $p(y; \psi)$ where ψ is a vector of parameters. It will often be abbreviated to $p(y)$. The *expectation* of y is the mean of this distribution; if y is continuous over the range $-\infty < y < \infty$,

$$E(y) = \mu = \int_{-\infty}^{\infty} y p(y) dy \tag{3.3}$$

The variance is defined as

$$\text{Var}(y) = \sigma^2 = E[(y - \mu)^2] = \int_{-\infty}^{\infty} (y - \mu)^2 p(y) dy \tag{3.4}$$

For some purposes it is only necessary, or desirable, to specify the mean and variance of a distribution. When this is the case, the abbreviation $y \sim WS(\mu, \sigma^2)$ is sometimes used; *WS* stands for wide sense.

A linear combination of two random variables is defined by $w_1 y_1 + w_2 y_2$ where w_1 and w_2 are fixed weights. If

$$y_1 \sim WS(\mu_1, \sigma_1^2) \qquad \text{and} \qquad y_2 \sim WS(\mu_2, \sigma_2^2)$$

then

$$E(w_1 y_1 + w_2 y_2) = w_1 \mu_1 + w_2 \mu_2 \tag{3.5}$$

while

$$\text{Var}(w_1 y_1 + w_2 y_2) = w_1 \sigma_1^2 + w_2 \sigma_2^2 + 2 w_1 w_2 \cdot \text{Cov}(y_1, y_2) \tag{3.6}$$

If the two variables are uncorrelated, $\text{Cov}(y_1, y_2) = 0$ and the last term in (3.6) disappears.

More generally we may consider the vector of random variables $y = (y_1, \ldots, y_n)'$. This has a multivariate distribution with mean vector

$$\mu = E(y) = \{E(y_1), E(y_2), \ldots, E(y_n)\}' \tag{3.7}$$

and variance–covariance matrix

$$V = E\{(y - \mu)(y - \mu)'\} \tag{3.8}$$

The matrix V, which is usually known simply as the *covariance matrix*, is p.d. provided that individual variances, which are the elements on the main diagonal, are strictly positive. Results (3.5) and (3.6) may be extended by defining an $m \times n$ matrix of fixed weights, W, and considering the properties of the $m \times 1$ vector of random variables $z = Wy$. If $y \sim WS(\mu, V)$, then

$$E(z) = W \cdot E(y) = W\mu \tag{3.9}$$

while

$$\text{Var}(z) = E\{W(y - \mu)(y - \mu)'W'\} = WVW' \tag{3.10}$$

The Normal Distribution

The density function for the normal distribution is

$$p(y; \mu, \sigma^2) = \frac{1}{\sigma^2 \sqrt{2\pi}} \exp\left\{ -\frac{1}{2} \frac{(y - \mu)^2}{\sigma^2} \right\} \tag{3.11}$$

The distribution is completely specified once μ and σ^2 are given, and it is usual to summarise (3.11) by writing $y \sim N(\mu, \sigma^2)$. The *standard* normal distribution has $\mu = 0$ and $\sigma^2 = 1$.

If the $n \times 1$ vector $y = (y_1, \ldots, y_n)'$ has a multivariate normal distribution, the joint density of the variables depends on the mean vector, μ, and the covariance matrix, V, as defined in (3.7) and (3.8). The p.d.f. is

$$p(y; \mu, V) = (2\pi)^{-n/2} \, |\, V \,|^{-1/2} \cdot \exp\{-\tfrac{1}{2}(y - \mu)' V^{-1}(y - \mu)\}$$

$$(3.12)$$

where V must be p.d. As in the univariate case, the abbreviation, $y \sim N(\mu, V)$ can be adopted. If V is diagonal, $y_1, \ldots, y_n$ are mutually uncorrelated. Furthermore, the joint density function (3.12) can be decomposed as $p(y_1, \ldots, y_n) = p(y_1) \cdot p(y_2) \ldots p(y_n)$. Thus, if normal variables are uncorrelated, they are also independent.

A set of independent normally distributed random variables, $y_1, \ldots, y_T$ with mean μ and variance σ^2 will be indicated by writing $y_t \sim NID(0, \sigma^2)$, $t = 1, \ldots, T$.

By using the moment generating function of the normal distribution it can be shown that any linear combination of normal variables is itself normal. More generally, if $y \sim N(\mu, V)$ and $z = Wy$, where W is an $m \times n$ matrix, it follows that $z \sim N(W\mu, WVW')$. Note that W must have full row rank if WVW' is to be p.d.

A number of distributions are associated with the normal. If $y_1, \ldots, y_n$ are independent standard normal variates then $z = y_1^2 + y_2^2 + \cdots + y_n^2$ has a *chi-square* distribution with n degrees of freedom; i.e. $z \sim \chi_n^2$. Furthermore, if z and x are independent χ^2 variates with n and m degrees of freedom respectively, then $z + x \sim \chi_{n+m}^2$.

An important result links the χ^2 distribution to the multivariate normal. If y is an $n \times 1$ vector such that $y \sim N(\mu, V)$, then

$$(y - \mu)' V^{-1}(y - \mu) \sim \chi_n^2 \qquad\qquad (3.13)$$

The proof is as follows. Since V is p.d., V^{-1} is also p.d. and there exists a non-singular matrix, L, such that $L'L = V^{-1}$. The n variables defined by the transformation $z = L(y - \mu)$ are $NID(0, 1)$ since $E(z) = 0$ and

$$\text{Var}(z) = LVL' = L(L'L)^{-1}L' = LL^{-1}(L')^{-1}L' = I$$

The result then follows immediately since the quadratic form in (3.13) is equal to the sum of squares, $z'z$.

The F-distribution is defined in terms of the ratio of two independent χ^2 variates. If $z \sim \chi_n^2$ and $x \sim \chi_m^2$, then

$$F = \frac{z/n}{x/m} \qquad\qquad (3.14)$$

is an F-distribution with (n, m) degrees of freedom.

Finally, if $z \sim N(0, 1)$ and $x \sim \chi_m^2$, and z and x are independent, the ratio

$$t = z\sqrt{m}/\sqrt{x} \tag{3.15}$$

is distributed as Student's 't' with m degrees of freedom. Note that squaring (3.15) yields a variate which follows an F-distribution with $(1, m)$ degrees of freedom.

Properties of Estimators

Suppose, for simplicity, that a model contains a single unknown parameter, ψ. If this is to be estimated, we need a rule telling us how to process the observations, $y_1, \ldots, y_n$, once they become available. Such a rule defines an *estimator* of ψ, and this estimator is a function of the observations. Once the estimator has been evaluated for a particular set of observations, it becomes an *estimate*.

A 'good' estimator should have a 'reasonably' high probability of yielding estimates which are 'reasonably' close to the true parameter value. One possible criterion is the demand that it should have minimum mean square error. The *mean square error* (MSE) of an estimator, $\hat{\psi}$, is defined by

$$\text{MSE}(\hat{\psi}) = E[(\hat{\psi} - \psi)^2] \tag{3.16}$$

If $\hat{\psi}^*$ denotes any other estimator of ψ, $\hat{\psi}$ can be said to minimise the MSE if $\text{MSE}(\hat{\psi}) \leqslant \text{MSE}(\hat{\psi}^*)$ for all ψ. Unfortunately this is usually an unrealistic demand since $\hat{\psi}^*$ can always be set equal to an arbitrary value within the parameter space and its MSE will then be zero for that particular value. An estimator of this kind is hardly sensible, but it can be eliminated by imposing the condition of unbiasedness. An estimator is said to be *unbiased* if

$$E(\hat{\psi}) = \psi \qquad \text{for all } \psi \tag{3.17}$$

The MSE is equal to the variance for an unbiased estimator and this leads to the concept of a *minimum variance unbiased estimator* (MVUE).

Unfortunately, the MVUE criterion has severe drawbacks as a method of assessing different estimators. It contains an element of arbitrariness, particularly with regard to unbiasedness. Thus, while the concept of a MVUE is relevant for a static regression model such as (2.1), it is of little value in (2.10), which represents one of the simplest examples of a dynamic model. The least squares estimator in (2.1) is *linear* in the observations since it may be written in the form

$$b = \sum_{t=1}^{T} w_t y_t \qquad (3.18)$$

where $w_t = x_t/\Sigma x_t^2$. The weights, $w_1, \ldots, w_T$ are non-stochastic, and so the mean and variance of b may be obtained using (3.5) and (3.6). It is not difficult to show that b is unbiased. On the other hand, regressing y_t on y_{t-1} in (2.10) yields an estimator of ϕ which is biased. To insist on applying the MVUE criterion here would rule out the least squares estimator, even though it is perfectly sensible. In fact it is difficult to obtain exact expressions for both the mean and the variance of this estimator and so no criterion based on small sample considerations is particularly helpful.

4. Asymptotic Theory

For many estimators exact small sample results cannot be established. However, it is usually possible to determine at least the properties of an estimator *approximately*. Such approximations generally become more accurate as the sample size increases and they may be established formally by deriving a result which holds exactly in the limit as $T \to \infty$. Properties of estimators obtained on this basis are said to be asymptotic. This section sketches out these properties; a more detailed treatment will be found in Theil (1971, Chapter 8).

Consistency

As the sample size increases, it is desirable for an estimator to become more accurate in the sense that the probability of its being 'close' to the true value of the parameter increases. This notion is captured by the property of *consistency* which may be formally stated as follows. Consider a sequence of random variables (a_T), i.e. $a_1, a_2, \ldots, a_T \ldots$ This sequence is said to *converge in probability* to a constant, a, if

$$Pr(|a_T - a| < \epsilon) > 1 - \eta \qquad \text{for all } T > T_0, \qquad (4.1)$$

where ϵ and η are arbitrarily small numbers. In other words, given ϵ and η, it is always possible to find a sample size such that the probability that a_T is outside the range $a \pm \epsilon$ is less than η.

An alternative way of expressing (4.1) is to write

$$\lim_{T \to \infty} Pr(|a_T - a| > \epsilon) = 0 \qquad \text{for any } \epsilon > 0 \qquad (4.2)$$

The notation in (4.1) and (4.2) is rather awkward. If a sequence does converge in probability to a it is more common to write

$$\text{plim } a_T = a \tag{4.3}$$

This states that the *probability limit* (or 'plim') of the sequence (a_T) is a.

If (a_T) and (b_T) are sequences of random variables with probability limits a and b respectively, it is not difficult to show that

$$\text{plim}(a_T + b_T) = a + b \tag{4.4}$$

This property is also shared by the expectation operator. However, probability limits have a number of properties which are not shared by expectations. In particular

$$\text{plim}(a_T b_T) = ab \tag{4.5}$$

irrespective of whether or not a_T and b_T are independent. Furthermore, if $g(.)$ is a continuous function, then

$$\text{plim } g(a_T) = g(\text{plim } a_T) = g(a) \tag{4.6}$$

The concept of a probability limit may be extended to vectors and matrices, and the properties (4.4) to (4.6) hold in these cases also. One result which is particularly important follows from the generalisation of (4.6). If A_T is a square matrix of random variables such that

$$\text{plim } A_T = A$$

where A is p.d., then

$$\text{plim } A_T^{-1} = (\text{plim } A_T)^{-1} = A^{-1} \tag{4.7}$$

In the context of estimation theory, an estimator of a single parameter, ψ, based on T observations may be denoted by $\hat{\psi}_T$. The estimator is then said to be a *consistent* estimator of ψ if

$$\text{plim } \hat{\psi} = \psi \tag{4.8}$$

As an example consider a random sample of T observations, $y_1, \ldots, y_T$, drawn from a probability distribution with mean μ and variance σ^2. The sample mean, $\bar{y}$, is unbiased and its variance is σ^2/T. As T gets bigger, $\text{Var}(\bar{y})$ tends to zero, indicating that the distribution of $\bar{y}$ becomes more and more concentrated around the true parameter value μ.

Consistency may be shown formally by means of *Chebyshev's inequality*. This states that for any random variable, z, with finite mean and variance, μ and ω^2 respectively, the probability of a deviation from the mean equal to k or more times the standard deviation is at most equal to $1/k^2$, i.e.

$$\text{Prob}[|z - \mu| \geqslant k\omega] \leqslant 1/k^2, \qquad \text{for any } k > 0 \qquad (4.9)$$

In the case of the sample mean, $\bar{y}$, the mean is μ and the variance σ^2/T, so that the inequality in square brackets becomes $|\bar{y} - \mu| \geqslant k\sigma/\sqrt{T}$. Setting $k = \epsilon\sqrt{T}/\sigma$ gives

$$\text{Prob}[|\bar{y} - \mu| \geqslant \epsilon] \leqslant \sigma^2/\epsilon^2\sqrt{T}, \qquad \text{for any } \epsilon > 0.$$

Since $\sigma^2/\epsilon^2\sqrt{T}$ converges to zero as $T \to \infty$ for any $\epsilon > 0$, it follows from the definition (4.2), that $\bar{y}$ is a consistent estimator of μ.

As a second example, consider the estimation of σ^2 in the above example. A suitable estimator is

$$s^2 = (T - 1)^{-1}\Sigma(y_t - \bar{y})^2 \qquad (4.10)$$

and if the observations are normally distributed it can be shown that

$$(T - 1)s^2/\sigma^2 \sim \chi^2_{T-1} \qquad (4.11)$$

Since $\text{Var}(\chi^2_f) = 2f$, it follows that $\text{Var}(s^2) = 2\sigma^4/(T - 1)$. Proceeding as before it may be shown that

$$\text{plim } s^2 = \sigma^2 \qquad (4.12)$$

An alternative estimator of the variance in this situation is

$$\tilde{\sigma}^2 = T^{-1}\Sigma(y_t - \bar{y})^2 \qquad (4.13)$$

Although this estimator is biased in small samples, it is asymptotically unbiased in the sense that $\lim E(\tilde{\sigma}^2) = \sigma^2$. Given this result it is relatively straightforward to show that the estimator is consistent.

The consistency of an estimator can often be demonstrated under much weaker conditions than are implied by the above discussion. In fact, if a random sample is drawn from a parent distribution with a finite mean, μ, it can be shown that the sample mean will be a consistent estimator of μ even if the variance, σ^2, is infinite.

A final point about consistent estimators concerns the implications of (4.6). For example if $y_t \sim NID(\mu, \sigma^2)$, $\bar{y}^{-1}$ is a consistent estimator of μ^{-1}. It is *not* true, however, that $E(\bar{y}^{-1}) = \mu^{-1}$, for the simple reason that the expectation cannot be evaluated in this case.

Asymptotic Distributions

Suppose a sample of T independent observations, $y_1, \ldots, y_T$ is drawn from a normal distribution with mean μ and variance σ^2. Assume for simplicity that σ^2 is known. From the properties of the normal distribution, the sample mean, $\bar{y}$, is normally distributed with mean μ and variance, σ^2/T. Now suppose we have exactly the same situation, except that the observations no longer come from a normal distribution. In these circumstances it will not usually be possible to derive the exact distribution of $\bar{y}$. However, for a 'reasonably large' value of T, $\bar{y}$ can be taken to be *approximately* normally distributed with mean μ and variance, σ^2/T. What constitutes a 'reasonably large' sample in these circumstances depends on the form of the distribution of y_t, and the degree of approximation which can be tolerated. The *central limit theorem*, the result which forms the basis of our assertion regarding the approximate normality of $\bar{y}$, tells us nothing on this point. What it does prove formally is that $\sqrt{T} \cdot \bar{y}$ *converges in distribution* to a normal variate with mean $\sqrt{T} \cdot \mu$ and variance σ^2. In other words, as T becomes larger, the distribution of $\sqrt{T} \cdot \bar{y}$ approaches the distribution $N(\sqrt{T}\mu, \sigma^2)$. This is usually expressed by writing

$$\sqrt{T}(\bar{y} - \mu) \xrightarrow{L} N(0, \sigma^2) \tag{4.14}$$

the notation indicating that $\sqrt{T} \cdot \bar{y}$ has a *limiting normal distribution*.

Expressing the result in terms of $\sqrt{T} \cdot \bar{y}$ rather than $\bar{y}$ sometimes causes confusion. However, this is necessary in order to establish an asymptotic distribution. Suppose we have the original case where $y_t \sim N(\mu, \sigma^2)$. The variance of $\bar{y}$ is σ^2/T, but in the limit as $T \to \infty$, this goes to zero. Thus the asymptotic distribution is degenerate, all its mass being concentrated at the point $\bar{y} = \mu$. This is clearly no guide to the kind of distributional results which might be expected in finite samples. Nevertheless, although the asymptotic results do formally refer to the limiting distribution of $\sqrt{T} \cdot \bar{y}$, it is the approximation to the distribution of $\bar{y}$ which is of interest. Thus, when (4.14) holds, $\bar{y}$ is said to be *asymptotically normally distributed* and this may be written

$$\bar{y} \sim AN(\mu, \sigma^2/T) \tag{4.15}$$

The *asymptotic variance* is

$$\mathrm{Avar}(\bar{y}) = \sigma^2/T \tag{4.16}$$

Cramér's Theorem

The following result, which is sometimes known as *Cramér's theorem* plays an important role in the derivation of the asymptotic properties of certain estimators.

Suppose that $\eta_T = H_T \xi_T$, where ξ_T is an $n \times 1$ vector of random variables with a limiting multivariate normal distribution with mean

μ and covariance Ω, i.e. $\xi_T \xrightarrow{L} N(\mu, \Omega)$, and H_T is an $m \times n$ matrix with the property

$$\text{plim } H_T = H$$

where H is p.d. Then

$$\eta_T \xrightarrow{L} N(H\mu, H\Omega H') \tag{4.17}$$

Example　　Consider the statistic

$$\frac{\bar{y} - \mu}{s/\sqrt{T}} \tag{4.18}$$

constructed from a random sample from a non-normal distribution. The standard deviation of the parent distribution, σ, is assumed to be unknown and so is replaced by the consistent estimator, s, defined by (4.10). From the central limit theorem $\sqrt{T}(\bar{y} - \mu) \xrightarrow{L} N(0, \sigma^2)$. If $\sqrt{T}(\bar{y} - \mu)$ is taken to be the random variable, ξ_T, in Cramér's theorem, then $H_T = s$. Since plim $s = \sigma$, it follows from (4.17) that

$$\frac{\bar{y} - \mu}{s/\sqrt{T}} \xrightarrow{L} N(0, 1) \tag{4.19}$$

This is the justification for regarding (4.18) as an *asymptotic t-ratio* which can be treated as a standard normal variate in large samples. The square root of the estimator of the asymptotic variance, (4.16),

$$\text{avar}(\bar{y}) = s^2/T \tag{4.20}$$

is an *asymptotic standard error*.

5.　Time Series Analysis

Autoregressive-moving average processes were introduced in Section 2. The aim of this section is to set out the basic properties of these models, to extend them to cover non-stationarity, and to comment briefly on the model selection procedures which have been proposed in the time series literature. A more detailed discussion of these matters will be found in the companion volume, 'Time Series Models', (TSM).

Stationarity

The disturbance term, ϵ_t, which featured in all the ARMA models of Section 2 is the simplest example of a stationary stochastic process.

For any value of t, ϵ_t has zero mean and constant variance, and is uncorrelated with any other variable in the sequence. For a sequence of T terms, $\epsilon_1, \epsilon_2, \ldots, \epsilon_T$, these properties may be stated formally as:

$$E(\epsilon_t) = 0, \qquad t = 1, \ldots, T \tag{5.1a}$$

$$E(\epsilon_t^2) = \sigma^2, \qquad t = 1, \ldots, T \tag{5.1b}$$

$$E(\epsilon_t \epsilon_s) = 0, \qquad t \neq s \text{ and } t, s = 1, \ldots, T \tag{5.1c}$$

The expectation operator denotes the expected value over all possible realisations.

The properties in (5.1) define a (zero mean) 'white noise' process. The process is weakly, or covariance, stationary because it satisfies the following criteria: (i) the mean is independent of t; (ii) the variance is independent of t; and (iii) each *autocovariance*, $E(\epsilon_t \epsilon_s)$, depends only on the difference between t and s. Had a normal distribution been specified for ϵ_t, the conditions in (5.1) would have been sufficient to ensure that the process was strongly stationary. However, the distinction between weak and strong stationarity is not important in this book and the term stationarity will be employed whenever the criteria for weak stationarity are satisfied.

Now consider the first order moving average process, usually referred to as an MA(1) process, defined in (2.13). It follows from (5.1a) that

$$E(y_t) = E(\epsilon_t) + \theta E(\epsilon_{t-1}) = 0 \tag{5.2a}$$

while from (5.1b) and (5.1c)

$$E(y_t^2) = E(\epsilon_t^2) + 2\theta E(\epsilon_t \epsilon_{t-1}) + \theta^2 E(\epsilon_{t-1}^2) = \sigma^2(1 + \theta^2) \tag{5.2b}$$

Thus the mean and variance of the process are independent of t. Furthermore,

$$E(y_t y_{t-1}) = E\{(\epsilon_t + \theta \epsilon_{t-1})(\epsilon_{t-1} + \theta \epsilon_{t-2})\} = \theta \sigma^2 \tag{5.2c}$$

and

$$E(y_t y_{t-\tau}) = 0, \qquad \tau = \pm 2, \pm 3, \ldots \tag{5.2d}$$

The autocovariances therefore depend only on $\tau = t - s$ and so the model satisfies the requirements for stationarity.

In a similar way it can be shown that the MA(q) process,

$$y_t = \theta_0 + \epsilon_t + \theta_1 \epsilon_{t-1} + \cdots + \theta_q \epsilon_{t-q} \tag{5.3}$$

is stationary. No restrictions are needed either on θ_0 or on the moving average parameters, $\theta_1, \ldots, \theta_q$. Restrictions must, however, be imposed on the parameters of an autoregressive model. For the

AR(p) process, (2.12), the condition for stationarity is that the roots of the characteristic equation

$$x^p - \phi_1 x^{p-1} - \cdots - \phi_p = 0 \tag{5.4}$$

should be less than one in absolute value. The same condition must be imposed on the general ARMA(p, q) process defined by (2.15).

For the AR(1) model, (2.10), the condition for stationarity is $|\phi| < 1$. When this condition holds, repeated substitution for lagged values of y_t yields an infinite moving average,

$$y_t = \sum_{j=0}^{\infty} \phi^j \epsilon_{t-j} \tag{5.5}$$

This is a reflection of a more general result, namely that any stationary AR or ARMA process may be expressed as an infinite MA process.

It follows immediately from (5.1a) that $E(y_t) = 0$. As regards the variance,

$$E(y_t^2) = \sum_{j=0}^{\infty} \sum_{i=0}^{\infty} \phi^{i+j} E(\epsilon_{t-j}\epsilon_{t-i}) = \sigma^2 \sum_{j=0}^{\infty} \phi^{2j}$$

and summing the sequence $1, \phi^2, \phi^4, \ldots$ as an infinite geometric progression yields

$$\mathrm{Var}(y_t) = \sigma^2/(1 - \phi^2) \tag{5.6}$$

The autocovariances may be obtained by a similar argument, although the derivation given in the next sub-section is slightly easier.

The Autocorrelation Function

The autocovariances of a stationary stochastic process are defined by

$$\gamma_\tau = E(y_t y_{t-\tau}), \qquad \tau = 0, \pm 1, \pm 2, \ldots \tag{5.7}$$

The plot of γ_τ, the autocovariance at lag τ, against τ is known as the *autocovariance* function. Since $\gamma_\tau = \gamma_{-\tau}$ negative values of τ can be ignored. In a similar way, the autocorrelations at lag τ,

$$\rho_\tau = \gamma_\tau/\gamma_0, \qquad \tau = 0, 1, 2, \ldots \tag{5.8}$$

can be plotted against τ to yield the *autocorrelation function*. Both functions contain exactly the same information on the correlation structure of the process, the only difference being that the auto-correlations are dimensionless with $\rho_0 = 1$.

The autocorrelation function for an MA(1) process exhibits a sharp cut-off after $\tau = 1$. For an AR(1) process, however, the auto-correlations decay exponentially as shown in figure 1.3. The form of the autocorrelation function may be obtained by multiplying both sides of (2.10) by $y_{t-\tau}$ and taking expectations. Thus:

$$E(y_t y_{t-\tau}) = \phi E(y_{t-1} y_{t-\tau}) + E(\epsilon_t y_{t-\tau}) \tag{5.9}$$

becomes

$$\gamma_\tau = \phi \gamma_{\tau-1}, \qquad \text{for } \tau = 1, 2, \ldots \tag{5.10}$$

and this implies a similar expression for the autocorrelations.

In general an MA(q) process has a cut-off in the autocorrelation function beyond lag q. On the other hand, the autocorrelation function for an AR(p) process decays gradually towards zero. It has the form of a pth-order difference equation and the question of whether the decay is exponential, cyclical or oscillatory depends on the roots of the characteristic equation, (5.4).

The properties of a stochastic process can also be examined in the frequency domain using the power spectrum. The spectrum is basically a transformation of the autocovariance function and it contains exactly the same information as the autocovariance function. However, the information is presented in a different way, and this can be particularly useful if interest is centred on cyclical movements. Frequency domain methods will not be employed in this book, but a full discussion of the subject will be found in TSM.

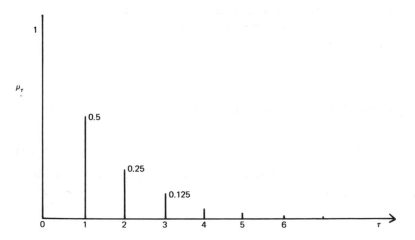

Figure 1.3   *Autocorrelation Function for AR(1) Process with $\phi = 0.5$*

The Lag Operator

The lag operator, L, is defined by the transformation

$$Ly_t = y_{t-1} \qquad (5.11)$$

Applying L to y_{t-1} yields $Ly_{t-1} = y_{t-2}$. Substituting from (5.11) gives $L(Ly_t) = L^2 y_t = y_{t-2}$ and so, in general,

$$L^\tau y_t = y_{t-\tau}, \qquad \tau = 1, 2, 3, \ldots \qquad (5.12)$$

It is logical to complete the definition by letting L^0 have the property $L^0 y_t = y_t$ so that (5.12) holds for all non-negative integers.

The lag operator can be manipulated in a similar way to any algebraic quantity. Thus (5.5) may be expressed in the form

$$y_t = \sum_{j=0}^{\infty} (\phi L)^j \epsilon_t$$

and if $|\phi L| < 1$ summing $1, \phi L, (\phi L)^2, \ldots$ as an infinite geometric progression yields

$$y_t = \epsilon_t / (1 - \phi L) \qquad (5.13)$$

Multiplying both sides of (5.13) by $(1 - \phi L)$ gives

$$y_t - \phi y_{t-1} = \epsilon_t$$

which on re-arrangement becomes the original form of the model, (2.10).

The first difference operator,

$$\Delta y_t = y_t - y_{t-1}$$

can be used in the same way as the lag operator as $\Delta = 1 - L$. The relationship between the two operators can often be usefully exploited. For example,

$$\Delta^2 y_t = (1 - L)^2 y_t = (1 - 2L + L^2) y_t = y_t - 2y_{t-1} + y_{t-2} \qquad (5.14)$$

The general ARMA(p, q) model, (2.15), can be written in the form

$$\phi(L) y_t = \theta(L) \epsilon_t \qquad (5.15)$$

where $\phi(L)$ and $\theta(L)$ are polynomials in the lag operator defined by

$$\phi(L) = 1 - \phi_1 L - \cdots - \phi_p L^p \qquad (5.16)$$

and

$$\theta(L) = 1 + \theta_1 L + \cdots + \theta_q L^q \qquad (5.17)$$

The stationarity condition can be expressed in terms of (5.16) by requiring that the roots of $\Phi(L)$ lie *outside* the unit circle, i.e. are greater than one in absolute value, cf. (5.4). A similar condition may be imposed on the roots of $\theta(L)$ in order to ensure *invertibility*. An invertible process is one which can be expressed as an infinite autoregression; cf. (2.14).

Model Building

The attraction of the ARMA(p, q) model is that it provides the most parsimonious representation of a stationary stochastic process. It may be extended to encompass a much wider class of nonstationary models by differencing. If the difference operator must be applied d times before an ARMA(p, q) representation is appropriate, the variable is said to follow an *autoregressive-integrated-moving average* process of order (p, d, q). This is abbreviated as ARIMA(p, d, q). Thus (2.11) is an ARIMA$(1, 1, 0)$ process.

   An ARMA model may be estimated by maximum likelihood, a nonlinear optimisation procedure being necessary except in the case of pure autoregression. However, this is a mere technicality. The most difficult part of ARIMA model building is the selection of suitable values for d, p and q. The strategy suggested by Box and Jenkins (1976) consists of three elements: identification[1], estimation and diagnostic checking.

   At the identification stage, tentative values of d, p and q are chosen, primarily from a study of the sample autocorrelation function or *correlogram*. The sample autocorrelation at lag τ is defined as

$$r_\tau = c_\tau/c_0, \qquad \tau = 1, 2, 3, \ldots \tag{5.18}$$

where c_τ is the sample autocovariance,

$$c_\tau = T^{-1} \sum_{t=\tau+1}^{T} (y_t - \bar{y})(y_{t-\tau} - \bar{y}), \qquad \tau = 0, 1, \ldots \tag{5.19}$$

and $\bar{y}$ is the sample mean. If the process is stationary, the correlogram should have similar properties to the theoretical autocorrelation function of an ARMA model of a given order. If it is not stationary the correlogram will fail to damp down as τ increases and differencing must be carried out.

---

[1]Used in this context, the term 'identification' is unconnected with the concept of 'identification' or 'identifiability' referred to in Section 1.1 and defined in Section 3.6.

After estimating the chosen model, its adequacy can be assessed by testing whether the residuals are approximately random. This is diagnostic checking. The main test statistic employed is the *Box–Pierce Q-statistic*, which is defined as

$$Q = T \sum_{\tau=1}^{P} \tilde{r}_\tau^2 \qquad\qquad (5.20)$$

where $\tilde{r}_\tau$ is the τth sample autocorrelation in the residuals. If the model is correctly specified, Q has a χ^2 distribution with $P - p - q$ degrees of freedom. High values of Q lead to a rejection of the hypothesis of correct specification and if the model is rejected the whole sequence of identification, estimation and diagnostic checking is repeated until a satisfactory model is obtained.

Prediction

The main reason for building an ARIMA time series model is to make predictions of future observations. These predictions are built up recursively. Consider then the ARMA(1, 1) process,

$$y_t = \phi y_{t-1} + \epsilon_t + \theta \epsilon_{t-1}, \qquad t = 1, \ldots, T \qquad (5.21)$$

and suppose that ϕ and θ are given, and that the disturbances are known up to time T. The minimum mean square linear estimator of y_{T+1} is then

$$\tilde{y}_{T+1/T} = \phi y_T + \theta \epsilon_T \qquad\qquad (5.22)$$

This follows directly from writing down (5.21) for $t = T + 1$ and setting the unobserved disturbance, ϵ_{T+1}, equal to its expected value of zero. Further predictions are built up from the difference equations,

$$\tilde{y}_{T+l/T} = \phi \tilde{y}_{T+l-1/T}, \qquad l = 2, 3, \ldots \qquad (5.23)$$

with (5.22) providing the starting value. In practice, ϕ and θ will be estimated, and ϵ_t will be replaced by a residual. However, the principles upon which predictions are made remain the same.

Multivariate Time Series

The above concepts generalise to multivariate time series. A (zero mean) multivariate white noise process is defined as an $N \times 1$ vector, ϵ_t, with the following properties:

$$E(\epsilon_t) = 0, \qquad t = 1, \ldots, T \qquad (5.24a)$$

$$E(\epsilon_t \epsilon_t') = \Omega, \qquad t = 1, \ldots, T \qquad\qquad (5.24b)$$

$$E(\epsilon_t \epsilon_s') = 0, \qquad t \neq s \quad \text{and} \quad t, s = 1, \ldots, T \qquad (5.24c)$$

Such a process will be denoted by writing $\epsilon_t \sim WN(0, \Omega)$ cf. (5.1).

The multivariate, or *vector*, ARMA(p, q) process (2.16) may be written in the form

$$\Phi(L) y_t = \Theta(L) \epsilon_t \qquad\qquad (5.25)$$

where $\Phi(L)$ and $\Theta(L)$ are polynomial matrices in the lag operator defined in the same way as (5.16) and (5.17). The disturbance term, ϵ_t, is multivariate white noise and the variables in y_t are *jointly stationary if the roots of the determinantal polynomial, $|\Phi(L)|$,* lie outside the unit circle. Any stationary multivariate ARMA process may be written as an infinite MA process:

$$y_t = \Phi^{-1}(L) \Theta(L) \epsilon_t \qquad\qquad \checkmark \qquad (5.26)$$

The relationship between a pair of jointly stationary stochastic processes, y_{1t} and y_{2t}, is characterised by the *cross-covariance function*. If both processes have zero mean this is defined by

$$\gamma_\tau(y_1, y_2) = E(y_{1t} y_{2,t-\tau}), \qquad \tau = 0, \pm 1, \pm 2, \ldots \qquad (5.27)$$

The *cross-correlation function* is a dimensionless quantity obtained by dividing the cross-covariances function by the standard deviations of y_1 and y_2. Neither the cross-correlation function, nor the cross-covariance function is, in general, symmetric about $\tau = 0$.

Predictions can be made in the same way as for univariate processes. Thus for the vector AR(1) model,

$$y_t = \Phi y_{t-1} + \epsilon_t \qquad\qquad (5.28)$$

with $\epsilon_t \sim WN(0, \Omega)$, predictions are built up from the recursion

$$\tilde{y}_{t+l/T} = \Phi \tilde{y}_{T+l-1/T}, \qquad l = 1, 2, \ldots \qquad (5.29)$$

with $\tilde{y}_{T/T} = y_T$.

6. Econometric Models

Figure 1.4 shows the links between time series models and regression models and highlights the way in which they come together to produce dynamic econometric models. In univariate time series analysis a variable is modelled in terms of its own past, while in econometrics a model is set up on the basis of a behavioural relationship suggested by economic theory. The static regression models described in Chapter 2 provide the basic framework for handling behavioural relationships. However, although such models are appropriate when the system is in equilibrium, they are unable to capture the dynamic effects associated with the adjustment from one equilibrium position to another. Hence the introduction of concepts and modelling techniques derived from univariate time series analysis.

Chapters 7 and 8 contain a detailed treatment of dynamic single equation models. There are a number of different ways of constructing such models, but whatever approach is adopted, the main problem is to determine a suitable dynamic specification. Chapter 6 is a link between Chapters 7 and 8 and Chapter 2. The models here are static, but the disturbances are assumed to follow ARMA processes. Such models provide an introduction to estimation and testing in dynamic models and so it is possible to justify studying them on purely pedagogic grounds. However, the view taken here is that the choice between static and dynamic models is dictated partly by the level of temporal aggregation in the data. If observations are available annually, rather than on a quarterly or monthly basis, it is very difficult to model dynamic effects directly. In these circumstances, a static model may be the best approximation. The ARMA disturbance term not only reflects movements in omitted variables, it also captures the effects of aggregation in what is basically a dynamic relationship.

Systems of equations are also examined in Chapters 2, 6 and 8. The static seemingly unrelated regression equation (SURE) model is

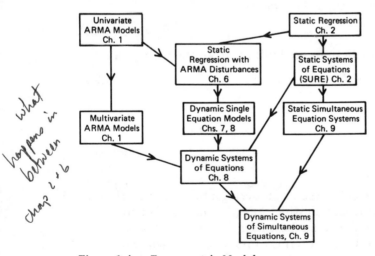

Figure 1.4   *Econometric Models*

introduced in Chapter 2. This can be used for modelling a set of observations over time. The observations may be on different variables or on the same variables measured for different individuals. The latter case represents a time series of cross-sections. The SURE model is extended in Chapters 6 and 8. Chapter 9 deals with simultaneous equation systems, both static and dynamic models being considered.

The relationship between time series analysis and econometrics may be viewed in a somewhat different light once the idea of multivariate time series modelling is introduced. A dynamic econometric model may be expressed as a multivariate ARMA model in which *a priori* restrictions have been imposed on the parameters on the basis of economic theory.

Example   Consider a simple dynamic Keynesian model in which consumption, c, is assumed to depend on national income, y, lagged one period, and investment, i, is determined by an 'accelerator' mechanism. Adding the national income accounting identity yields:

$$c_t = \alpha y_{t-1} + \epsilon_t \tag{6.1a}$$

$$i_t = \beta(c_{t-1} - c_{t-2}) + \eta_t \tag{6.1b}$$

$$y_t = c_t + i_t \tag{6.1c}$$

where ϵ_t and η_t are disturbance terms. Substituting for c_t and i_t in (6.1c) gives

$$y_t = \alpha y_{t-1} + \beta(c_{t-1} - c_{t-2}) + \zeta_t$$

where $\zeta_t = \epsilon_t + \eta_t$; and the three equations may now be written as a multivariate ARMA process (2.16), i.e.

$$\begin{pmatrix} c_t \\ i_t \\ y_t \end{pmatrix} = \begin{bmatrix} 0 & 0 & \alpha \\ \beta & 0 & 0 \\ \beta & 0 & \alpha \end{bmatrix} \begin{pmatrix} c_{t-1} \\ i_{t-1} \\ y_{t-1} \end{pmatrix} + \begin{bmatrix} 0 & 0 & 0 \\ -\beta & 0 & 0 \\ \beta & 0 & 0 \end{bmatrix} \begin{pmatrix} c_{t-2} \\ i_{t-2} \\ y_{t-2} \end{pmatrix} + \begin{pmatrix} \epsilon_t \\ \eta_t \\ \zeta_t \end{pmatrix} \tag{6.2}$$

Unrestricted, the matrices of AR coefficients, Φ_1, and Φ_2, would contain eighteen unknown parameters. In (6.2) there are only two.

The restrictions placed on (6.2) are fairly tight, and it could be argued that economic theory has no business to be so specific about the form of the dynamics. Thus in (6.1a) there is no *a priori* economic reason why c_{t-1}, c_{t-2}, y_t and y_{t-2} should not appear as explanatory variables. However, investment, whether current or lagged, could still be excluded quite reasonably on economic grounds.

Restrictions are also imposed on econometric models by the assumption that certain variables are exogenous. In a multivariate ARMA model all the variables are effectively endogenous. The advantage of assuming that some of the variables are exogenous is that the remaining variables can be modelled independently without any loss of information.

If the sample size is small, *a priori* restrictions are rather important, since it may be difficult to arrive at a suitable specification for a multivariate ARMA process on the basis of the data alone. Provided the restrictions are a reasonably accurate reflection of those in the real world, incorporating them in the model will yield a more powerful tool for forecasting. This theme is taken up and examined in the final chapter of the book. The forecasts of the endogenous variables in an econometric model are assumed to be based on predictions of the exogenous variables made on the basis of a suitable multivariate ARMA time series model. It is then argued that if a behavioural model is correctly specified, it will always forecast at least as well as a rival time series model. Indeed, this proposition is fairly obvious, although it appears that it is not always properly understood. The reason for the confusion, as well as for the quite legitimate controversy, concerns the validity of the economic theory used in constructing the model.

In the early seventies, there appeared a number of studies comparing the forecasts from econometric models with those from 'naive' time series models. A number of these were favourable to the time series approach. One explanation for such a finding could be the shaky foundation of the economic theory upon which the econometric models were built. A second explanation concerns inadequate dynamic modelling. Most of the economic theory brought to bear on such problems is static, rather than dynamic, and this, coupled with data and computing considerations, tended to result in the construction of models which were essentially static. One of the key contributions of time series analysis has been to focus attention on the dynamic aspects of these models, and this is why the incorporation of dynamics into econometric models forms one of the principal themes of this book.

Notes

Section 2   The concept of exogeneity introduced in the section on data generation is based on the definition of 'weak exogeneity' in Engle *et al.* (1979).

Section 3   Useful references for matrix algebra include Theil (1971, Ch. 1) and Johnston (1972, Ch. 4). Theil (1971, Ch. 2) also presents some of the basic statistical theory.

Section 4   The classic reference is Cramér (1946). Theil (1971, pp. 357–378) provides a review.

Section 6   Prior beliefs in parameter values can be incorporated into an econometric model by adopting a Bayesian approach; see, for example, Leamer (1978).

Exercises

1. Consider a random sample of size T from a distribution with mean μ and variance σ^2. Show that the estimator $\hat{\mu} = 0$ has a smaller MSE than the sample mean if $T \cdot \mu^2 < \sigma^2$.

2. By substituting repeatedly for lagged values of y_t, find the coefficients in the infinite MA representation of the ARMA(1, 1) process, (5.21). Check your answer using the lag operator. State any assumptions you make.

3. Derive expressions for the variance and autocorrelations of an MA(2) model. Explain how you would predict 1, 2 and 3 periods ahead.

4. What is the effect of taking first differences in the model,

$$y_t = \alpha + \beta t + \epsilon_t$$

where α and β are parameters? Can the resulting model be expressed as an AR process?

5. Find the mean of y_t in the model,

$$y_t = \alpha + \phi y_{t-1} + \epsilon_t$$

where α is a parameter and $|\psi| < 1$.

6. Is an AR(2) process stationary if $\phi_1 = 1.7$ and $\phi_2 = -0.72$?

7. Derive the autocorrelation function for the seasonal AR model

$$(1 - \phi L)(1 - \phi^* L^4) y_t = \epsilon_t$$

where $|\phi| < 1$ and $|\phi^*| < 1$.

2
Regression

1. Linear Regression Models

The starting point in the development of econometric theory is the linear multiple regression model. This may be written

$$y_t = \beta_1 x_{1t} + \beta_2 x_{2t} + \cdots + \beta_k x_{kt} + u_t, \qquad t = 1, \ldots, T \qquad (1.1)$$

where y_t is the tth observation on the *dependent* variable, x_{it} is the tth observation on the ith *independent* variable, u_t is a disturbance term and $\beta_1, \ldots, \beta_k$ are unknown parameters. The first independent variable, x_{1t}, will usually be a constant, i.e. $x_{1t} = 1$ for all t. In this case any systematic variation in y_t depends on the *explanatory* variables, $x_2, \ldots, x_k$.

The explanatory variables are assumed to be *exogenous*. For statistical purposes they can therefore be regarded as being *fixed* in repeated samples. As a result, the stochastic properties of the disturbances can be defined unconditionally rather than conditional on $x_2, \ldots, x_k$.

The disturbance term cannot be observed directly. By assuming that $E(u_t) = 0$ for all t, it follows that

$$E(y_t) = \beta_1 x_{1t} + \cdots + \beta_k x_{kt}, \qquad t = 1, \ldots, T \qquad (1.2)$$

This is the systematic part of the model.

Vector and Matrix Notation

Let $x_t = (x_{1t} x_{2t} \ldots x_{kt})'$ be the $k \times 1$ vector of observations on the independent variables at time t, and let $\beta = (\beta_1, \ldots, \beta_k)'$ be the corresponding $k \times 1$ vector of regression parameters. A more concise formulation of the model in terms of *vector notation* is given by

$$y_t = x_t'\beta + u_t, \qquad t = 1, \ldots, T \qquad (1.3)$$

Rather than consider each of the T equations defined by (1.1) separately, all T sets of observations on the dependent and independent variables may be written together using *matrix notation*. Let $y = (y_1, \ldots, y_T)'$ be the $T \times 1$ vector of observations on the dependent variable, and define the $T \times k$ matrix

$$X = \begin{pmatrix} 1 & x_{21} & \cdots & x_{k1} \\ 1 & x_{22} & \cdots & x_{k2} \\ \vdots & \vdots & & \vdots \\ 1 & x_{2T} & \cdots & x_{kT} \end{pmatrix} = \begin{pmatrix} x_1' \\ x_2' \\ \vdots \\ x_T' \end{pmatrix}$$

If $u = (u_1, \ldots, u_T)'$ denotes the $T \times 1$ vector of disturbances, the complete model may be expressed as

$$y = X\beta + u \tag{1.4}$$

The Classical Assumptions

The classical linear regression model satisfies the following assumptions:

   (i) the explanatory variables are fixed;
  (ii) the rank of X is equal to k;
 (iii) the disturbances are uncorrelated, each having zero mean and a constant, but finite, variance.

The last assumption will be denoted explicitly by writing the disturbance term as ϵ_t. In matrix notation, (1.4) becomes

$$y = X\beta + \epsilon, \tag{1.5}$$

where $\epsilon = (\epsilon_1, \ldots, \epsilon_T)'$. The properties of the disturbance term may then be expressed by writing:

$$E(\epsilon) = 0, \quad \text{Var}(\epsilon) = E(\epsilon\epsilon') = \sigma^2 I_T. \tag{1.6}$$

Note that condition (iii) does not specify the full distribution of the disturbances, nor does it assume that they are independent over time. Taking them to be uncorrelated is a weaker requirement than independence, although it does imply independence under normality.

Functional Form

The assumption of linearity may be less restrictive than appears at first sight. Consider, for example, the Cobb–Douglas production function

$$Q = AK^\beta L^\gamma \tag{1.7}$$

where Q is quantity produced, K is capital, L is labour, and A, β and γ are parameters. From the point of view of economic theory (1.7) has a number of attractive properties, which is why it is often chosen as the basis for applied work. In particular β and γ have a direct interpretation as the elasticities of output with respect to capital and labour respectively.

Suppose that the parameters in (1.7) are to be estimated on the basis of a set of observations on Q, K and L. A stochastic formulation of the exact relationship in (1.7) must be set up and one way of doing this is to posit a disturbance term that enters *multiplicatively*. The model to be estimated is then

$$Q_t = AK_t^\beta L_t^\gamma \exp(\epsilon_t), \qquad t = 1, \ldots, T \tag{1.8}$$

where ϵ_t is the disturbance term. This is *intrinsically linear* since a *double-log transformation* yields

$$\log Q_t = \alpha + \beta \log K_t + \gamma \log L_t + \epsilon_t, \qquad t = 1, \ldots, T \tag{1.9}$$

where $\alpha = \log A$. Although the model is nonlinear in variables, it is linear in parameters. It may be cast in the form (1.1) by defining $x_{2t} = \log K_t$, $x_{3t} = \log L_t$ and $y_t = \log Q_t$.

On the other hand, suppose there were strong *a priori* grounds for imposing an additive disturbance term on (1.7), i.e.,

$$Q_t = AK_t^\beta L_t^\gamma + \epsilon_t, \qquad t = 1, \ldots, T, \tag{1.10}$$

There is no way in which (1.10) may be transformed into a linear model, and in these circumstances estimation must be carried out by the techniques of nonlinear regression. This topic is pursued in Chapters 3 and 4.

The constant elasticity of substitution (CES) production function is rather more general than the Cobb–Douglas form. It is specified by the exact functional relationship

$$Q = \exp(\alpha)\{(1 - \delta)L^{-\gamma} + \delta K^{-\gamma}\}^{-\beta/\gamma} \tag{1.11}$$

where α, β, γ and δ are parameters. In this case, the model to be estimated will be nonlinear irrespective of whether the disturbance term is multiplicative or additive. However, multiplicative disturbances are somewhat easier to handle, since taking logarithms of both sides of the equation yields

$$\log Q_t = \alpha - (\beta/\gamma)\log\{(1 - \delta)L^{-\gamma} + \delta K^{-\gamma}\} + \epsilon_t \tag{1.12}$$

The model is still nonlinear when viewed as a whole, but for given values of γ and δ it is linear with respect to α and β. The implications for estimation are discussed in Section 4.4.

Although the logarithmic transformation is widely used, it is not the only transformation available. It is sometimes useful to let a variable enter in reciprocal form, i.e. $x_t = 1/X_t$, where X_t is the original variable. In fact, the same variable may enter an equation in several different forms, a classic example being polynomial regression where

$$y_t = \alpha + \beta_1 X_t + \beta_2 X_t^2 + \cdots + \beta_p X_t^p + \epsilon_t \qquad (1.13)$$

Qualitative Variables

The explanatory variables in a regression model may be qualitative or quantitative. Suppose that a production function is being fitted to a set of time series observations as in (1.9), but that at a certain point in time, t^*, it is known that a technological innovation was introduced. This may be handled by means of a *dummy* variable, D_t, which takes the value zero for $t < t^*$ and unity thereafter. Equation (1.9) therefore becomes

$$\log Q_t = \log A + \delta D_t + \beta \log K_t + \gamma \log L_t + \epsilon_t, \qquad t = 1, \ldots, T$$
$$(1.14)$$

For $t < t^*$, the parameter associated with the constant term in the model is $\log A$, while for $t \geqslant t^*$ it is $\log A + \delta$. The parameter δ therefore measures the effect of the technological innovation on production.

Note that only one dummy variable is needed in this example. As a general rule, if a qualitative variable can fall in any one of m classes, $m - 1$ dummies are needed in addition to the constant term. Alternatively the constant may be dropped and m dummy variables defined.

From the point of view of regression theory, dummy variables may be treated in exactly the same way as quantitative variables. No new issues are raised. This is not the case with qualitative dependent variables. A typical example might concern a model to explain car ownership, and the dependent variable is then unity when a family owns a car and zero otherwise. Least squares will not usually be an appropriate technique for estimating a model of this kind, and a discussion of what should be done is deferred until Section 3.4.

2. Least Squares Estimation

The linear model (1.1) contains k unknown regression parameters, $\beta_1, \ldots, \beta_k$, and these are to be estimated from the T sets of observations. The model contains one other unknown parameter, namely σ^2. However, the estimation of σ^2 is of secondary importance. To some extent this is also true of β_1, the parameter associated with the constant term, but for the moment all the regression parameters will be viewed in exactly the same way.

Suppose $\hat{\beta}$ is an estimator of the parameter vector β. Corresponding to this estimator are a set of T *residuals*, defined by $y_t - x_t'\hat{\beta}$ for $t = 1, \ldots, T$. The *ordinary least squares* estimator is obtained by choosing $\hat{\beta}$ such that the residual sum of squares, $\Sigma(y_t - x_t'\hat{\beta})^2$, is a minimum.

In order to derive an expression for the OLS estimator, the *sum of squares function*

$$S(\beta) = \sum_{t=1}^{T} (y_t - x_t'\beta)^2 \qquad (2.1)$$

is defined. In the context of (2.1), β is regarded not as a vector of fixed values, but as a vector of variables. The observations, $y_1, \ldots, y_T$, are given, and the object of the exercise is to minimise $S(\beta)$ with respect to the elements of β. Although the symbol β is also used to denote the true set of parameter values, its use as an argument in $S(.)$ should not lead to any confusion.

The sum of squares function may also be expressed in matrix form as

$$S(\beta) = (y - X\beta)'(y - X\beta) \qquad (2.2)$$

Differentiating with respect to β gives

$$\frac{\partial S}{\partial \beta} = -2X'(y - X\beta)$$

Since

$$\frac{\partial S}{\partial \beta \, \partial \beta'} = 2X'X$$

is a p.d. matrix, the least squares estimator, b, is obtained by solving the k *normal equations*,

$$X'Xb = X'y \qquad (2.3)$$

Pre-multiplying both sides of (2.3) by the inverse of $X'X$ leads to the expression:

$$b = (X'X)^{-1}X'y \qquad (2.4)$$

The relevance of the second of the classical assumptions should be apparent from this formula. Unless the matrix X is of full rank, i.e. of rank k, the cross-product matrix will be singular and there will be no unique solution to the normal equations. Having $T \geqslant k$ is a necessary, but not a sufficient condition for X to be of full rank.

The Residual Sum of Squares and the Coefficient of Multiple Correlation

The OLS residuals are defined by

$$e_t = y_t - \hat{y}_t = y_t - x_t'b, \qquad t = 1, \ldots, T. \tag{2.5}$$

These residuals are orthogonal to the independent variables. If $e = (e_1, \ldots, e_T)'$, then $X'e = X'(y - Xb) = X'y - X'Xb = 0$, the last step following directly from (2.3). Using this result,

$$y'y = (\hat{y} + e)'(\hat{y} + e) = (Xb + e)'(Xb + e)$$

$$= b'X'Xb + e'e + 2b'X'e$$

$$= b'X'Xb + e'e \tag{2.6}$$

This may be re-written as

$$\Sigma y_t^2 = \Sigma \hat{y}_t^2 + \Sigma e_t^2 \tag{2.7}$$

where the summation is from $t = 1$ to $t = T$. Thus the total sum of squares may be decomposed into two parts, the explained sum of squares and the residual sum of squares. This last term, which will be denoted by SSE, may be expressed as

$$SSE = e'e = (y - Xb)'(y - Xb)$$

$$= y'y + b'X'Xb - 2b'X'y$$

$$= y'y - b'X'y \tag{2.8}$$

Formula (2.8) is convenient for calculation.

Similar results may be derived by working with the observations in deviation from the mean form. If the first element in x_t is the constant term, x_t may be partitioned as $x_t = (1\ x_{*t}')'$. Similarly β and b may be partitioned as $(\beta_1, \beta_*')'$ and $(b_1, b_*')'$, where β_* and b_* are the $(k - 1) \times 1$ vectors of coefficients associated with the explanatory variables. Thus

$$y_t = x_t'b + e_t = b_1 + x_{*t}'b_* + e_t, \qquad t = 1, \ldots, T \tag{2.9}$$

Summing over all observations and dividing by T gives

$$\bar{y} = b_1 + \bar{x}_{*t}'b_* \tag{2.10}$$

the mean of the residuals, $\bar{e}$, being equal to zero in view of the orthogonality of X and e. Subtracting (2.10) from (2.9) leaves

$$y_t - \bar{y} = (x_{*t} - \bar{x}_*)'b_* + e_t, \qquad t = 1, \dots, T \qquad (2.11)$$

The residuals therefore remain the same when the fitted equation is written in deviation from the mean form.

If the tth element in y is taken to be $y_t - \bar{y}$, and X is redefined as a $T \times (k-1)$ matrix in which the tth row is $(x_{*t} - \bar{x}_*)'$, formulae (2.6) and (2.8) remain valid when b is replaced by b_*. This follows immediately from (2.11) which may be written in matrix form as $y = X'b_* + e$. It can also be shown that (2.4) yields the estimator b_* when X and y are in deviation from the mean form.

Expression (2.6) can now be interpreted as

$$\sum (y_t - \bar{y})^2 = \sum (\hat{y}_t - \bar{y})^2 + \sum e_t^2 \qquad (2.12)$$

The decomposition in (2.12) is rather more useful than that in (2.7). The constant term in the regression equation cannot be regarded as 'explaining' the behaviour of y_t in any meaningful sense. It is therefore more sensible to measure 'goodness of fit' in terms of the deviations of the observations from their sample mean. This is reflected in the *coefficient of multiple correlation*

$$R^2 = 1 - \frac{\sum e_t^2}{\sum (y_t - \bar{y})^2} = \frac{\sum (\hat{y}_t - \bar{y})^2}{\sum (y_t - \bar{y})^2} \qquad (2.13)$$

which lies in the range $0 \leqslant R^2 \leqslant 1$, and may be interpreted as measuring the proportion of the variance of y_t explained by the regression. The closer R^2 is to unity, the better the fit of the regression model.

One objection to the use of R^2 is that it cannot decrease, and will usually increase, if additional variables are introduced into the set of regressors. The corrected R^2, $\bar{R}^2$, makes some allowance for the number of independent variables in the model. It is defined in the following way;

$$\bar{R}^2 = 1 - \frac{\sum e_t^2/(T-k)}{\sum (y_t - \bar{y})^2/(T-1)}$$

Therefore

$$\bar{R}^2 = 1 - \{(T-1)/(T-k)\}(1 - R^2) \qquad (2.14)$$

Restricted Least Squares

It is sometimes desirable to impose constraints on the coefficients in a regression model. Thus suppose we have a Cobb—Douglas production function, (1.8), and it is felt reasonable, on *a priori* grounds, to assume constant returns to scale. This being the case, $\beta + \gamma = 1$, and such a constraint may easily be incorporated in the model by setting $\gamma = 1 - \beta$, and re-arranging (1.9) to give

$$\log(Q_t/L_t) = \alpha + \beta \log(K_t/L_t) + \epsilon_t \tag{2.15}$$

It is not usually possible to solve the problem of imposing constraints on a model quite so easily. However, a general solution may be obtained by the use of Lagrange multipliers. Suppose there are $m < k$ linear constraints. These may be written in the form

$$R\beta = r \tag{2.16}$$

where R is an $m \times k$ matrix of rank m, and r is an $m \times 1$ vector. Thus if the constraints are $\beta_1 + \beta_2 = 3$ and $\beta_2 - \beta_3 + \beta_k = 0$, they may be expressed as

$$\begin{bmatrix} 1 & 1 & 0 & 0 & \dots & 0 & 0 \\ 0 & 1 & -1 & 0 & \dots & 0 & 1 \end{bmatrix} \begin{bmatrix} \beta_1 \\ \beta_2 \\ \vdots \\ \beta_k \end{bmatrix} = \begin{bmatrix} 3 \\ 0 \end{bmatrix} \tag{2.17}$$

The *restricted least squares* estimator of β is obtained by minimising the function

$$S(\beta, \lambda) = (y - X\beta)'(y - X\beta) - \lambda(R\beta - r) \tag{2.18}$$

with respect to β and the $m \times 1$ vector of Lagrange multipliers, λ. This yields

$$\frac{\partial S}{\partial \beta} = -2X'y + 2X'X\beta + R'\lambda = 0 \tag{2.19}$$

$$\frac{\partial S}{\partial \lambda} = -r + R\beta = 0 \tag{2.20}$$

Premultiplying both sides of (2.19) by $R(X'X)^{-1}$ gives

$$-2R(X'X)^{-1}X'y + 2R\beta + R(X'X)^{-1}R'\lambda = 0$$

Since, from (2.20), $R\beta = r$, and $R(X'X)^{-1}R'$ is nonsingular,

$$\lambda = -2[R(X'X)^{-1}R']^{-1}[r - R(X'X)^{-1}X'y]$$

Substituting this expression into (2.19) and premultiplying by $(X'X)^{-1}$ gives the restricted least squares estimator of β,

$$b\dagger = b + (X'X)^{-1}R'[R(X'X)^{-1}R']^{-1}(r - Rb) \tag{2.21}$$

If the OLS estimator actually satisfied the constraints exactly, it can be seen from (2.21) that b and $b^\dagger$ would be identical.

Except when $b = b^\dagger$, the sum of squares of the residuals associated with $b^\dagger$ must exceed the sum of squares of the OLS residuals. The following result may be shown to hold

$$(y - Xb^\dagger)'(y - Xb^\dagger) = (y - Xb)'(y - Xb)$$
$$+ (r - Rb)'[R(X'X)^{-1}R']^{-1}(r - Rb);$$

$$(2.22)$$

see Theil (1971, Ch. 1).

3. Properties of the Ordinary Least Squares Estimator

The mean and variance of the OLS estimator may be derived by substituting for y in (2.4). This yields

$$b = (X'X)^{-1}X'(X\beta + u) = \beta + (X'X)^{-1}X'u \tag{3.1}$$

If $E(u) = 0$, b is unbiased since, with X fixed,

$$E(b) = \beta + (X'X)^{-1}X'E(u) = \beta \tag{3.2}$$

The variance is then

$$\mathrm{Var}(b) = E[(b - \beta)(b - \beta)'] = (X'X)^{-1}X'E(uu')X(X'X)^{-1}$$

$$(3.3)$$

If the disturbance term obeys the classical assumptions, $E(uu') = E(\epsilon\epsilon') = \sigma^2 I$ and (3.3) becomes

$$\mathrm{Var}(b) = \sigma^2(X'X)^{-1} \tag{3.4}$$

If the disturbances are normally distributed it can be shown that b is the MVUE of β; see Section 3.4. However, if only the first two moments of the distribution of ϵ are specified, as in (1.6), a somewhat weaker result can be shown to hold. The OLS estimator is linear in the observations, and within the class of *linear* unbiased estimators it has minimum variance. The proof of this result, which is known as the Gauss—Markov theorem, is set out below.

Without knowing the form of the distribution of ϵ, very little can be said about the distribution of b in small samples. Nevertheless an asymptotic distribution for b may be derived under suitable regularity conditions. The theory extends to dynamic models. The key results are presented in the last two sub-sections.

The Gauss–Markov Theorem

When a regression model satisfies the classical assumptions, the OLS estimator is the best linear unbiased estimator (BLUE) of β in the sense that the *covariance matrix of any other linear unbiased estimator exceeds that of b by a positive semi-definite (p.s.d.) matrix*. The proof of the theorem is as follows. Let $\hat{b} = D^*y$, where D^* is a $k \times T$ matrix which does not depend on y. Hence $\hat{b}$ is a linear estimator of β. Define $D = D^* - (X'X)^{-1}X'$, so that the estimator may be written

$$\hat{b} = [D + (X'X)^{-1}X']y$$
$$= [D + (X'X)^{-1}X'](X\beta + \epsilon)$$
$$= (DX + I)\beta + [D + (X'X)^{-1}X']\epsilon$$

If $\hat{b}$ is to be unbiased, it is necessary to have $DX = 0$. With this condition imposed, its covariance matrix becomes

$$\text{Var}(\hat{b}) = E[(b - \beta)(b - \beta)']$$
$$= [D + (X'X)^{-1}X']E(\epsilon\epsilon')[D' + X(X'X)^{-1}]$$
$$= \sigma^2[DD' + DX(X'X)^{-1} + (X'X)^{-1}X'D' + (X'X)^{-1}]$$

Since $DX = 0$, this expression simplifies to

$$\text{Var}(\hat{b}) = \sigma^2 DD' + \sigma^2(X'X)^{-1} \tag{3.5}$$

Comparing (3.5) with (3.4) shows that $\text{Var}(\hat{b})$ exceeds $\text{Var}(b)$ by $\sigma^2 DD'$, which is a p.s.d. matrix.

The implications of the Gauss–Markov theorem are brought out more clearly by considering the estimation of an arbitrary linear combination of the elements of β. If w is a $k \times 1$ vector of fixed weights, the quantity to be estimated is $w'\beta$. The corresponding combination of the elements of b is unbiased, and its variance is given by

$$\text{Var}(w'b) = w'E[(b - \beta)(b - \beta)']w = \sigma^2 w'(X'X)^{-1}w \tag{3.6}$$

Any other linear, unbiased estimator of β, $\hat{b}$, may be used to form an unbiased estimator of $w'\beta$ but, from (3.5), its variance will be

$$\text{Var}(w'\hat{b}) = \sigma^2 w'DD'w + \sigma^2 w'(X'X)^{-1}w \tag{3.7}$$

Since DD' is p.s.d., the variance of $w'b$ cannot exceed the variance of any other linear, unbiased estimator.

As a special case, w may be chosen so that all of its elements are zero, except the ith which is unity. The above result then shows that the variance of any element in the OLS estimator cannot exceed the variance of the corresponding element in any other linear unbiased estimator of β.

It is important to appreciate that the Gauss–Markov theorem only proves that OLS is best within the class of *linear* estimators. A non-linear estimator will invariably be more time consuming to compute, but the gain in statistical efficiency may justify the extra expense. As an example, suppose that the disturbance term is known to follow a double exponential, or Laplace distribution. This distribution is symmetric and unimodal, and has a finite variance, but in large samples the variance of the OLS estimator of a given element in β is twice as large as the variance of the most efficient (nonlinear) estimator; see Section 3.7.

Distribution of the OLS Estimator

The assumption that the disturbances are normally distributed enables the exact small sample distribution of the OLS estimator to be obtained. From (3.1) it will be seen that each element of b is a linear combination of normally distributed variables. Hence b has a multivariate normal distribution. Since it has already been shown that b is unbiased with covariance matrix $\sigma^2(X'X)^{-1}$ under the classical assumptions, we may now write

$$b \sim N[\beta, \sigma^2(X'X)^{-1}] \tag{3.8}$$

If the disturbances are non-normal, the small sample distribution of b is, in general, unknown. However, if the sample size is large, b may be treated as being approximately normally distributed. This result is essentially a generalisation of the central limit theorem for a sample mean.

Consider a random sample of T observations, $y_1, \ldots, y_T$, drawn from a distribution with a mean, μ, and finite variance, σ^2. The sample mean, $\bar{y}$, is actually the OLS estimator of μ, and in Section 1.4, it was stated that the central limit theorem implies that $\sqrt{T}(\bar{y} - \mu)$ converges in distribution to a normal variate with mean zero and variance σ^2. A similar result may be demonstrated for the OLS estimator in the classical regression model, (1.5), provided that

$$\lim_{T \to \infty} T^{-1}X'X = Q \tag{3.9}$$

where Q is a p.d. matrix. This being the case, it can be shown that $\sqrt{T}(b - \beta)$ has a limiting normal distribution with a mean of 0 and a covariance matrix, $\sigma^2 Q^{-1}$. Following on from the definition in (1.4.14), b is said to be asymptotically normal with a mean of β and a covariance matrix, $\sigma^2 T^{-1}Q^{-1}$. An alternative way of expressing this result is to write

$$b \sim AN[\beta, \sigma^2(X'X)^{-1}] \tag{3.10};$$

cf. (3.8). In doing this, it is to be understood that $\sigma^2(X'X)^{-1}$ is an approximation to the asymptotic covariance matrix,

$$\text{Avar}(b) = \sigma^2 T^{-1} Q^{-1} \tag{3.11}$$

The conditions embodied in (3.9) will be referred to as the *standard regularity conditions* for the explanatory variables. They are somewhat restrictive for time series regression in that they do not permit the variables to exhibit trends. As a simple illustration consider a model in which a time trend is fitted through the origin, i.e.

$$y_t = \beta x_t + \epsilon_t \tag{3.12}$$

with $x_t = t$ for $t = 1, \ldots, T$. Since

$$T^{-1}X'X = T^{-1}\sum x_t^2 = T^{-1}\sum t^2 = (T+1)(2T+1)/6$$

the limiting distribution of $\sqrt{T}(b - \beta)$ is degenerate. Fortunately, cases like this pose no problem from the theoretical point of view. The limiting distribution of $\sqrt{\sum x_t^2}(b - \beta)$ is $N(0, \sigma^2)$ and this provides the basis for the assertion that b will be approximately normal with a mean of β and a variance of $\sigma^2/\sum x_t^2$. The conditions which must be imposed on the explanatory variables for this device to be used are relatively weak, and are known as Grenander's conditions. One important feature of these conditions is that they ensure that b is consistent.

From the practical point of view, the regularity conditions are not particularly important. Unless the explanatory variables are known to have been generated in a particular way, it is impossible to check whether or not they hold for a particular sample. Imposing regularity conditions is a formal requirement needed to derive the appropriate limiting distribution. Stating that a particular estimator has a certain limiting distribution then acts as a guide indicating the approximate results which can be expected in finite samples. That the conditions could be conceived as holding as $T \to \infty$, however, is neither necessary nor sufficient for the asymptotic approximation to be a reasonable one in practice.

Stochastic Regressors: The Mann–Wald Theorem

The discussion of OLS up to this point has been based on the assumption that the regressors are non-stochastic. As pointed out in Section

1, relaxing this assumption makes no difference to small sample properties if the disturbances are distributed independently of X. Furthermore if the standard regularity conditions are modified to

$$\text{plim } T^{-1}X'X = Q$$

the asymptotic theory goes through exactly as before. However, in many models the disturbance term is distributed independently of the current observations on the explanatory variables, but is not independent of past and/or future observations. If the disturbances are generated as an independent sequence, the asymptotic properties of the OLS estimator may be derived using a result due to Mann and Wald. This result is an extremely powerful one, and it will become apparent later that it is applicable in a much wider range of situations than the context of this section would suggest.

Mann—Wald Theorem   Let z_t denote a $k \times 1$ vector of random variables with the property that

$$\text{plim } T^{-1} \sum_{t=1}^{T} z_t z_t' = Q \tag{3.13}$$

where Q is a p.d. matrix. Let ϵ_t be a sequence of independently and identically distributed random variables with zero mean, finite variance, σ^2, and finite moments of every order. Finally assume that $E(z_t \epsilon_t) = 0$ for all $t = 1, \ldots, T$. Then

$$\text{(i)} \quad \text{plim } T^{-1} \sum z_t \epsilon_t = 0 \tag{3.14a}$$

and

$$\text{(ii)} \quad \sqrt{1/T} \sum z_t \epsilon_t \xrightarrow{L} N(0, \sigma^2 Q) \tag{3.14b}$$

Now consider a specific application of the Mann—Wald theorem. In

$$y_t = \alpha y_{t-1} + \beta x_t + \epsilon_t, \qquad t = 1, \ldots, T \tag{3.15}$$

it is no longer the case that the disturbances are independent of all the explanatory variables since $E(y_t \epsilon_{t-\tau}) \neq 0$ for $\tau \geq 1$. The OLS estimators of α and β will, in general, be biased in small samples, and it becomes very difficult to say anything positive about small sample properties. However, if the model is written in the form

$$y_t = z_t' \gamma + \epsilon_t, \qquad t = 1, \ldots, T \tag{3.16}$$

where $z_t = (y_{t-1} x_t)'$ and $\gamma = (\alpha, \beta)'$, the Mann—Wald result may be applied directly. Cramér's theorem, (1.4.17) may then be invoked to derive the asymptotic distribution of the OLS estimator, $c = (a, b)'$. Since $\sqrt{T}(c - \gamma)$ has a limiting multivariate normal distribution with mean 0 and variance $Q^{-1}\sigma^2 Q \cdot Q^{-1} = \sigma^2 Q^{-1}$, it follows that

$$c \sim AN[\gamma, \sigma^2(\textstyle\sum z_t z_t')^{-1}] \quad \text{approximately} \tag{3.17}$$

multivariate

The regularity conditions, (3.13), can be shown to hold if $|\alpha| < 1$ and both $\sum x_t^2/T$ and $\sum x_t x_{t-1}/T$ converge to finite positive numbers.

The implication of this result is that the OLS estimator may be treated as having the usual properties provided that the sample size is large. Furthermore, this conclusion generalises to any dynamic model of the form,

$$y_t = \alpha_1 y_{t-1} + \cdots + \alpha_r y_{t-r} + x_t'\beta + \epsilon_t \tag{3.18}$$

4. Generalised Least Squares

Under the classical assumptions, the disturbances are uncorrelated and have constant variance. These assumptions are fairly restrictive, and a natural extension is to allow the covariance matrix of u to have a more general form. Thus (1.6) is modified to

$$E(u) = 0, \quad E(uu') = \sigma^2 V \tag{4.1}$$

where σ^2 is an unknown positive, finite parameter, and V is a $T \times T$ symmetric, p.d. matrix. The assumption of a zero mean for u is retained, as are the classical assumptions relating to the non-stochastic nature of the explanatory variables and the rank of X. Splitting up the covariance matrix into two terms, σ^2 and V, is essentially arbitrary. However, it is a natural division for many models.

The GLS Estimator and its Properties

If V is *known*, obtaining a BLUE of β is not difficult in principle. Since V is symmetric and p.d., its inverse is also symmetric and p.d., and so there exists a nonsingular $T \times T$ matrix L such that $L'L = V^{-1}$. Pre-multiplying both sides of (1.4) by L gives

$$Ly = LX\beta + Lu \tag{4.2}$$

The transformed observations on the dependent and independent variables are given by the vector Ly and the matrix LX respectively. The new disturbance vector is Lu, and since L is fixed, the expectation of Lu is zero, while its covariance matrix is

$$\text{Var}(Lu) = E(Luu'L') = \sigma^2 LVL'$$
$$= \sigma^2 L(L'L)^{-1}L' = \sigma^2 I \qquad (4.3)$$

The transformed disturbance vector therefore satisfies the classical assumptions, and so OLS applied to the transformed observations will yield the BLUE of β.

Replacing y and X by Ly and LX respectively in the OLS formula, (2.4), gives

$$\tilde{b} = (X'L'LX)^{-1}X'L'Ly = (X'V^{-1}X)^{-1}X'V^{-1}y \qquad (4.4)$$

This is known as the *generalised least squares* (GLS) estimator of β. Since the model obeys the classical assumptions it follows from (3.4) that

$$\text{Var}(\tilde{b}) = \sigma^2(X'L'LX)^{-1} = \sigma^2(X'V^{-1}X)^{-1} \qquad (4.5)$$

Furthermore, because $\tilde{b}$ is a linear estimator, it will have a multi-variate normal distribution when the disturbances are normal.

Example 1   Consider the bivariate regression model,

$$y_t = \alpha + \beta x_t + u_t, \qquad t = 1, \ldots, T \qquad (4.6)$$

in which $E(u_t) = 0$ for all t, $E(u_t u_s) = 0$ for all $t \neq s$, and $\text{Var}(u_t) = \sigma_t^2 = \sigma^2 x_t^2$ for all t. Since the variances are unequal, the disturbances are said to be heteroscedastic. However, because the disturbances are serially uncorrelated, V is diagonal. The transformation matrix, L, is also diagonal with the tth diagonal element given by $1/x_t$. The GLS estimators of α and β are therefore obtained by applying OLS to the model

$$y_t/x_t = \alpha x_t^{-1} + \beta + u_t/x_t$$

i.e. y_t/x_t is regressed on $1/x_t$.

In the above example, GLS reduces to *weighted least squares* (WLS), with the criterion to be minimised given by

$$S(\beta) = \sum_{t=1}^{T} \sigma_t^{-2}(y_t - x_t'\beta)^2$$

More generally, $S(\beta)$ is the quadratic form

$$S(\beta) = (Ly - LX\beta)'(Ly - LX\beta) = (y - X\beta)'V^{-1}(y - X\beta) \qquad (4.7)$$

Properties of the OLS Estimator

The assumption that $E(u) = 0$ ensures that the OLS estimator, b, is unbiased. However, from (3.3), its covariance matrix is:

$$\mathrm{Var}(b) = \sigma^2(X'X)^{-1}X'VX(X'X)^{-1} \qquad (4.8)$$

Since the GLS estimator is the BLUE of β, it follows from the Gauss–Markov theorem, that (4.8) will exceed (4.5) by a p.s.d. matrix. In general, therefore, the OLS estimator will be less efficient than GLS. The extent of the inefficiency depends on the structure of both V and X.

Example 2   Consider the model, (4.6), with $x_t = t$, for $t = 1,\ldots,$ T. Figures for the variance of the GLS estimator of β divided by the variance of the corresponding OLS estimator are shown in table 2.1 for two types of heteroscedasticity. Details of the calculations will be found in Geary (1966).

Table 2.1 $\mathrm{Var}(\tilde{b})/\mathrm{Var}(b)$ for Model (4.6)

T	10	20	30	40	50
$\sigma_t^2 = \sigma^2 x_t$	0.72	0.64	0.60	0.58	0.57
$\sigma_t^2 = \sigma^2 x_t^2$	0.41	0.33	0.30	0.28	0.27

Feasible GLS

In (4.6) the covariance matrix of the disturbances was completely specified up to a factor of proportionality. In other words V was known. However, setting $\sigma_t^2 = \sigma^2 x_t^2$ imposes a rather special structure on the variances, and if heteroscedasticity is to be handled in a more general fashion, greater flexibility is required. One possibility is to go to the opposite extreme and simply let V be a diagonal matrix. Unfortunately this approach is not viable, since the covariance matrix then contains T unknown parameters and these cannot be sensibly estimated from T observations. Some structure is therefore necessary, and the usual solution is to let σ_t^2 depend on a small number of unknown parameters. Estimators of these parameters can then be used to construct a *feasible* GLS estimator of β.

Example 3   Consider model (4.6) with

$$\sigma_t^2 = \delta_1 + \delta_2 x_t, \qquad t = 1, \ldots, T \qquad (4.9)$$

where δ_1 and δ_2 are parameters. Suitable estimators, $\hat{\delta}_1$ and $\hat{\delta}_2$, may be formed from the OLS residuals by regressing e_t^2 on x_t. A feasible GLS estimator may then be based on $\hat{\sigma}_t^2 = \hat{\delta}_1 + \hat{\delta}_2 x_t$, and computed by WLS.

Serial correlation may be handled in a similar way, with the V matrix made to depend on a limited number of unknown parameters.

If the parameters in a V matrix can be estimated consistently, the feasible GLS estimator will, under suitable regularity conditions, have the same asymptotic distribution as the GLS estimator; see Theil (1971, Ch. 8).

5. Prediction

In the context of a regression model, a *prediction* is an estimate of a future value of the dependent variable made *conditional* on the corresponding future values of the independent variables. Thus, having estimated the parameters of the model using the observations at times $t = 1, \ldots, T$, the problem is to estimate y_{T+l} given x_{T+l} for $l = 1, 2, 3, \ldots$.

Prediction in the Classical Model

If the model obeys the classical assumptions, the OLS estimator, b, is the BLUE of β. An obvious predictor is therefore

$$\tilde{y}_{T+l/T} = x'_{T+l}b, \qquad l = 1, 2, \ldots \tag{5.1}$$

This is the BLUE of the *fixed part* of y_{T+l}, namely $x'_{T+l}\beta$. The proof follows as a straightforward application of the Gauss–Markov theorem by setting the vector of weights, w, equal to x_{T+l}. However, $\tilde{y}_{T+l}$ cannot be regarded as a BLUE of y_{T+l} itself; it is not an unbiased estimator of y_{T+l}, since $E(\tilde{y}_{T+l/T}) = x'_{T+l}\beta$, and this will not, in general, be equal to y_{T+l}.

Since y_{T+l} is stochastic, the properties of $\tilde{y}_{T+l/T}$ as an estimator of y_{T+l} must be found in terms of the prediction error

$$e_{T+l} = y_{T+l} - \tilde{y}_{T+l/T}$$
$$= x'_{T+l}(\beta - b) + \epsilon_{T+l}, \qquad l = 1, 2, \ldots \tag{5.2}$$

It follows immediately from (5.2) that the expectation of e_{T+l} is zero, while its variance is given by

$$\text{Var}(e_{T+l}) = \sigma^2[1 + x'_{T+l}(X'X)^{-1}x_{T+l}] \tag{5.3}$$

In evaluating (5.3), the cross-product terms have zero expectation as b depends only on $\epsilon_1, \ldots, \epsilon_T$ and these are assumed to be uncorrelated with the future disturbance, ϵ_{T+l}, as well as with each other.

Bearing the above results in mind, $\tilde{y}_{T+l/T}$ may be regarded as an unbiased estimator of y_{T+l} in the sense that the expectation of the associated estimation error is zero. An estimator having this property is sometimes referred to as being *unconditionally unbiased* or simply *u-unbiased*. Within the class of linear, *u*-unbiased estimators of y_{T+l}, $\tilde{y}_{T+l/T}$ is best in the sense that the variance of its estimation error cannot exceed that of any other estimator. The proof is as follows. Let $\hat{y}_{T+l/T}$ denote an estimator of y_{T+l}. If this is to be linear in the observations it must be of the form

$$\hat{y}_{T+l/T} = [x'_{T+l}(X'X)^{-1}X' + d']y, \qquad (5.4)$$

where d is a $T \times 1$ vector of fixed elements. The associated estimation error is

$$
\begin{aligned}
y_{T+l} - \hat{y}_{T+l/T} &= [x'_{T+l}(X'X)^{-1}X' + d'](X\beta + \epsilon) \\
&\quad - x'_{T+l}\beta - \epsilon_{T+l} \\
&= d'X\beta + [x'_{T+l}(X'X)^{-1}X' + d]\epsilon - \epsilon_{T+l} \quad (5.5)
\end{aligned}
$$

If this is to have zero expectation, the constraint $d'X = 0$ must be imposed. The variance of the estimation error is then

$$E[(y_{T+l} - \hat{y}_{T+l/T})^2] = \mathrm{Var}(e_{T+l}) + \sigma^2 d'd \qquad (5.6)$$

Unless $d = 0$, in which case $\hat{y}_{T+l/T} = \tilde{y}_{T+l/T}$, (5.6) will exceed (5.3).

The estimator $\tilde{y}_{T+l/T}$ is sometimes known as the *best linear unbiased predictor* (BLUP) of y_{T+l}. However, if the problem is taken to be one of minimising the MSE of the prediction error, $\tilde{y}_{T+l/T}$ is again obtained. The property of *u*-unbiasedness then emerges as a consequence of the criterion adopted, rather than as a condition imposed at the outset. The term minimum mean square linear estimator (MMSLE) may therefore be used to describe $\tilde{y}_{T+l/T}$, and in these circumstances it is often more natural to refer to (5.3) as the MSE of $\tilde{y}_{T+l/T}$. In fact this avoids any ambiguity since the MSE of the predictor can be written down in exactly the same way as the MSE of the prediction error.

When the disturbances in the model are serially correlated, (5.1) will no longer be the BLUP of y_{T+l}. If w is a $T \times 1$ vector in which the tth element is equal to $E(u_t u_{T+l})$, the BLUP of y_{T+l} is given by

$$\tilde{y}_{T+l/T} = x'_{T+l}\tilde{b} + w'V^{-1}(y - X\tilde{b}) \qquad (5.7)$$

The second term on the right hand side of (5.7) is the contribution which results from taking the correlation between the realised and future disturbances into account. The original derivation of (5.7) will be found in Goldberger (1962). In most time series models, however, the predictions are built up recursively and an expression of this form is rarely used explicitly.

6. Recursive Least Squares

If an estimator of β in a linear regression model is calculated from the first k observations, it can be updated as each subsequent observation, $y_{k+1}, y_{k+2}, \ldots, y_T$, is added to the data set. The updating may be based on recursive formulae which avoid repeated matrix inversions. Calculating the OLS estimator in this way has a number of attractions. Firstly, it enables the changes in the estimator to be tracked over time and secondly it produces a set of residuals which are uncorrelated when the model satisfies the classical assumptions. These residuals may be used in conjunction with the OLS residuals to assess the suitability of the model specification.

Updating Formulae

Consider the model in its vector formulation, (1.3). The OLS estimator based on the first t observations may be written as

$$b_t = (X_t'X_t)^{-1}X_t'y_t^*, \qquad t = k, \ldots, T \tag{6.1}$$

where $X_t = (x_1, x_2, \ldots, x_t)'$ and $y_t^* = (y_1, \ldots, y_t)'$. Now suppose that an estimator has been computed from the first $t - 1$ observations. This estimator is

$$b_{t-1} = (X_{t-1}'X_{t-1})^{-1}X_{t-1}'y_{t-1}^* \tag{6.2}$$

it being assumed that $t - 1 \geqslant k$ and that X_{t-1} is of full rank.

Once the tth observation becomes available, b_t may be obtained from b_{t-1} and y_t without the matrix inversion implied by (6.1). The cross-product matrix may be updated in a similar way. The relevant formulae are given by

$$b_t = b_{t-1} + (X_{t-1}'X_{t-1})^{-1}x_t(y_t - x_t'b_{t-1})/f_t \tag{6.3a}$$

and

$$(X_t'X_t)^{-1} = (X_{t-1}'X_{t-1})^{-1}$$
$$- (X_{t-1}'X_{t-1})^{-1}x_t x_t'(X_{t-1}'X_{t-1})^{-1}/f_t \tag{6.3b}$$

where

$$f_t = 1 + x_t'(X_{t-1}'X_{t-1})^{-1}x_t, \qquad t = k+1, \ldots, T \tag{6.3c}$$

Expression (6.3b) follows directly as a special case of a matrix inversion lemma given in TSM. If (6.3b) is substituted into (6.1), and $X_t'y_t$ is decomposed as

$$X_t'y_t^* = X_{t-1}'y_{t-1}^* + x_t y_t \tag{6.4}$$

a little algebraic manipulation gives (6.3a).

In the classical model, the explanatory variables are fixed and known for all time periods, and so the only random variables appearing in the updating formulae are b_{t-1} and y_t. Their role emerges more clearly if (6.3a) is rewritten as

$$b_t = b_{t-1} + (X'_{t-1}X_{t-1})^{-1}x_t\tilde{v}_t/f_t \tag{6.5}$$

where $\tilde{v}_t$ is the one-step ahead prediction error; cf. (5.2). This prediction error contains all the new information needed to update b_{t-1}.

A unique OLS estimator of β can only be obtained if at least k observations are available. If X_k is of full rank, b_k is given by

$$b_k = (X'_k X_k)^{-1}X'_k y^*_k = X_k^{-1}y^*_k \tag{6.6}$$

Once b_k has been computed, $b_{k+1}, \ldots, b_T$ may be obtained by repeated application of the updating formulae. No further matrix inversions are needed, and the final estimator, b_T, is identical to the OLS estimator defined by (2.4).

Recursive Residuals

A set of $T - k$ prediction errors are produced as a by-product of the OLS recursions. If the disturbances obey the classical assumptions, (1.6), these may be written

$$\tilde{v}_t = y_t - x'_t b_{t-1}$$
$$= x'_t(\beta - b_{t-1}) + \epsilon_t, \qquad t = k+1, \ldots, T \tag{6.7}$$

and it will be seen that each one has a mean of zero and variance $\sigma^2 f_t$. The prediction errors have a further property, however, which is not immediately apparent from (6.7). Substituting for $\beta - b_{t-1}$ in (6.7) yields

$$\tilde{v}_t = \epsilon_t - x'_t(X'_{t-1}X_{t-1})^{-1}\sum_{j=1}^{t-1} x_j \epsilon_j \tag{6.8}$$

Then for all $t > s = k+1, \ldots, T-1$,

$$E(\tilde{v}_t\tilde{v}_s) = E(\epsilon_t\epsilon_s) - x'_t(X'_{t-1}X_{t-1})^{-1}\sum_{j=1}^{t-1} x_j E(\epsilon_j\epsilon_s)$$

$$- x'_s(X'_{s-1}X_{s-1})^{-1}\sum_{j=1}^{s-1} x_j E(\epsilon_j\epsilon_t)$$

$$+ E\left[\left\{x'_t(X'_{t-1}X_{t-1})^{-1}\sum_{j=1}^{t-1} x_j \epsilon_j\right\}\left\{\left(\sum_{j=1}^{s-1} x_j \epsilon_j\right)'(X'_{s-1}X_{s-1})^{-1}\right\}x_s\right]$$

$$= 0 - \sigma^2 x_t'(X_{t-1}'X_{t-1})^{-1}x_s - 0$$
$$+ \sigma^2 x_t'(X_{t-1}'X_{t-1})^{-1}X_{s-1}'X_{s-1}(X_{s-1}'X_{s-1})^{-1}x_s$$
$$= 0$$

The one-step ahead prediction errors are therefore uncorrelated. An intuitive rationale for this result is provided by (6.5). Since $\tilde{v}_t$ contains all the new information required to update the estimator, it seems reasonable to expect it to be uncorrelated with the previous prediction errors as these are incorporated in the current estimator, b_{t-1}.

The *recursive residuals* are the standardised prediction errors,

$$v_t = \tilde{v}_t/f_t^{1/2}, \qquad t = k+1, \ldots, T \tag{6.9}$$

They are uncorrelated with zero mean and constant variance, σ^2. Furthermore, since they are linear in the observations they are normally distributed when the disturbances are normal.

The recursive residuals feature in the updating formula for the residual sum of squares. This is

$$SSE_t = SSE_{t-1} + v_t^2 \tag{6.10}$$

where

$$SSE_t = (y_t - X_t b_t)'(y_t - X_t b_t) \tag{6.11}$$

is the sum of the squares of the OLS residuals based on the first t observations; see Brown, Durbin and Evans (1975). Since $SSE_k = 0$, it follows from (6.10) that

$$SSE_T = \sum_{t=1}^{T} e_t^2 = \sum_{t=k+1}^{T} v_t^2 \tag{6.12}$$

Computational Aspects of the Recursions

An alternative to starting the recursions with an estimator of β based on the first k observations, is to begin at $t = 0$ with an arbitrary value of b_0, say $b_0 = 0$. The corresponding 'cross-product' matrix, which may be labelled $(X_0'X_0)^{-1}$, is set equal to κI, where κ is a large number, say 10^6. Once the first observation becomes available, the recursions can proceed. Although b_0 is arbitrary, setting κ equal to a large number means that its effect on estimates and prediction errors is negligible once k observations have been processed. In other words, b_k and $(X_k'X_k)^{-1}$ will be very close to the values which would have been obtained had these quantities been calculated directly. The 'estimates' $b_1, \ldots, b_{k-1}$ and the 'prediction errors', $\tilde{v}_1, \ldots, \tilde{v}_k$, are, however, completely arbitrary and should be discarded. Setting $b_0 = 0$ and $(X_0'X_0)^{-1} = \kappa I$ is similar to adopting a 'diffuse prior' in Bayesian analysis. However, the term is open to some confusion here since β is a fixed parameter and the 'diffuse prior' is simply a technical device for starting the recursions.

If multicollinearity is present in the independent variables, the cross product matrix, $X'X$, may be close to singularity. As a result the OLS estimator may be subject to inaccuracies in computation. If it is calculated recursively, the possibilities for numerical instability are even greater since rounding errors may tend to build up as the recursions proceed. One way of avoiding this problem is to adopt the QR decomposition algorithm of Gentleman (1973). This is an efficient and numerically stable method of calculating the OLS estimate. However, its construction is such that the recursive residuals emerge as an automatic by-product.

7. Residuals

The residuals from a regression play an important role in detecting possible misspecifications in the model. The nature of this role is described in Section 5.2, where both graphical procedures and formal test statistics are described. In this section attention is focused on the properties of the OLS residuals when the model obeys the classical assumptions. These properties are then contrasted with those of alternative sets of residuals which are available in linear regression, and the usefulness of such residuals is examined. The final sub-section is concerned with the distribution of the residual sum of squares and the estimation of the parameter σ^2.

OLS Residuals

The $T \times 1$ vector of OLS residuals, $e = (e_1, \ldots, e_T)'$, may be expressed as a linear combination of the observations. Substituting from (2.4),

$$e = y - Xb = [I - X(X'X)^{-1}X']y = My \tag{7.1}$$

Furthermore since the $T \times T$ matrix, M, is orthogonal to X, i.e. $MX = 0$, substituting for y yields

$$e = M(X\beta + \epsilon) = M\epsilon \tag{7.2}$$

Thus the OLS residuals are linear combinations of the true disturbances.

If the model obeys the classical assumptions, $E(e) = M \cdot E(\epsilon) = 0$, while

$$E(ee') = M \cdot E(\epsilon\epsilon')M' = \sigma^2 M \qquad (7.3)$$

Thus although the OLS residuals have zero mean, the fact that $M \neq I$ means that, in general, they will be neither homoscedastic nor serially uncorrelated. Hence their properties do not mirror those of the true disturbances even when the model is correctly specified. Fortunately this problem becomes less important in large samples. Under the standard regularity conditions, (3.9), it can be shown that M is closely approximated by the identity matrix when T is large, and that e converges in distribution to ϵ.

The OLS residuals are uncorrelated with the elements of b. From (3.1) and (7.2)

$$E[e(b - \beta)'] = E[Mee'X'(X'X)^{-1}] = MX'(X'X)^{-1} = 0 \qquad (7.4)$$

If the disturbances are normally distributed, e and b are independent.

LUS Residuals

It follows from (7.3) that in small samples the distribution of the OLS residuals will depend on the X matrix. This creates difficulties in constructing exact test procedures, since the distribution of any test statistic based on OLS residuals will vary with the data. It is therefore natural to ask whether it is possible to form a set of residuals which are homoscedastic and uncorrelated when the model satisfies the classical assumptions.

The OLS residuals are linear in the observations. This ensures that they are normally distributed when the disturbances are normal. If $e*$ denotes any other residual vector which is linear in the observations, it may be written as

$$e* = Cy \qquad (7.5)$$

where C is a matrix with T columns, in which the elements are fixed. If $CX = 0$, $e*$ will have zero expectation since $e* = C(X\beta + \epsilon) = C\epsilon$. Furthermore if $CC' = I$, the residuals can be said to have a *scalar* covariance matrix since

$$E(e*e*') = E[C\epsilon\epsilon'C'] = \sigma^2 CC' = \sigma^2 I$$

However, if this is to hold, C can have no more than $T - k$ rows; see Theil (1971, Ch. 5).

If C is a $(T - k) \times T$ matrix with the properties $CX = 0$ and $CC' = I$,

(7.5) defines a set of $T - k$ *LUS* residuals. These residuals are (i) *linear*; (ii) *unbiased*, in the sense that $E(e^* - \epsilon) = 0$; and (iii) have a *scalar* covariance matrix. If $\epsilon_t \sim NID(0, \sigma^2)$, the LUS residuals are distributed in exactly the same way, i.e. $e_t^* \sim NID(0, \sigma^2)$.

There are a number of different principles upon which the construction of LUS residuals may be based. The approach adopted by Theil (1971, Ch. 5) is to derive a vector of $T - k$ LUS residuals, which is likely to be a good estimator of a corresponding subset of $T - k$ disturbances, ϵ_1. His *BLUS* residuals are best in the sense that they minimise the expected sum of squares of the estimation errors, $E[e^* - \epsilon_1)'(e^* - \epsilon_1)]$. Unfortunately, the calculations needed to obtain a set of BLUS residuals are relatively heavy and this means that they are rarely used in practice. The recursive residuals are also LUS residuals, but are much easier to compute as they emerge directly from the updating formulae in (6.3). The numerically stable OLS routine of Gentleman also yields recursive residuals and if it is felt that this algorithm should be employed simply on the basis of its merits as an OLS computational procedure, the recursive residuals are essentially a free good. In a similar manner, other sets of LUS residuals may be produced as a by-product of different algorithms; see, for example, Groluh and Styan (1973).

Since $e^* = C(Xb + e) = Ce$, all the information contained in a set of LUS residuals is also contained in the corresponding OLS vector. This has important implications for testing, with a statistic based on OLS residuals being preferable, in most cases, to a corresponding statistic based on LUS residuals. The case for using a particular set of LUS residuals to detect misspecifications rests on their ability to present information in a manner which is complementary to that of the OLS residuals. Because the recursive residuals have a natural interpretation as standardised prediction errors they are particularly suitable for such a role.

The Residual Sum of Squares

The residual sum of squares, *SSE*, was defined in (2.8), and it was shown in (6.12) that it is identical to the sum of squares of the $T - k$ recursive residuals. It follows immediately that $E(SSE) = (T - k)\sigma^2$, and so

$$s^2 = SSE/(T - k) \tag{7.6}$$

is an unbiased estimator of σ^2.

Under the normality assumption it is apparent from (6.12) that SSE/σ^2 is the sum of $T - k$ independent χ_1^2 variates. Hence

$(T - k)s^2/\sigma^2 \sim \chi^2_{T-k}$. Furthermore, because the variance of a χ^2 distribution with f degrees of freedom is $2f$,

$$\text{Var}(s^2) = 2\sigma^4/(T - k) \tag{7.7}$$

8. Test Statistics and Confidence Intervals

If the disturbances in a classical linear regression model are normally distributed, small sample test statistics can be derived. Similarly, exact confidence intervals may be constructed.

This section merely presents the standard results on testing within the context of the classical regression model. As such the tests may seem to be *ad hoc*. However, in Chapter 5 it is shown that all the tests presented here may be derived from general principles associated with maximum likelihood estimation. This ensures that the tests have certain desirable statistical properties.

Individual Parameters: the t-statistic

Consider β_i, the ith element in the parameter vector β. If b_i denotes the corresponding element in the OLS vector, it follows from (3.8) that

$$b_i \sim N(\beta_i, \sigma^2 \zeta_i), \qquad i = 1, \ldots, k \tag{8.1}$$

where ζ_i denotes the ith diagonal element in $(X'X)^{-1}$.

Since σ^2 is unknown it must be estimated. Expression (7.6) gives a suitable estimator of σ^2 and it was shown in the previous section that

$$(T - k)s^2 \sim \sigma^2 \chi^2_{T-k} \tag{8.2}$$

Furthermore, in view of (7.4), s^2 and b_i are independent. Hence the statistic

$$\frac{(b_i - \beta_i)/\sigma\sqrt{\zeta_i}}{s/\sigma} = \frac{b_i - \beta_i}{s\sqrt{\zeta_i}}, \qquad i = 1, \ldots, k \tag{8.3}$$

follows a t-distribution with $T - k$ degrees of freedom, since it is the ratio of a standard normal variate to the square root of a χ^2 variate divided by its degrees of freedom. The estimated standard deviation of b, $s_i\sqrt{\zeta_i}$, is usually known as the *standard error*.

Hypotheses concerning specific values of the parameters may be tested using (8.3). Of particular importance is the *test of significance*, in which the null hypothesis is $H_0 : \beta_i = 0$. Failure to reject this hypothesis at a particular level of significance leads to the conclusion that b_i is 'insignificant'. This suggests one of two possibilities. Either

the sample evidence is consistent with $\beta_i = 0$ and so the corresponding variable should not appear in the model, or $\beta_i \neq 0$, but the size of
the sample is such that β_i cannot be estimated with the degree of
accuracy required to reject the null hypothesis.

Tests of significance are particularly appropriate when the data
come from a controlled experiment. Unfortunately this is rarely the
case in econometrics. In a regression based on time series data, tests
of significance should be interpreted with great care. Time series are
typically highly correlated even when there is no underlying relationship between them. This has two consequences. Firstly an
explanatory variable may be 'significant', even though it has no
effect on the dependent variable. Secondly, if two explanatory
variables are highly correlated it may be difficult to disentangle their
separate effects, even though both should properly be included in
the model. Hence it is quite possible that neither will be significant
on statistical grounds. The first of these points is a reflection of the
problem of *spurious correlation*. The second problem is known as
multicollinearity.

Confidence Intervals

The result (8.3) may be used to construct an exact confidence interval
for β_i. Let $t_{\alpha/2}$ denote the significance value for a two-sided test
based on a t-distribution with $T - k$ degrees of freedom. If
$Pr(t > t_{\alpha/2}) = \alpha/2$ it follows that $Pr(-t_{\alpha/2} < t < t_{\alpha/2}) = \alpha$. Hence the
probability that the interval

$$b_i \pm t_{\alpha/2} s \sqrt{\xi_i}, \qquad i = 1, \ldots, k \tag{8.4}$$

contains the true parameter, β_i, is $1 - \alpha$. Thus (8.4) is said to be the
$(1 - \alpha)\%$ *confidence interval* for β_i.

In practice α is usually set at 0.05 leading to a 95% confidence
interval. Unless T is very small, a useful rule of thumb is to set $t_{0.025}$
equal to two.

Linear Combinations of Parameters

In discussing the Gauss–Markov theorem, it was observed that
attention is often focused on a linear combination of the parameters,
$w'\beta$. From (3.6) it follows that the distribution of the unbiased
estimator, $w'b$, is given by

$$w'b \sim N[w'\beta, \sigma^2 w'(X'X)^{-1}w] \tag{8.5}$$

By the same argument used to obtain the t-statistic in (8.3), it follows that

$$\frac{w'b - w'\beta}{s\sqrt{w'(X'X)^{-1}w}} \tag{8.6}$$

also has a t-distribution with $T - k$ degrees of freedom. Hypotheses
of the form $H_0 : w'\beta = r_0$, where r_0 is a particular value, may therefore
be tested using (8.6). Furthermore, confidence intervals for $w'\beta$ may
be constructed along the lines of (8.4).

If w is set equal to x_{T+l} the above results may be applied to
testing hypotheses and constructing confidence intervals for future
expected values of y_t. However, a more relevant question concerns
inferences about y_t itself. As shown in Section 5, attention then
focuses on the prediction error, $y_{T+l} - x'_{T+l}b$, and it follows from
the results in Section 5 that

$$y_{T+l} - x'_{T+l}b \sim N[0, \sigma^2 \{1 + x'_{T+l}(X'X)^{-1}x_{T+l}\}],$$

$$l = 1, 2, \ldots \tag{8.7}$$

This result may be used to test whether a new observation on y_t
obeys the same model as the observations in the sample since

$$\frac{y_{T+l} - x'_{T+l}b}{s\sqrt{1 + x'_{T+l}(X'X)^{-1}x_{T+l}}}, \qquad l = 1, 2, \ldots \tag{8.8}$$

has a t-distribution with $T - k$ degrees of freedom under the null
hypothesis that the model is unchanged.

A $(1 - \alpha)\%$ *prediction interval* may be constructed from (8.8).
This has the same form as a confidence interval but the interpretation is slightly different because y_{T+l} is stochastic.

The Subset F-test

A test of the joint significance of a subset of $m \ (\leqslant k)$ variables in a
regression model may be based on the F-distribution. Without any
loss of generality the variables may be ordered so that the last m are
those under test, i.e.

$$H_0 : \beta_{k-m+1} = \cdots = \beta_k = 0 \tag{8.9}$$

Let SSE denote the residual sum of squares, $e'e$, when the model is
fitted with all k variables, and let SSE_0 be the residual sum of squares
when the last m variables are omitted. When the null hypothesis is
true, the statistic

$$\frac{(SSE_0 - SSE)/m}{SSE/(T - k)} = (SSE_0 - SSE)/ms^2 \tag{8.10}$$

has an F-distribution with $(m, T - k)$ degrees of freedom.

An interesting special case arises when the model contains a constant term and the null hypothesis is that the coefficients of the other regressors are all zero. Since the constant term has no explanatory power, this test may be regarded as an overall test on the significance of the relationship as a whole. The test may be cast in terms of the coefficient of multiple correlation, R^2, by noting that $SSE_0 = \Sigma (y_t - \bar{y})^2$ and substituting from (2.13). This yields the statistic

$$\frac{R^2/(k - 1)}{(1 - R^2)/(T - k)} \tag{8.11}$$

As with tests on individual coefficients, tests on subsets of parameters should be interpreted somewhat cautiously for non-experimental data. In particular an insignificant value of (8.11) tends to be the exception rather than the rule for time series data. Indeed it can be reasonably argued that the test is usually meaningless in such a context.

Analysis of Covariance

The available observations may sometimes fall naturally into two or more distinct groups. For cross-sectional data these divisions may depend on some qualitative characteristic associated with the observations. For example, the sample may be drawn from a number of different towns or countries. With time series data, the divisions may be associated with certain events, such as the outbreak of war or a change of government. However, irrespective of how the grouping arises, the basic statistical question is the same. Is the structure of the model different for different groups?

A general framework may be established by writing the complete set of observations as

$$y_j = Z_j \gamma + X_j \beta_j + \epsilon_j, \qquad j = 1, \ldots, G \tag{8.12}$$

where $[Z_j : X_j]$ is a $T_j \times k$ matrix of observations on the explanatory variables, and $(\gamma', \beta_j')'$ is the corresponding $k \times 1$ vector of regression parameters. The $n \times 1$ parameter vector β_j is subscripted, since it is the parameters in this vector which are to be tested for equality. Any test of a hypothesis of the form

$$H_0 : \beta_1 = \cdots = \beta_G = \beta \tag{8.13}$$

is an exercise in the *analysis of covariance*.

The disturbances are assumed to be normally distributed with a mean of zero and a variance, σ^2, which is the same for all groups. Furthermore the disturbances are independent both within and between groups. Under the restrictions imposed by (8.13) the model

$$
\begin{bmatrix} y_1 \\ \vdots \\ y_G \end{bmatrix} = \begin{bmatrix} Z_1 \\ \vdots \\ Z_G \end{bmatrix} \gamma + \begin{bmatrix} X_1 \\ \vdots \\ X_G \end{bmatrix} \beta + \begin{bmatrix} \epsilon_1 \\ \vdots \\ \epsilon_G \end{bmatrix}
\tag{8.14}
$$

is appropriate. Let the residual sum of squares be denoted by SSE_0. The unrestricted model is

$$
\begin{bmatrix} y_1 \\ y_2 \\ \vdots \\ y_G \end{bmatrix} = \begin{bmatrix} Z_1 \\ Z_2 \\ \vdots \\ Z_G \end{bmatrix} \gamma + \begin{bmatrix} X_1 & 0 & \cdots & 0 \\ 0 & X_2 & \cdots & 0 \\ & & \ddots & \\ 0 & 0 & & X_G \end{bmatrix} \begin{bmatrix} \beta_1 \\ \beta_2 \\ \vdots \\ \beta_G \end{bmatrix} + \begin{bmatrix} \epsilon_1 \\ \epsilon_2 \\ \vdots \\ \epsilon_G \end{bmatrix}
\tag{8.15}
$$

and the residual sum of squares is SSE. In the special case where $n = k$, SSE is given by adding up the residual sums of squares after applying OLS to each group in turn.

Model (8.14) embodies $n(G - 1)$ restrictions as compared with (8.15). If these restrictions are valid, the statistic

$$
\frac{(SSE_0 - SSE)/\{n(G-1)\}}{SSE/\{T - (k-n) - nG\}}
\tag{8.16}
$$

has an F-distribution with $(nG - n, T - k - nG + n)$ degrees of freedom. Note that $T = T_1 + \cdots + T_G$ is the total number of observations, while $(k - n) + nG = k + nG - n$ is the number of independent variables in the unrestricted model.

Example   A household budget survey carried out in Kenya in 1968/9 covered three towns: Nairobi, Mombasa and Kisumu. The following model was estimated for a number of commodities:

$$
y_t = \beta_1 + \beta_2 x_{2t} + \beta_3 x_{3t} + \epsilon_t = x_t'\beta + \epsilon_t, \qquad t = 1, \ldots, T
$$

where y_t is the logarithm of expenditure on a given commodity by the tth household, x_{2t} is the logarithm of household income, and x_{3t} is the logarithm of household size. (Although the data are cross-sectional, the subscript t is used to avoid any ambiguity.) A whole range of results is reported in the *Kenya Statistical Digest* of April, 1972, but the findings for rice are particularly interesting. Estimating the above equation for each town separately and adding up the total sum of squared residuals gave $SSE = 150.6$. Pooling the data gave $SSE_0 = 298.2$. The sample sizes in Nairobi, Mombasa and Kisumu were $T_1 = 100$, $T_2 = 86$ and $T_3 = 45$

respectively, and so the appropriate statistic for testing the equality of the regression coefficients in the three towns is

$$F = \frac{(298.2 - 150.6)/6}{150.6/222} = 36.27$$

Since the 5% significance point for an F-distribution with $(6,222)$ degrees of freedom is approximately 2.10, the calculated value is highly significant.

In terms of (8.16), $n = k$ in the above test. However, a rather different picture emerges if the test of equality is confined to β_2 and β_3. This means that when the data are pooled, the intercept terms for the three towns are allowed to differ. Introducing dummy variables into the pooled model gave the following results:

$$\hat{y}_t = -1.626 + 1.288 D_{2t} - 0.860 D_{3t} + 0.284 x_{2t} + 0.915 x_{3t},$$
$$(0.124) \quad (0.150) \quad (0.068) \quad (0.110)$$
$$t = 1, \ldots, T$$

where

$$D_{2t} = \begin{cases} 1 - \text{if the observation was from town 2 (Mombasa)} \\ 0 - \text{otherwise} \end{cases}$$

$$D_{3t} = \begin{cases} 1 - \text{if the observation was from town 3 (Kisumu)} \\ 0 - \text{otherwise.} \end{cases}$$

The figures in parentheses are standard errors and it will be observed that both dummy variables are highly significant.

The pooled model is of the form (8.14) with the tth rows of $Z = (Z_1', \ldots, Z_G')'$ and $X = (X_1', \ldots, X_G')$ given by $(1 \ D_{2t} \ D_{3t})$ and $(x_{2t} \ x_{3t})$ respectively. The residual sum of squares was 156.7 and so a test of the equality of the coefficients β_2 and β_3 in the three towns is based on

$$F = \frac{(156.7 - 150.6)/6}{150.6/222} = 2.25$$

This is insignificant at the 5% level.

These results can be interpreted as follows. The hypothesis that β_2, the income elasticity of demand for rice, and β_3, the household size elasticity, are the same in all three towns cannot be rejected. However, the level of consumption of rice in the three towns appears to be different. Since $\exp(1.288) = 3.42$, the expected household consumption of rice for a given level of income and household size is 342% higher in Mombasa than in Nairobi. On the other hand, since $\exp(-0.860) = 0.42$, rice consumption in Kisumu is only 42% of that in Nairobi.

In time series, a fundamental application of the analysis of covariance is in testing for *structural change*. Suppose that the sample period divides into two groups, with all the observations after a certain point in time falling into the second group. The hypothesis to be tested is that the complete parameter vector is the same in both periods. Stated explicitly the unrestricted model is

$$y_t = x_t'\beta_1 + \epsilon_t, \qquad t = 1, \ldots, T_1$$
$$y_t = x_t'\beta_2 + \epsilon_t, \qquad t = T_1 + 1, \ldots, T_1 + T_2$$

and the null hypothesis is $H_0 : \beta_1 = \beta_2$. Since $m = k$ and $G = 2$, the test statistic reduces to

$$\frac{(SSE_0 - SSE)/k}{SSE/(T - 2k)} \tag{8.17}$$

where SSE_0 is the sum of the residual sum of squares in each group. Under the null hypothesis of no structural change, this statistic has an F-distribution with $(k, T - 2k)$ degrees of freedom.

In order to construct (8.17) it is necessary that both T_1 and T_2 be greater than k. Indeed an assumption of this kind was implicitly made in setting up the general test statistic, (8.16). However, in testing for structural change it is not uncommon to find $T_2 \leqslant k$. In these circumstances a slightly different approach must be adopted. This is described in Section 5.7.

A General Test of Restrictions

In Section 2, the estimation of β was considered subject to a set of m restrictions defined by (2.16). The validity of these restrictions may be tested using the statistic

$$\frac{(SSE_0 - SSE)/m}{SSE/(T - k)} \tag{8.18}$$

where SSE_0 and SSE are the residual sums of squares for the restricted and unrestricted models respectively. Under the null hypothesis, this statistic has an F-distribution with $(m, T - k)$ degrees of freedom.

All the test statistics considered in the last sub-section are special cases of (8.18). In the case of the structural change test, for example, the constraints may be written in the form (2.16) by defining $R = [I, -I]$, $r = 0$, and $\beta = (\beta_1', \beta_2')'$. Thus if (8.18) can be shown to have an F-distribution under H_0 all the other results follow automatically. A proof is given in Theil (1971, pp. 143–144).

9. Systems of Equations: Seemingly Unrelated Regression Equations

When the model in (1.1) satisfies the classical assumptions, the OLS estimator is best linear unbiased. However it is sometimes possible to obtain a better estimator by viewing the equation as part of a *system*. This implies no contradiction with the BLUE property of OLS in a single equation, since extra information is being used in constructing the estimator.

There are basically two ways in which a gain in efficiency can occur when a single equation is considered as part of a system. The first arises when the disturbances in a particular equation are contemporaneously correlated with the disturbances in other equations. Such systems are known as *seemingly unrelated regression equations* (SURE). Secondly, models are sometimes formulated with constraints across different equations. The need to consider such equations together is rather more obvious in this case.

Seemingly Unrelated Regressions

A system of N equations may be written as

$$y_i = X_i \beta_i + \epsilon_i, \qquad i = 1, \ldots, N \tag{9.1}$$

The notation employed in (9.1) is analogous to that in (1.5) with y_i being a $T \times 1$ vector and so on. The number of independent variables for a given i in (9.1) will be denoted by k_i, and their sum, $k_1 + k_2 + \cdots + k_N$ will be denoted by k. None of the variables or parameters in the N equations need be related: the connection between the equations lies solely in the disturbance terms, which are correlated across different equations.

If the disturbances in each of the N equations obey the classical assumptions then

$$E(\epsilon_i) = 0, \quad E(\epsilon_i \epsilon_i') = \omega_{ii} I_N \tag{9.2}$$

The correlations between disturbances in different equations may be characterised by writing

$$E(\epsilon_{it} \epsilon_{jt}) = \omega_{ij}, \qquad \text{for } i, j = 1, \ldots, N \tag{9.3}$$

All non-contemporaneous disturbances are, however, uncorrelated and so

$$E(\epsilon_{it}\epsilon_{j\tau}) = 0, \qquad \text{for } t \neq \tau \tag{9.4}$$

The information concerning the correlation structure of the disturbances may therefore be summarised by an $N \times N$ matrix Ω, the ijth element of which is equal to ω_{ij}.

The complete system may be written out in matrix terms as

$$
\begin{pmatrix} y_1 \\ y_2 \\ \vdots \\ y_N \end{pmatrix}
=
\begin{pmatrix} X_1 & 0 & \cdots & 0 \\ 0 & X_2 & \cdots & 0 \\ \vdots & \vdots & & \vdots \\ 0 & 0 & \cdots & X_N \end{pmatrix}
\begin{pmatrix} \beta_1 \\ \beta_2 \\ \vdots \\ \beta_N \end{pmatrix}
+
\begin{pmatrix} \epsilon_1 \\ \epsilon_2 \\ \vdots \\ \epsilon_N \end{pmatrix}
\tag{9.5}
$$

or more compactly as

$$y = X\beta + u \tag{9.6}$$

where y is a vector of length TN, X is an $N \times k$ matrix and β is a vector of length k. The $N \times 1$ vector of disturbances has a covariance matrix

$$
V = E(uu') =
\begin{pmatrix}
E(\epsilon_1\epsilon_1') & E(\epsilon_1\epsilon_2') & \cdots & E(\epsilon_1\epsilon_N') \\
E(\epsilon_2\epsilon_1') & E(\epsilon_2\epsilon_2') & \cdots & E(\epsilon_2\epsilon_N') \\
\vdots & \vdots & & \vdots \\
E(\epsilon_N\epsilon_1') & E(\epsilon_N\epsilon_2') & \cdots & E(\epsilon_N\epsilon_N')
\end{pmatrix}
\tag{9.7}
$$

For a given value of Ω, the BLUE of β is given simply by applying GLS to (9.6). Formula (4.4) implies the inversion of the matrix V which is of order TN. However, this may be avoided by first noting that the individual terms in (9.7) are all scalar matrices since

$$E(\epsilon_i\epsilon_j') = \omega_{ij}I_N, \qquad i,j = 1, \ldots, N \tag{9.8}$$

Thus V may be written as

$$V = \Omega \otimes I \tag{9.9}$$

where $\otimes$ denotes a Kronecker or tensor product. The algebra of Kronecker products is discussed briefly in the appendix.

If ω^{ij} denotes the ijth element in the inverse of Ω, it follows from the properties of Kronecker products that

$$
\begin{aligned}
V^{-1} &= \Omega^{-1} \otimes I \\
&=
\begin{pmatrix}
\omega^{11}I & \cdots & \omega^{1N}I \\
\vdots & & \vdots \\
\omega^{N1}I & \cdots & \omega^{NN}I
\end{pmatrix}
\end{aligned}
\tag{9.10}
$$

Substituting (9.10) into the GLS formula (4.4) then yields a more tractable expression for the SURE estimator,

$$
\tilde{b} =
\begin{pmatrix}
\omega^{11} X_1' X_1 & \omega^{12} X_1' X_2 & \cdots & \omega^{1N} X_1' X_N \\
\vdots & \vdots & & \vdots \\
\omega^{N1} X_N' X_1 & \omega^{N2} X_N' X_2 & \cdots & \omega^{NN} X_N' X_N
\end{pmatrix}^{-1}
\begin{pmatrix}
\sum_{j=1}^{N} \omega^{1j} X_1' y_j \\
\vdots \\
\sum_{j=1}^{N} \omega^{Nj} X_N' y_j
\end{pmatrix}
$$

$$(9.11)$$

The covariance matrix of $\tilde{b}$ is simply the inverted matrix in expression (9.11); cf. (4.5).

Efficiency of the SURE Estimator

The potential gain in efficiency from using a SURE estimator depends on a number of factors. However, when the explanatory variables are pairwise orthogonal and $\omega_{ij} = \sigma^2 \rho$ for all $i \neq j$, Zellner (1963, p. 354) has shown that the relative efficiency of the OLS estimator of a particular coefficient is given by

$$
\frac{\mathrm{Var}(\tilde{b})}{\mathrm{Var}(b)} = \frac{(1-\rho)(1-\rho+\rho N)}{1-2\mu+\rho N}
\tag{9.12}
$$

As might be expected the OLS estimator becomes more inefficient as $\rho \to 1$. On the other hand when $\rho = 0$, OLS is fully efficient. Indeed it is easy to see from (9.11) that the SURE and OLS estimators are identical when $\omega_{ij} = 0$ for $i \neq j$. This is only to be expected since when the disturbances are uncorrelated across equations there is nothing to connect them.

A second result to emerge from (9.12) is that for $\rho > 0$ the gain in efficiency of the SURE estimator relative to OLS increases with the number of equations. The Monte Carlo evidence in Kmenta and Gilbert (1968) suggests that this is the case even if the explanatory variables are not orthogonal, provided that T is large.

As already noted, OLS is fully efficient when the disturbances are uncorrelated across equations. There is, however, another instance when OLS yields efficient estimators. This arises when the regressors in all N equations are identical, i.e. $X_i = \bar{X}, i = 1, \ldots, N$. The proof is as follows. In terms of (9.6), $X = I \otimes \bar{X}$ and so, using (A.8),

$$
X' V^{-1} X = (I \otimes \bar{X})'(\Omega^{-1} \otimes I)(I \otimes \bar{X}) = \Omega^{-1} \otimes (\bar{X}'\bar{X})
\tag{9.13}
$$

The GLS estimator is therefore

$$
\begin{aligned}
\tilde{b} &= \{\Omega \otimes (\bar{X}'\bar{X})^{-1}\}(I \otimes \bar{X}')(\Omega^{-1} \otimes I)y \\
&= \{I \otimes (\bar{X}'\bar{X})^{-1}\bar{X}'\}y
\end{aligned}
$$

On examining this expression it will be seen that the SURE estimator in the ith equation is

$$\tilde{b}_i = (\bar{X}'\bar{X})^{-1}\bar{X}'y \tag{9.14}$$

for all $i = 1, \ldots, N$. In other words the OLS and SURE estimators are identical.

An Alternative Formulation of the SURE Model

An alternative way of developing the SURE estimator — which does not involve Kronecker products — is to write the N equations together as

$$y_t = X_t \beta + \epsilon_t, \qquad t = 1, \ldots, T \tag{9.15}$$

where y_t is an $N \times 1$ vector and X_t is an $N \times k$ matrix

$$X_t = \begin{pmatrix} x'_{1t} & 0 & \cdots & 0 \\ 0 & x'_{2t} & \cdots & 0 \\ \vdots & \vdots & & \vdots \\ 0 & 0 & \cdots & x'_{Nt} \end{pmatrix} \tag{9.16}$$

where x_{it} is the $k_i \times 1$ vector of observations on the independent variables in the ith equation. The $k \times 1$ vector β is exactly as defined in (9.5), while the ϵ_t's are serially uncorrelated with mean zero and covariance matrix, $E(\epsilon_t \epsilon'_t) = \Omega$.

If the T equations in (9.15) are stacked in the usual way, the covariance matrix of the disturbances $u^* = (\epsilon'_1, \ldots, \epsilon'_T)'$ is

$$E(u^* u^{*'}) = \begin{pmatrix} \Omega & 0 & \cdots & 0 \\ 0 & \Omega & \cdots & 0 \\ \vdots & \vdots & & \vdots \\ 0 & 0 & \cdots & \Omega \end{pmatrix}$$

This may be written more concisely as $E(u^* u^{*'}) = I \otimes \Omega$. However, it is not necessary to use the algebra of Kronecker products in order to see that the GLS estimator of β is

$$\tilde{b} = \left(\sum_{t=1}^{T} X'_t \Omega^{-1} X_t \right)^{-1} \sum_{t=1}^{T} X'_t \Omega^{-1} y_t \tag{9.17}$$

From the definition of X_t in (9.16) it is not difficult to see that (9.17) is identical to the SURE estimator given by (9.11).

Constraints across Equations

The formulation in the previous sub-section is particularly appropriate for incorporating constraints in a model. For example,
suppose that in the two equation system

$$y_{1t} = \beta_1 x_{1t} + \beta_2 x_{2t} + \epsilon_{1t}$$
$$y_{2t} = \beta_3 x_{3t} + \beta_4 x_{4t} + \epsilon_{2t}$$
(9.18)

it is known that $\beta_2 = \beta_4$. Writing the model in the form (9.15) yields

$$\begin{pmatrix} y_{1t} \\ y_{2t} \end{pmatrix} = \begin{pmatrix} x_{1t} & x_{2t} & 0 \\ 0 & x_{4t} & x_{3t} \end{pmatrix} \beta + \begin{pmatrix} \epsilon_{1t} \\ \epsilon_{2t} \end{pmatrix}$$
(9.19)

where $\beta = (\beta_1, \beta_2, \beta_3)'$. Estimating the coefficients with an allowance
made for the constraint will give more efficient estimators of the
elements in β even if the disturbances in the two equations are
uncorrelated.

A more general method of allowing for constraints is to write them
in the form (2.16). Adopting the notation of (9.6) then leads to a
natural generalisation of the restricted least squares estimator, *viz.*

$$\tilde{b}^* = \tilde{b} + (X'V^{-1}X)^{-1}R'[R(X'V^{-1}X)^{-1}R']^{-1}(r - R\tilde{b})$$
(9.20)

where $\tilde{b}$ is the unconstrained GLS estimator.

A Feasible SURE Estimator

In general, Ω will be unknown and so (9.11) cannot be used directly.
However, it may be estimated by applying OLS separately to each
equation. If e_i denotes that $T \times 1$ vector of OLS residuals in the ith
equation, an estimator of the ijth element in Ω is given by

$$\hat{\omega}_{ij} = e_i'e_j/T, \qquad i,j = 1, \dots, N$$
(9.21)

Substituting for Ω in (9.11) or (9.17) then gives a *feasible* SURE
estimator. The discussion at the end of Section 4 suggests that the
resulting estimator of β will have the same asymptotic distribution as
(9.11) under suitable conditions. The estimator of the asymptotic
covariance matrix of the feasible SURE estimator is

$$\text{avar}(\tilde{b}) = (\sum X_t'\hat{\Omega}^{-1}X_t)^{-1}$$
(9.22)

Example 1   The application of Grunfeld's investment theory to
data from two corporations, General Electric and Westinghouse, is
a classic example of the use of the SURE technique; see Theil
(1971) where the data are reproduced on p. 296. The theory
states that next year's investment, y, depends on x_1 the 'market

value of the firm', defined as the total value of the outstanding stock at the end-of-year stock market quotations, and the capital stock, x_2. Estimating each equation separately by OLS gives the following results

$$\text{General Electric} \quad \hat{y}_1 = -10.0 + 0.027x_{11} + 0.152x_{21}$$
$$\text{Westinghouse} \quad \hat{y}_2 = -0.5 + 0.053x_{12} + 0.092x_{22}$$
$$(9.23)$$

where the second subscripts on x_1 and x_2 indicate the equation and the time subscript has been dropped for convenience.

The estimate of Ω from the OLS residuals is

$$\hat{\Omega} = \begin{bmatrix} 660.83 & 176.46 \\ 176.46 & 88.67 \end{bmatrix} \tag{9.24}$$

The correlation between the disturbances in the two equations is 0.53, which suggests that joint estimation would be appropriate. Indeed this makes sense from the theoretical point of view, since one would expect investment to be governed, to some extent, by considerations related to the economy as a whole. No such variables are specifically included in the model and so their effect must be felt through the disturbance terms.

Using the estimate of Ω in (9.24), the feasible SURE method yields

$$\hat{y}_1 = -27.7 + 0.038x_{11} + 0.139x_{21}$$
$$\hat{y}_2 = -1.3 + 0.058x_{12} + 0.064x_{22}$$
$$(9.25)$$

Incorporating the constraint that the coefficients of the explanatory variables are the same for each firm gives

$$\hat{y}_1 = -23.0 + 0.36x_{11} + 0.139x_{21}$$
$$\hat{y}_2 = 6.9 + 0.36x_{12} + 0.139x_{22}$$
$$(9.26)$$

cf. (9.19).

Example 2   Hart, Hutton and Sharot (1975) propose the following model for explaining the attendance at an English Football League match:

$$\log y_t = \alpha + \beta_1 \log x_{1t} + \beta_2 \log x_{2t} + \beta_3 \log x_{3t}$$
$$+ \beta_4 \log x_{4t} + \beta_5 \log x_{5t} + \epsilon_t \tag{9.27}$$

where y_t is attendance at the tth match of the season, x_1 and x_2 are the league positions of the home and away teams prior to the match, x_3 is the population of the home team's catchment area, x_4 is distance travelled by away support and x_5 is a time trend.

Equation (9.25) was estimated for four teams, Leeds, Newcastle, Nottingham Forest and Southampton. The results for Newcastle are typical:

$$\log \hat{y}_t = 8.85 + 0.06x_{1t} - 0.08x_{2t} + 0.31x_{3t} - 0.22x_{4t} - 0.80x_{5t}$$
$$\quad\;\; (0.84)\,(0.12)\quad (0.04)\quad (0.08)\quad (0.06)\quad (0.33)$$

$$(9.28)$$

However, the OLS residuals had the following correlation matrix:

$$\begin{pmatrix} 1 & & & \\ 0.32 & 1 & & \\ 0.13 & 0.32 & 1 & \\ 0.14 & 0.42 & 0.12 & 1 \end{pmatrix}$$

with the teams in the order previously cited. This suggested that it would be appropriate to treat the four equations as a system. The Newcastle equation estimated by the feasible SURE method then became

$$\log \hat{y}_t = 8.92 + 0.08x_{1t} - 0.08x_{2t} + 0.31x_{3t} - 0.24x_{4t} - 0.07x_{5t}$$
$$\quad\;\; (0.64)\,(0.09)\quad (0.03)\quad (0.06)\quad (0.05)\quad (0.03)$$

$$(9.29)$$

There are some differences in the two sets of estimates. However, the important point to note is that the estimated standard errors in (9.29) are a good deal smaller than those in (9.28). Although the standard errors quoted in (9.29) only have large sample validity they nevertheless suggest a gain in efficiency.

10. Multivariate Regression

Systems of equations are sometimes written in the form

$$y_t = Bx_t + \epsilon_t, \qquad t = 1, \ldots, T \tag{10.1}$$

where y_t and ϵ_t have the same interpretation as in (9.15), but B is an $N \times K$ matrix of parameters and x_t is a $K \times 1$ vector of regressors. However, unless restrictions are placed on the elements of B, a fully efficient set of estimators will be obtained simply by regressing each element of y_t on the vector x_t. Taking account of the covariances between the disturbances in different equations cannot lead to an improved estimator since each equation has exactly the same set of explanatory variables; cf. (9.14).

The *multivariate least squares* estimator of B may be written as

$$\tilde{B}' = \left[\sum_{t=1}^{T} x_t x_t' \right]^{-1} \sum_{t=1}^{T} x_t y_t' \tag{10.2}$$

The ith column of (10.2) gives the vector of OLS estimates from regressing y_{it} on x_t. Defining the $N \times T$ matrix $Y' = (y_1, \ldots, y_T)$ and the $K \times T$ matrix $X' = (x_1, \ldots, x_T)$ leads to a more compact matrix representation of (10.2), as

$$\tilde{B}' = (\bar{X}'\bar{X})^{-1}\bar{X}'Y \tag{10.3}$$

or

$$\tilde{B} = Y'\bar{X}(\bar{X}'\bar{X})^{-1} \tag{10.4}$$

In terms of (9.14), $\tilde{B} = (b_1, \ldots, b_N)'$.

Model (10.1) may be written in the form (9.15) by defining $\beta = \text{vec}(B')$. The operator vec(.) indicates that the columns of the appropriate matrix are stacked one above the other. Thus if β_i' denotes the ith row of B, $\text{vec}(B') = (\beta_1', \ldots, \beta_N')'$. Corresponding to (9.15) therefore

$$y_t = (I \otimes x_t')\beta + \epsilon_t, \qquad t = 1, \ldots, T \tag{10.5}$$

Thus, suppose that $N = K = 2$, so that the model is

$$\begin{aligned} y_{1t} &= \beta_{11}x_{1t} + \beta_{12}x_{2t} + \epsilon_{1t} \\ y_{2t} &= \beta_{21}x_{1t} + \beta_{22}x_{2t} + \epsilon_{2t} \end{aligned}, \qquad t = 1, \ldots, T \tag{10.6}$$

This may be written in the form (10.5) as follows

$$\begin{bmatrix} y_{1t} \\ y_{2t} \end{bmatrix} = \begin{bmatrix} x_{1t} & x_{2t} & 0 & 0 \\ 0 & 0 & x_{1t} & x_{2t} \end{bmatrix} \begin{pmatrix} \beta_{11} \\ \beta_{12} \\ \beta_{21} \\ \beta_{22} \end{pmatrix} + \begin{bmatrix} \epsilon_{1t} \\ \epsilon_{2t} \end{bmatrix} \tag{10.7}$$

Hypothesis Testing

Hypotheses concerning elements in different rows of B may be framed in terms of the vector β in (10.5). The covariance matrix of b, the OLS estimator of β, is then given by the inverse of (9.13), i.e.

$$\text{Var}(b) = \Omega \otimes (\bar{X}'\bar{X})^{-1} \tag{10.8}$$

Suppose we wish to test $H_0 : \beta_{11} = \beta_{22}$. Any hypothesis which imposes linear restrictions on the parameters may be written in the form $w'\beta = r$; cf. Section 3. Thus

$$\text{Var}(w'\tilde{b}) = w'[\Omega \otimes (\bar{X}'\bar{X})^{-1}]w \tag{10.9}$$

For $N = 1$ this expression reduces to (3.6).

If ϵ_t follows a multivariate normal distribution, then

$$(w'b - r)/\sqrt{\mathrm{Var}(w'b)} \sim N(0, 1)$$

under H_0. However, Ω will usually be unknown, and so must be replaced by an estimator such as (9.21). In general any tests carried out on this basis will only have asymptotic validity, although if an hypothesis involves only coefficients on the *same* explanatory variable in each equation, an exact test is available. This is Hotelling's T^2-test, details of which will be found in Malinvaud (1970, pp. 236–237).

Example   Consider the hypothesis $H_0 : \beta_{11} = \beta_{22}$ in (10.6), and suppose that the sums of squares and cross-products for all variables are given by

	y_1	y_2	x_1	x_2
y_1	250			
y_2	100	500		
x_1	100	50	50	
x_2	75	150	50	100

Thus, for example, $\Sigma x_{1t} y_{2t} = 50$.

The multivariate least squares estimator may be computed from formula (10.4) as

$$\tilde{B} = \begin{bmatrix} 100 & 75 \\ 50 & 150 \end{bmatrix} \begin{bmatrix} 50 & 50 \\ 50 & 100 \end{bmatrix}^{-1} = \begin{bmatrix} 100 & 75 \\ 50 & 150 \end{bmatrix} \begin{bmatrix} \frac{2}{50} & -\frac{1}{50} \\ -\frac{1}{50} & \frac{1}{50} \end{bmatrix}$$

$$= \begin{bmatrix} 2.5 & -0.5 \\ -1 & 2 \end{bmatrix} \tag{10.10}$$

and so $b = \mathrm{vec}(\tilde{B}') = (2.5\ -0.5\ -1\ 2)'$.

The hypothesis to be tested may be framed as $H_0 : w'\beta = 0$, where $w' = (1\ 0\ 0\ -1)$. Expression (10.9) may be written out as

$$\mathrm{Var}(w'b) = \frac{1}{50} \cdot w' \left[\begin{array}{c|c} \omega_{11}\begin{pmatrix} 2 & -1 \\ -1 & 1 \end{pmatrix} & \omega_{12}\begin{pmatrix} 2 & -1 \\ -1 & 1 \end{pmatrix} \\ \hline \omega_{21}\begin{pmatrix} 2 & -1 \\ -1 & 1 \end{pmatrix} & \omega_{22}\begin{pmatrix} 2 & -1 \\ -1 & 1 \end{pmatrix} \end{array} \right] w$$

$$= \frac{1}{50}(\omega_{11} + 2\omega_{12} + 2\omega_{22}) \tag{10.11}$$

as $\omega_{12} = \omega_{21}$.

The estimate of Ω, (9.21), may be obtained from the formula

$$e'e = Y'Y - \tilde{B}X'Y \tag{10.12}$$

cf. (2.8). Dividing through by T gives the following estimate of Ω:

$$\hat{\Omega} = \frac{1}{50}\left[\begin{pmatrix} 250 & 100 \\ 100 & 500 \end{pmatrix} - \begin{pmatrix} 2.5 & -0.5 \\ -1 & 2 \end{pmatrix}\begin{pmatrix} 100 & 50 \\ 75 & 150 \end{pmatrix}\right]$$

$$= \begin{bmatrix} 0.75 & 1 \\ 1 & 5 \end{bmatrix} \tag{10.13}$$

Substituting for the ω_{ij}'s in (10.11) gives

$$\text{var}(w'b) = (0.75 + 10 + 2)/50 = 12.75/50 = 0.255$$

and so the required test statistic is

$$z = \frac{w'b - w'\beta}{\sqrt{\text{var}(w'b)}} = \frac{0.5}{0.505} = 0.99$$

Under $H_0: z \sim AN(0, 1)$ and since $-1.96 < 0.99 < 1.96$, the null hypothesis cannot be rejected at a (nominal) 5% level of significance.

Prediction

The BLUP of y_{T+l} is given by

$$\tilde{y}_{T+l/T} = \tilde{B}x_{T+l} = (I \otimes x_{T+l})b \tag{10.14}$$

and so the prediction error is

$$y_{T+l} - \tilde{y}_{T+l/T} = (I \otimes x_{T+l})(\beta - b) + u_{T+l} \tag{10.15}$$

The MSE of the prediction is therefore

$$\begin{aligned} \text{MSE}(\tilde{y}_{T+l/T}) &= (I \otimes x_{T+l})\text{Var}(b)(I \otimes x_{T+l})' + \Omega \\ &= (I \otimes x'_{T+l})[\Omega \otimes (\bar{X}'\bar{X})^{-1}](I \otimes x_{T+l}) + \Omega \end{aligned} \tag{10.16}$$

$$\begin{aligned} &= \Omega + \Omega \otimes x'_{T+l}(\bar{X}'\bar{X})^{-1}x_{T+l} \\ &= \Omega[1 + x'_{T+l}(\bar{X}'\bar{X})^{-1}x_{T+l}] \end{aligned} \tag{10.17}$$

since $x'_{T+l}(\bar{X}'\bar{X})^{-1}x_{T+l}$ is a scalar.

Constraints

The above example included a test of a restriction across the equations. Once restrictions across equations are taken to hold, the parameters in the model may no longer be efficiently estimated by OLS. The constraints must be incorporated in the estimation procedure in the manner suggested by (9.19) or (9.20).

Similarly, once different restrictions are introduced within equations, a SURE estimator will be appropriate unless, of course, Ω is diagonal.

11. The Method of Instrumental Variables

Suppose we have a linear model of the form

$$y_t = z_t'\delta + u_t, \qquad t = 1, \ldots, T \tag{11.1}$$

where z_t is a $k \times 1$ vector of observations, and δ is a corresponding $k \times 1$ vector of parameters. The models considered until now have been of a similar form to (11.1), but the change in notation in (11.1) as compared with, say, (1.3) serves to indicate that the explanatory variables may be stochastic. If the assumption that the disturbances are distributed independently of the explanatory variables is relaxed, the standard least squares estimators may lose some of their optimal properties. More seriously, if the disturbance at time t is correlated with some of the explanatory variables at time t, OLS will not, in general, be consistent.

Dynamic regression provides an important example of a situation where OLS may be inconsistent. Suppose that the disturbance term in (1.2.3) is serially correlated, i.e. $E(u_t u_{t-1}) \neq 0$. In these circumstances both y_{t-1} and u_t depend on u_{t-1}, and so $E(y_{t-1}u_t) \neq 0$. The consequences of this are examined further in Chapter 8.

A second example arises in the context of a simultaneous equations model. If an endogenous variable appears as a regressor in a particular equation, it will not in general be independent of the disturbance term, since it is itself partly determined by the dependent variable in that equation.

The Instrumental Variable Estimator

It is convenient to write (11.1) in matrix form as

$$y = Z\delta + u \tag{11.2}$$

where Z is a $T \times k$ matrix of explanatory variables. Since these variables may be stochastic, the standard regularity condition is

$$\text{plim } T^{-1}Z'Z = Q_{zz} \tag{11.3}$$

where Q_{zz} is p.d.

The OLS estimator of δ is

$$d = (Z'Z)^{-1}Z'y = \delta + (Z'Z)^{-1}Z'u \qquad (11.4)$$

and if z_t and u_t are correlated it will normally be the case that

$$\text{plim } T^{-1}Z'u \neq 0 \qquad (11.5)$$

Therefore d will be inconsistent, since

$$\text{plim}(d - \delta) = \text{plim}(Z'Z)^{-1}Z'u = Q_{zz} \text{ plim } T^{-1}Z'u \neq 0 \qquad (11.6)$$

Note that, as a general rule, the vector plim $T^{-1}Z'u$ need only contain one non-zero element in order to render all the estimators in d inconsistent.

The consequences of (11.5) on the OLS estimator are therefore rather serious. However, a general approach to estimation problems of this kind is provided by the method of instrumental variables. Let w_t denote a $k \times 1$ vector of variables at time t. If these variables are to provide a suitable set of *instruments* for the variables in z_t, they must have two properties. Firstly they should be uncorrelated with the disturbance term in (11.1) i.e.

$$E(w_t u_t) = 0, \qquad t = 1, \ldots, T \qquad (11.7)$$

Secondly, they should be highly correlated with the explanatory variables.

Let W denote a $T \times k$ matrix in which the tth row is w_t. The standard regularity conditions, (11.3), will be replaced by two similar sets of conditions. These are

$$\text{plim } T^{-1}W'Z = Q_{wz} \qquad (11.8)$$

and

$$\text{plim } T^{-1}W'W = Q_{ww} \qquad (11.9)$$

where Q_{wz} and Q_{ww} are p.d. matrices.

Now consider the instrumental variable (IV) estimator,

$$\tilde{d} = (W'Z)^{-1}W'y \qquad (11.10)$$

If

$$\text{plim } T^{-1}W'u = 0 \qquad (11.11)$$

the IV estimator is consistent since

$$\text{plim}(\tilde{d} - \delta) = \text{plim}(W'Z)^{-1}W'u = Q_{wz} \cdot 0 = 0 \qquad (11.12)$$

Note that (11.11) is weaker than (11.7); it may be interpreted as saying that u_t and w_t are uncorrelated in large samples. If it is assumed that the disturbances in (11.1) are independently and identically distributed with mean zero and constant variance, σ^2, the asymptotic distribution of $\tilde{d}$ may be derived from the results of Section 1.4. Given (11.7), it follows immediately from the first part of the Mann–Wald theorem that (11.11) holds. Furthermore, the second part of the theorem implies that

$$T^{-1/2}W'u \xrightarrow{\ L\ } N(0, \sigma^2 Q_{ww})$$

Now consider

$$\sqrt{T}(\tilde{d} - \delta) = \left(\frac{W'Z}{T}\right)^{-1}\frac{W'u}{\sqrt{T}}$$

In view of (11.8) it follows from Cramér's theorem that

$$\sqrt{T}(\tilde{d} - \delta) \xrightarrow{\ L\ } N(0, \sigma^2 Q_{zw}^{-1} Q_{ww} Q_{wz}^{-1}) \tag{11.13}$$

In other words $\tilde{d}$ is asymptotically normally distributed with a mean of δ and a covariance matrix

$$\mathrm{Avar}(\tilde{d}) = \sigma^2(Z'W)^{-1}W'W(W'Z)^{-1} \tag{11.14}$$

Although instrumental variables would never be used in a classical regression model (except in the trivial sense that W can be set equal to Z), this case nevertheless provides some insight into the properties of the IV estimator in more general situations. To simplify matters further suppose that $k = 1$, so that the model contains a single, non-stochastic explanatory variable, x. If an IV estimator is based on a non-stochastic instrument, it is straightforward to show that it will be unbiased with a variance given by

$$\mathrm{Var}(\tilde{d}) = \sigma^2 \cdot \sum w_t^2 / (\sum x_t w_t)^2 \tag{11.15}$$

The variance of the OLS estimator, d, is

$$\mathrm{Var}(d) = \sigma^2 / \sum z_t^2 \tag{11.16}$$

and if the correlation between z_t and w_t is

$$r_{zw}^2 = \frac{(\sum z_t w_t)^2}{\sum z_t^2 \cdot \sum w_t^2} \tag{11.17}$$

it follows that the efficiency of $\tilde{d}$ is

$$\frac{\mathrm{Var}(d)}{\mathrm{Var}(\tilde{d})} = r_{zw}^2 \tag{11.18}$$

Thus the higher the correlation between z and its instrument, w, the more efficient the IV estimator. This result essentially carries over to the more general cases where IV estimation is applied: the more highly correlated are the elements of z_t and w_t the more effective is the IV estimation procedure. On the other hand if plim $T^{-1}W'Z$ is close to a null matrix, the technique will yield very poor results.

The GIVE Estimator

The above analysis of a single variable regression model suggests that the best instrument for an explanatory variable is the one which is most highly correlated with it. However, there is no reason to restrict the choice to a single instrument. A more efficient procedure would be to take a weighted average of the possible instruments, with the weights chosen so as to maximise the correlation between this composite variable and z. Such an instrument is obtained simply by regressing z on all possible instruments. This yields the required weights. Thus if K instruments, w_j, $j = 1, \ldots, K$, are available, the most efficient choice of instrument is given by

$$w_t = \hat{z}_t = w_t'(\textstyle\sum w_t w_t')^{-1} \sum w_t z_t, \qquad t = 1, \ldots, T \qquad (11.19)$$

where $w_t = (w_1, \ldots, w_K)'$. As formula (11.19) makes clear, the optimal instrument is given by the best linear unbiased predictor of z_t based on the set of all possible instruments.

In the general case when the model contains $k > 1$ explanatory variables, and $K > k$ possible instruments are available, the *generalised instrumental variable estimator* (GIVE) is obtained by selecting the best predictor for each element of z_t in the manner suggested above. If W^* denotes the $T \times K$ matrix of all possible instruments, this may be reduced to the $T \times k$ matrix of optimal instruments, W, by multivariate regression. Thus

$$W = W^*(W^{*'}W^*)^{-1}W^{*'}Z \qquad (11.20)$$

The GIVE procedure may be extended to cases where the disturbances are correlated. Suppose that the disturbance vector has a covariance matrix given by (4.1). By defining a matrix L, such that $L'L = V^{-1}$, the observations may be transformed in the manner of (4.2) to yield

$$Ly = LZ\gamma + Lu$$

The matrix of optimal instruments for LZ is

$$W = W^*(W^{*'}W^*)^{-1}W^{*'}LZ \qquad (11.21)$$

and so the GIVE estimator is

$$\tilde{d} = (Z'L'PLZ)^{-1}Z'L'PLy \qquad (11.22)$$

where

$$P = W*(W*'W*)^{-1}W*'$$

Note that

$$\text{Avar}(\tilde{d}) = \sigma^2 (Z'L'PLZ)^{-1} \tag{11.23}$$

Notes

Sections 1–5, 8   Johnston (1972), Theil (1971), Malinvaud (1970), and Mann and Wald (1943).

Section 6   Plackett (1950), Brown, Durbin and Evans (1975) and Phillips and Harvey (1974). The link between the OLS recursions and the Kalman filter is discussed in TSM.

Section 7   Theil (1971, Ch. 5), Koerts and Abrahamse (1969) and Phillips and Harvey (1974).

Sections 9, 10   Theil (1971, Ch. 7) and Phillips and Wickens (1978, Ch. 3).

Section 11   Sargan (1958, 1961).

Exercises

1. Consider model (3.12) with $x_t = \kappa x_{t-1}$ for $t = 2, \ldots, T$, with $|\kappa| < 1$ and $x_1 \neq 0$. Is condition (3.9) satisfied? Is the OLS estimator consistent?

2. How would you estimate α and β from a set of observations on y_t and x_t if the underlying functional form is $y = \exp(\alpha + \beta/x)$? State any assumptions you make in writing down an appropriate stochastic model.

Graph y against x for $\beta < 0$ and comment on the usefulness of this particular functional form.

3. Derive an expression for the covariance matrix of the restricted least squares estimator (2.21).

4. Show that $R^2 = b'X'Xb/y'y$, where X and y are in deviation from the mean form.

5. Does $T^{3/2}(b - \beta)$ have a limiting distribution in (3.12)?

6. If, in (1.3), $\text{Var}(u_t) = \sigma^2 h_t$ where h_t is known, derive recursions similar to those in (6.3) for the weighted least squares estimator.

7. Prove (10.12).

8. Use the Mann–Wald theorem to find the asymptotic distribution of the OLS estimator of ϕ in the AR(1) model, (1.2.10).

9. Prove (11.23).

3
The Method of Maximum Likelihood

1. Introduction

The method of least squares, and related techniques, such as instrumental variables, use only the first two moments of the observations. However, if, in setting up a model, the form of the distribution is specified, restricting attention to the first two moments may be statistically inefficient. The method of maximum likelihood (ML) attempts to incorporate *all* the information in a model by working with the complete distribution of the observations.

In a linear model, the Gauss—Markov theorem provides a justification for the use of ordinary least squares. The only assumptions made about the distribution of the observations concern the first two moments, and provided that attention is confined to linear estimators OLS can be shown to be best. As noted in Section 1.3, however, the same criterion cannot be applied to a dynamic model. Nor can it be applied to a nonlinear model, such as (2.1.12), even though minimising the residual sum of squares has an obvious intuitive appeal, just as it does for the dynamic model (2.3.15).

If the observations in (2.1.12) and (2.3.15) are assumed to be normally distributed, the least squares estimators can be viewed as maximum likelihood estimators. Their use can then be justified in terms of statistical efficiency, provided, that is, that the sample is reasonably large. The assumption of normality also leads to stronger grounds for using OLS in the linear model. The estimator is now best within the class of all estimators, not simply those which are linear in the observations. However, once the question of specifying the complete distribution of the observations is brought into the picture, it may become more difficult to justify using OLS in certain situations. The logic of the model may be such that a normal distribution is inappropriate, even as an approximation. As an illustration, consider fitting a regression model in which the dependent variable is nominal and is defined to be zero or one. Least squares is statistically inefficient in this case, since it ignores the information that the observations are restricted to two values. The maximum likelihood estimator, on the other hand, takes the full structure of the model into account.

Viewing estimation within the least squares framework derived from classical regression has even more severe limitations. Consider the time series models introduced in Section 1.2. Although (1.2.10) and (1.2.12) can be estimated by least squares, it is not immediately apparent how to treat (1.2.13). The lagged disturbance term is unobservable, and so the unknown parameters cannot simply be estimated by regressing y_t on ϵ_{t-1}. This is typical of the kind of problem encountered in modelling time series and a more general approach is clearly needed.

The remaining parts of this section set out the principle of maximum likelihood estimation, and discuss the implications for computation. Section 2 investigates whether there are general rules for constructing minimum variance unbiased estimators of the parameters in any given model. It turns out that this is only possible in certain special cases. However, a minimum variance bound can be derived, and in Section 3 it is shown that this bound is attained by the ML estimator in large samples. This result is a fundamental one and it provides the main justification for the use of maximum likelihood in terms of statistical efficiency.

The application of ML techniques to regression models is considered in Section 4. In Section 5 it is shown how the likelihood function for a time series model may be decomposed in terms of one-step ahead prediction errors. This decomposition is the basis of ML estimation in all dynamic models.

Section 6 introduces the concept of identifiability. This is concerned with the uniqueness of estimates obtained from a particular model. Identifiability is of fundamental importance since if a model is not identified, it may not make sense. The final section considers the extent to which the ML estimator is robust to the assumption of normality. Since the discussion is carried out within the context of a linear regression model, it is also an examination of the robustness of ordinary least squares.

The Principle of Maximum Likelihood

Suppose that the sample consists of a set of observations on T random variables, $y_1, \ldots, y_T$. The statistical model specifies a distribution for $y_1, \ldots, y_T$, known as the *joint density function*.

This depends on n unknown parameters in a vector $\psi = (\psi_1, \ldots, \psi_n)'$. The joint density may be denoted by $L(y_1, \ldots, y_T; \psi)$, and interpreted as the probability of obtaining particular values of $y_1, \ldots, y_T$.

Once the sample has been drawn, $y_1, \ldots, y_T$ become a set of fixed numbers. The expression for the joint density can then be reinterpreted as a function of ψ, where ψ is any admissible value of the parameter vector, rather than the true value. It therefore indicates the plausibility of different values of ψ, given the sample. Viewed in this way, the expression for the joint density function is called the *likelihood function*, and is denoted by $L(\psi)$.

The maximum likelihood approach to the problem of estimating ψ is to ask what value of ψ is most likely, given the observations. This leads to the following principle: the ML estimate of ψ is given by $\tilde{\psi}$, where $\tilde{\psi}$, is a value such that

$$L(\tilde{\psi}) \geqslant L(\hat{\psi}) \tag{1.1}$$

$\hat{\psi}$ being any other admissible estimate of ψ. The rules by which an ML *estimator*, is constructed must then be such as to ensure that for any particular sample, the resulting ML *estimate* satisfies (1.1).

The classical theory of maximum likelihood estimation is based on a situation in which the T observations are drawn from the same distribution. Furthermore they are drawn independently of each other, and so the joint density function is given by

$$L(y_1, \ldots, y_T; \psi) = \prod_{t=1}^{T} p(y_t; \psi) \tag{1.2}$$

where $p(y_t; \psi)$ is the p.d.f. of y_t for $t = 1, \ldots, T$. The models encountered in econometrics rarely conform to this pattern, but the main results on the properties of ML estimators carry over in almost all cases. A full discussion of this issue is beyond the scope of this book, however, and the proofs sketched out in Sections 2 and 3 are limited to independent and identically distributed random variables.

The Likelihood Equations

In almost all the applications considered here, $L(\psi)$ is a continuous function of ψ and the ML estimator can be found by the differential calculus. As a rule, it is easier to work with the (natural) logarithm of $L(\psi)$. This raises no difficulties, since if $\tilde{\psi}$ satisfies the condition that $\log L(\tilde{\psi}) \geqslant \log L(\hat{\psi})$, it will also satisfy (1.1).

The $n \times 1$ vector of first derivatives, $\partial \log L(\psi)/\partial \psi$, will often be written using the differential operator, D. The advantage of this notation, $D \log L(\psi)$, is that it is easy to indicate if the derivative is to be evaluated at a certain point. Writing $D \log L(\hat{\psi})$ denotes the vector of first derivatives evaluated at $\psi = \hat{\psi}$, i.e.

$$D \log L(\hat{\psi}) = \frac{\partial \log L(\psi)}{\partial \psi} \bigg|_{\psi = \hat{\psi}} \tag{1.3}$$

Similar notation may be used for the $n \times n$ matrix of second derivatives. Thus

$$D^2 \log L(\tilde{\psi}) = \frac{\partial^2 \log L(\psi)}{\partial \psi \, \partial \psi'} \bigg|_{\psi = \tilde{\psi}} \tag{1.4}$$

indicates the Hessian of $\log L$ evaluated at the ML estimate, $\tilde{\psi}$.

The ML estimator is obtained as a solution to the likelihood equations,

$$D \log L(\psi) = 0 \tag{1.5}$$

Since there may be more than one solution to (1.5), it is important to check that a maximum has been obtained. The Hessian will be a negative definite matrix if the solution corresponds to a maximum of the likelihood function, although this does not guarantee that a global, as opposed to a local, maximum has been reached.

Example 1   The log–likelihood function for a sample of T independent observations from a normal distribution with mean μ and variance σ^2 is

$$\log L(\mu, \sigma^2) = -\frac{T}{2} \log 2\pi - \frac{T}{2} \log \sigma^2 - \frac{1}{2\sigma^2} \sum_{t=1}^{T} (y_t - \mu)^2 \tag{1.6}$$

In terms of the general notation, $\psi = (\mu, \sigma^2)'$ and the elements of $D \log L(\psi)$ are

$$\frac{\partial \log L}{\partial \mu} = \frac{1}{\sigma^2} \sum (y_t - \mu) \tag{1.7a}$$

and

$$\frac{\partial \log L}{\partial \sigma^2} = \frac{-T}{2\sigma^2} + \frac{1}{2\sigma^4} \sum (y_t - \mu)^2 \tag{1.7b}$$

Setting these derivatives equal to zero and solving yields

$$\tilde{\mu} = \sum_{t=1}^{T} y_t / T = \bar{y} \tag{1.8a}$$

and

$$\tilde{\sigma}^2 = \sum_{t=1}^{T} (y_t - \bar{y})^2/T \tag{1.8b}$$

These expressions are the only solutions to the likelihood equations and an examination of the matrix of second derivatives shows that they do indeed maximise $\log L(\psi)$. Specifically,

$$D^2 \log L(\psi) = \begin{bmatrix} \dfrac{\partial^2 \log L}{\partial \mu^2} & \dfrac{\partial^2 \log L}{\partial \mu \, \partial \sigma^2} \\ \dfrac{\partial^2 \log L}{\partial \sigma^2 \, \partial \mu} & \dfrac{\partial^2 \log L}{\partial (\sigma^2)^2} \end{bmatrix}$$

$$= \begin{bmatrix} \dfrac{-T}{\sigma^2} & -\dfrac{1}{\sigma^4} \Sigma(y_t - \mu) \\ -\dfrac{1}{\sigma^4} \Sigma(y_t - \mu) & \dfrac{T}{2\sigma^4} - \dfrac{1}{\sigma^6} \Sigma(y_t - \mu)^2 \end{bmatrix} \tag{1.9}$$

Substituting from (1.8) yields

$$D^2 \log L(\tilde{\psi}) = -\begin{bmatrix} \dfrac{T}{\tilde{\sigma}^2} & 0 \\ 0 & \dfrac{T}{2\tilde{\sigma}^4} \end{bmatrix} \tag{1.10}$$

Since $\tilde{\sigma}^2 > 0$, (1.10) is clearly negative definite and so the expressions (1.8a) and (1.8b) do indeed maximise the likelihood function.

Example 2   The p.d.f. for a standardised Cauchy variate is

$$p(y; \psi) = \pi^{-1} \{1 + (y_t - \psi)^2\}^{-1} \tag{1.11}$$

and so the log–likelihood function for a sample of T independent observations is

$$\log L = -\sum_{t=1}^{T} \log[\pi\{1 + (y_t - \psi)^2\}] \tag{1.12}$$

The ML estimator of the parameter, ψ, is given by the solution to the likelihood equation,

$$2\sum(y_t - \psi)/\{1 + (y_t - \psi)^2\} = 0 \qquad (1.13)$$

However, since this equation is nonlinear, it is impossible to write down an explicit solution for $\tilde{\psi}$, as was done in the previous example.

Computational Aspects of Maximum Likelihood

In the second example above, the estimate must be computed by some sort of iterative procedure. This is typical of maximum likelihood estimation, since a direct solution to the likelihood equations tends to be the exception rather than the rule. The need to resort to iterative procedures is sometimes regarded as a drawback to maximum likelihood, and an alternative strategy is to look for simpler estimators which are easier to compute. In econometrics such estimators are often based on the least squares principle. However, there is no guarantee that an estimator constructed along these lines will be efficient, or even consistent.

In the last twenty years or so, the developments in computing facilities and in algorithms for carrying out nonlinear optimisation have been remarkable. This has led to a shift in emphasis towards ML estimation. Since computational considerations are so important in this context, the next chapter is devoted to a review of nonlinear optimisation techniques and their application in econometrics. A good grasp of this material is vital to anyone actually interested in carrying out ML estimation in practice.

The stress on maximum likelihood methodology should not be taken to imply that other methods have become redundant. The amount of computer time which can be saved by the use of a simpler estimator can often be considerable, and there is evidence to suggest that in many cases the small sample performance of a carefully chosen alternative to the ML estimator is perfectly satisfactory. Furthermore, a cruder estimation technique may be more robust when the specification is inappropriate. Nevertheless, while accepting these points, a consideration of maximum likelihood is still fundamental to any discussion of estimation. Indeed, a maximum likelihood framework provides the basis for deriving simpler estimators which retain the property of large sample efficiency. The various two-step estimators described in this book all fall within this category. By following ML principles, it is usually possible to avoid the pitfalls which have often been encountered by those attempting to construct simple, but efficient, estimators within a least squares framework.

2. Sufficiency and the Cramér–Rao Lower Bound

The concept of minimum variance unbiased estimation was introduced in Section 1.3. It was pointed out that it suffers from severe limitations as a criterion for assessing different estimation procedures. Nevertheless, it is a useful ideal to begin with in examining the merits of various approaches to estimation. The first part of this section considers the notion of sufficiency and the way in which it is related to minimum variance unbiased estimation. Although a link can be established, it turns out that the existence of an MVUE can be deduced from sufficiency only in certain special cases. An alternative approach to the problem is to ask whether there is a lower bound to the variance of an unbiased estimator. Showing that a particular estimator attains the lower bound is then tantamount to demonstrating that the estimator is an MVUE.

Sufficiency

Heuristically, a sufficient statistic is one which contains all the information in the sample which is relevant for the estimation of a particular parameter. To state the property formally, suppose that $\hat{\psi}$ is a statistic and that ψ^* is any other statistic which is not a function of $\hat{\psi}$. If, for each ψ^*, the conditional distribution of ψ^*, given $\hat{\psi}$, does not involve ψ, then $\hat{\psi}$ is a *sufficient statistic* for ψ. In other words if $p(\psi^*/\hat{\psi})$ does not involve ψ, $\hat{\psi}$ is sufficient.

As it stands, this definition is not particularly helpful. However, if the likelihood function can be factorised as

$$L(y_1, \ldots, y_T; \psi) = g(\hat{\psi}; \psi) \cdot k(y_1, \ldots, y_T) \qquad (2.1)$$

where $g(\hat{\psi}; \psi)$ is a function of $\hat{\psi}$ and ψ alone, and $k(y_1, \ldots, y_T)$ is independent of ψ, then $\hat{\psi}$ is sufficient. In fact, under fairly general conditions, (2.1) is both a necessary and a sufficient condition for sufficiency.

A sufficient statistic, if it exists, is not unique. Any one-to-one function of a statistic which satisfies (2.1) will itself satisfy (2.1).

Example 1   For a Poisson distribution,

$$p(y; \psi) = e^{-\psi} \psi^y / y!$$

and the likelihood function for T observations may be factorised as

$$L(\psi) = \{e^{-T\psi} \psi^{\Sigma y}\} \cdot \left\{ \prod_{t=1}^{T} y_t! \right\}^{-1} \qquad (2.2)$$

Hence Σy or, equivalently, the sample mean is a sufficient statistic for ψ.

Example 2   The likelihood function for a random sample from an $N(0, \sigma^2)$ distribution is

$$L(\sigma^2) = (2\pi\sigma^2)^{-T/2} \cdot \exp(-\tfrac{1}{2}\Sigma y_t^2/\sigma^2) \qquad (2.3)$$

This factors trivially into a function of σ^2 and Σy_t^2, and a term $k(y_1, \ldots, y_T)$ which is simply unity. Hence Σy_t^2 is a sufficient statistic for σ^2.

When the model contains $n > 1$ unknown parameters, the factorisation criterion generalises to define sets of *jointly* sufficient statistics.

Example 3   The exponent in the likelihood function for a random sample from an $N(\mu, \sigma^2)$ distribution may be split into two parts to yield:

$$L(\mu, \sigma^2) = (2\pi\sigma^2)^{-T/2} \cdot \exp[-\tfrac{1}{2}\sigma^{-2}\{\Sigma(y_t - \bar{y})^2$$
$$+ T(\bar{y} - \mu)^2\}] \quad (2.4)$$

As in the previous example, $k(y_1, \ldots, y_T) = 1$, and $\bar{y}$ and $\Sigma(y_t - \bar{y})^2$ are jointly sufficient for μ and σ^2.

If the family of distributions for a sufficient statistic is *complete*, the Rao–Blackwell theorem may be used to show that a function of the sufficient statistic is an MVUE of the parameter in question; see Silvey (1975, pp. 28–33). Unfortunately there are a large number of cases where this condition does not hold.

The Cramér–Rao Lower Bound

When $n = 1$, the amount of *information* in the sample is defined by

$$I(\psi) = -E[D^2 \log L(\psi)]$$

$$= -\int D^2 \log L(\psi) \cdot L(y_1, \ldots, y_T; \psi) dy_1, \ldots, dy_T \quad (2.5)$$

Note that although $L(\psi)$ is regarded as a likelihood function for the purpose of differentiation, the expectation is taken with respect to the joint density function. The parameter, ψ, is then interpreted as the true value.

In the next sub-section it is shown that $1/I(\psi)$ is the *minimum variance bound* (MVB) for an unbiased estimator of ψ. This result extends to the case of n parameters. If $\hat{\psi}$ is an unbiased estimator of

ψ, the Cramér–Rao inequality states that the covariance matrix of $\hat{\psi}$ must exceed the inverse of the information matrix by a p.s.d. matrix.

If the variance of an unbiased estimator is equal to the MVB, the estimator is said to be efficient. However, failure to attain the lower bound does not necessarily rule out an estimator on the grounds of efficiency, since, in small samples, it is quite possible that no unbiased estimator will actually reach the lower bound. In some cases it can actually be shown that an estimator is the MVUE, even though it is not efficient in the sense defined above.

Example 4 If the sample consists of T independent observations from an $N(\mu, \sigma^2)$ distribution, the information matrix is obtained by taking expectations in (1.9); i.e.

$$I(\mu, \sigma^2) = -E[D^2 \log L(\mu, \sigma^2)] = \begin{pmatrix} \dfrac{T}{\sigma^2} & 0 \\ 0 & \dfrac{T}{2\sigma^4} \end{pmatrix} \qquad (2.6)$$

The MVB for μ is therefore σ^2/T and this is also the variance of the sample mean, $\bar{y}$. An unbiased estimator of σ^2, s^2, was defined in (1.4.10). It follows, as a special case of the result derived in Section 2.7, that the variance of s^2 is $2\sigma^4/(T-1)$. This is greater than the MVB of $2\sigma^4/T$. However, it is impossible to find an unbiased estimator of σ^2 which does attain the lower bound, and s^2 turns out to be the MVUE.

Derivation of the Cramér–Rao Inequality*

To simplify matters assume a single unknown parameter, ψ. The joint density of T independent observations is given by $L = L(y_1, \ldots, y_T; \psi)$ as defined in (1.2). This must, by definition, satisfy the condition

$$\int \cdots \int L \cdot dy_1, \ldots, dy_T = 1 \qquad (2.7)$$

Differentiating both sides of (2.7) with respect to ψ and interchanging the operations of differentiation and integration gives

$$\int \cdots \int \frac{\partial L}{\partial \psi} dy_1, \ldots, dy_T = 0$$

which may be written as

$$E[D \log L(\psi)] = \int \cdots \int \left(\frac{1}{L} \frac{\partial L}{\partial \psi} \right) L \, dy_1, \ldots, dy_T = 0 \qquad (2.8)$$

*Indicates a more difficult section which may be omitted without loss of continuity.

Differentiating again and interchanging operations yields

$$\int \cdots \int \left(\left(\frac{1}{L} \frac{\partial L}{\partial \psi} \right) \frac{\partial L}{\partial \psi} + L \frac{\partial}{\partial \psi} \left(\frac{1}{L} \frac{\partial L}{\partial \psi} \right) \right) dy_1, \ldots, dy_T = 0$$

which becomes

$$\int \cdots \int \left(\left(\frac{1}{L} \frac{\partial L}{\partial \psi} \right)^2 + \frac{\partial^2 \log L}{\partial \psi^2} \right) L \, dy_1, \ldots, dy_T = 0 \qquad (2.9)$$

Rearranging (2.9) then gives

$$E[\{D \log L(\psi)\}^2] = -E[D^2 \log L(\psi)] = I(\psi) \qquad (2.10)$$

If $\hat{\psi}$ denotes an unbiased estimator of ψ, then by definition

$$E(\hat{\psi}) = \int \cdots \int \hat{\psi} L \, dy_1, \ldots, dy_T = \psi \qquad (2.11)$$

Differentiating (2.11) yields

$$\int \cdots \int \hat{\psi} D \log L(\psi) L \, dy_1, \ldots, dy_T = 1 \qquad (2.12)$$

From (2.8), $E[D \log L(\psi)] = 0$ and so

$$E[\psi D \log L(\psi)] = \psi \cdot E[D \log L(\psi)] = 0$$

Incorporating this result into (2.10) gives

$$1 = \int \cdots \int (\hat{\psi} - \psi) D \log L(\psi) \cdot L \, dy_1, \ldots, dy_T$$

and so, by the Cauchy–Schwartz inequality,

$$1 \leqslant \left[\int \cdots \int (\hat{\psi} - \psi)^2 L \, dy_1, \ldots, dy_T \right]$$
$$\cdot \left[\int \cdots \int \{D \log L(\psi)\}^2 L \, dy_1, \ldots, dy_T \right] \qquad (2.13)$$

The first term on the r.h.s. of (2.13) is $E[(\hat{\psi} - \psi)^2]$. The second is the information, $I(\psi)$, defined in (2.10). Rearranging (2.13) gives the *Cramér–Rao inequality*

$$\boxed{\mathrm{Var}(\hat{\psi}) \geqslant 1/I(\psi)} \qquad (2.14)$$

3. Properties of the Maximum Likelihood Estimator

It follows from the factorisation theorem, (2.1), that the ML estimator of ψ is a function of the sufficient statistic. This means that the ML estimator depends only on relevant information in the sample, although it does not necessarily mean that it uses that information in the best possible way. Nevertheless, when there exists an unbiased estimator whose variance attains the lower bound, this estimator is identical to the ML estimator.

Unfortunately the MVB is often unattainable. This was illustrated in the last example, in connection with the estimation of σ^2. Indeed in this case, the ML estimator, (1.8b), is biased. However, since $E(\tilde{\sigma}^2) = \sigma^2(T-1)/T$, the bias becomes negligible as T becomes large. Furthermore the variance of $\tilde{\sigma}^2$ tends towards the MVB of $2\sigma^4/T$. Taken together, these results are indicative of a more general result, namely that ML estimators normally attain the Cramér—Rao lower bound in large samples.

Suppose that the information matrix, $I(\psi)$, divided by T converges to a p.d. matrix $IA(\psi)$ as $T \to \infty$, i.e.

$$IA(\psi) = \lim T^{-1}I(\psi) \tag{3.1}$$

Then, subject to certain regularity conditions $\sqrt{T}(\tilde{\psi} - \psi)$ has a limiting multivariate normal distribution with mean vector zero and covariance matrix, $IA^{-1}(\psi)$. The same result is implied by the statement that $\tilde{\psi}$ is asymptotically normal with mean ψ and covariance matrix,

$$\text{Avar}(\tilde{\psi}) = T^{-1}IA^{-1}(\psi) \tag{3.2}$$

Compare the discussion in Section 2.3. The proof is set out below for a single parameter.

Because ML estimators attain the MVB in large samples, they are said to be asymptotically efficient. Suppose that $\tilde{\psi}$ is the ML estimator of a single parameter, ψ, and that $\hat{\psi}$ is another estimator, which is asymptotically normal with mean ψ and variance, $\text{Avar}(\hat{\psi})$. The *asymptotic efficiency* of $\hat{\psi}$ is given by

$$\text{Eff}(\hat{\psi}) = \text{Avar}(\tilde{\psi})/\text{Avar}(\hat{\psi}) = T^{-1}IA^{-1}(\psi)/\text{Avar}(\hat{\psi}) \tag{3.3}$$

In general, asymptotic efficiency is the only theoretical guide to the relative merits of different estimators in econometrics and time series analysis. Exact finite sample distributions are difficult to obtain, and even when they are available, a measure analogous to (3.3) is only really applicable if both estimators are normal.

Derivation of the Asymptotic Distribution of the Maximum Likelihood Estimator*

Consider a sample of T independent observations drawn from a p.d.f. containing a single unknown parameter, ψ. The likelihood function, which is of the form (1.2), may be expanded around the true value of

ψ using Taylor's theorem. This yields:

$$D \log L(\tilde{\psi}) \simeq D \log L(\psi) + (\tilde{\psi} - \psi)D^2 \log L(\psi) \tag{3.4}$$

The leading term in the remainder of the expansion involves $(\tilde{\psi} - \psi)^2$, and it may be shown to be of a smaller order than $(\tilde{\psi} - \psi)D^2 \log L(\psi)$ when the regularity conditions are satisfied.

By definition, $D \log L(\tilde{\psi})$ is equal to zero, and so (3.4) may be rearranged to give

$$\sqrt{T}(\tilde{\psi} - \psi) \simeq \frac{-\sqrt{T} D \log L(\psi)}{D^2 \log L(\psi)} \tag{3.5}$$

From (1.2) it will be seen that $D \log L(\psi)$ is actually the sum of T identically distributed independent random variables, since

$$D \log L(\psi) = \sum_{t=1}^{T} \frac{\partial \log p(y_t; \psi)}{\partial \psi} = \sum_{t=1}^{T} D \log p(y_t; \psi)$$

The random variable, $D \log p(y_t; \psi)$, may be regarded as the derivative of a log–likelihood for a sample of size one, and from (2.8) and (2.10) it follows that it has zero expectation and variance $i(\psi)$ for $t = 1, \ldots, T$. The lower case notation, $i(\psi)$, indicates the information for a single observation. For the full sample

$$I(\psi) = T \cdot i(\psi) \tag{3.6}$$

and so $D \log L(\tilde{\psi})$ has zero expectation and variance $I(\psi)$. Under the regularity conditions (3.1) will hold, and so, from the Central Limit Theorem, $\xi = T^{-1/2} D \log L(\psi)/\sqrt{I(\psi)/T}$ has a limiting standardised normal distribution, i.e., $\xi \xrightarrow{L} N(0, 1)$.

Substituting for $D \log(\psi)$ in (3.5) yields

$$\sqrt{T}(\tilde{\psi} - \psi) \simeq \frac{\sqrt{I(\psi)/T}}{D^2 \log L(\psi)/T} \cdot \xi$$

Since

$$\mathrm{plim}\left\{\frac{\sqrt{I(\psi)/T}}{D^2 \log L(\psi)/T}\right\} = \{IA(\psi)\}^{-1/2}$$

it follows from Cramér's theorem, (1.4.17), that $\sqrt{T}(\tilde{\psi} - \psi)$ has a limiting normal distribution with mean zero and variance $IA^{-1}(\psi)$.

4. Maximum Likelihood Estimation of Regression Models

This section begins by examining the relationship between ML and OLS in a classical linear regression model. The results are then extended to cover the nonlinear case, and in the last sub-section the estimation of a model with a dichotomous dependent variable is considered.

The ML estimation of a linear regression model with hetero-scedastic disturbances is also examined. A model of this kind represents something of a departure from the usual ML framework in that although the observations are independently distributed, they are no longer identically distributed. However, the results on asymptotic efficiency continue to hold and the properties of the ML estimators may be determined exactly as before. In a similar way, ML theory can be extended to cover systems of equations. The SURE model of Section 2.9 provides an illustration.

The Classical Linear Regression Model

The log–likelihood function for the classical linear regression model (2.1.5) with normally distributed disturbances is

$$\log L(\beta, \sigma^2) = -\frac{T}{2} \log 2\pi - \frac{T}{2} \log \sigma^2$$

$$-\frac{1}{2\sigma^2} (y - X\beta)'(y - X\beta) \tag{4.1}$$

Maximising (4.1) with respect to β and σ^2 yields the estimators

$$\tilde{\beta} = b = (X'X)^{-1}X'y \tag{4.2}$$

and

$$\tilde{\sigma}^2 = e'e/T \tag{4.3}$$

The estimator of β is identical to the OLS estimator, (2.2.4). If the log–likelihood function is written in the form

$$\log L(\beta, \sigma^2) = -\frac{T}{2} \log 2\pi - \frac{T}{2} \log \sigma^2 - \frac{1}{2\sigma^2} S(\beta) \tag{4.4}$$

where $S(\beta)$ is the sum of squares function defined in (2.2.1), the reason for this correspondence becomes clear: $\partial S(\beta)/\partial \beta = 0$ implies that $\partial \log L/\partial \beta = 0$.

The information matrix is

$$I(\beta, \sigma^2) = \begin{pmatrix} \sigma^{-2}X'X & 0 \\ 0 & T/2\sigma^4 \end{pmatrix} \tag{4.5}$$

which is an obvious generalisation of (2.6), the information matrix associated with the estimator of a population mean. It was shown in

Section 2.3 that the OLS estimator of β is normally distributed with mean β and covariance matrix, $\sigma^2(X'X)^{-1}$. Hence it attains the lower bound and so is the MVUE of β. However, as in the location model, $\tilde{\sigma}^2$ is biased and s^2, the unbiased estimator, fails to attain the lower bound.

Although the MVB for σ^2 is unattainable in finite samples, both $\tilde{\sigma}^2$ and s^2 are MVB estimators in large samples. In the case of s^2, it was shown in Section 2.7 that $\text{Var}(s^2) = 2\sigma^4/(T-k)$. While this variance exceeds the MVB of $2\sigma^4/T$ in small samples, it obviously tends towards it as $T \to \infty$.

Nonlinear Regression

Some examples of nonlinear regression models were given in Section 2.1. A general formulation is as follows:

$$y_t = f(x_t; \beta) + \epsilon_t, \qquad t = 1, \ldots, T \tag{4.6}$$

where x_t is a $k \times 1$ vector of exogenous variables, β is an $n \times 1$ vector of parameters, and $f(x_t; \beta)$ denotes a function of x_t and β. If the disturbances are normally distributed the log–likelihood function can be written as in (4.4), with

$$S(\beta) = \sum_{t=1}^{T} \{y_t - f(x_t; \beta)\}^2 \tag{4.7}$$

defining the sum of squares function. The nonlinear (ordinary) least squares estimator of β, b, is therefore identical to the ML estimator.

Differentiating the log–likelihood function with respect to β yields

$$\frac{\partial \log L}{\partial \beta} = -\sigma^{-2} \sum z_t \{y_t - f(x_t; \beta)\} \tag{4.8}$$

where

$$z_t = \frac{\partial f_t(x_t; \beta)}{\partial \beta} = -\frac{\partial \epsilon_t}{\partial \beta}, \qquad t = 1, \ldots, T \tag{4.9}$$

Differentiating a second time leads to the expression

$$E\left\{-\frac{\partial^2 \log L}{\partial \beta \, \partial \beta'}\right\} = \sigma^{-2} \sum z_t z_t' \tag{4.10}$$

The off-diagonal blocks in the information matrix, $I(\beta, \sigma^2)$, are null since

$$E\left\{-\frac{\partial^2 \log L}{\partial \beta \, \partial \sigma^2}\right\} = -E\{\sigma^{-4}\textstyle\sum z_t \epsilon_t\} = 0 \qquad (4.11)$$

This result follows immediately because the elements of z_t are fixed.
    Under suitable regularity conditions, therefore,

$$b \sim AN\{\beta, \sigma^2(\textstyle\sum z_t z_t')^{-1}\} \qquad (4.12)$$

The regularity conditions are similar to those imposed in the linear model. In particular the condition that $\lim T^{-1}\sum z_t z_t'$ converges to a p.d. matrix is a natural extension of (2.3.9). The large sample properties of the ML estimator of σ^2 are exactly as in the linear model with

$$\tilde{\sigma}^2 \sim AN(\sigma^2, 2\sigma^4/T) \qquad (4.13)$$

    If normality of the disturbances is not assumed, the nonlinear least squares estimator will, in general, still have the large sample distribution indicated by (4.12); cf. the discussion in Section 2.3.

Heteroscedasticity

Two heteroscedastic regression models will be examined in this subsection. Apart from extending the ML results, these two models provide some insight into the kind of results which may be expected from more general regression models. In both cases the regression model is of the form

$$y_t = x_t'\beta + u_t, \qquad t = 1, \ldots, T \qquad (4.14)$$

with the disturbances independently distributed about a mean of zero.

Multiplicative Heteroscedasticity   Suppose that the variance of the disturbance term is given by the expression

$$\sigma_t^2 = \exp(w_t'\delta), \qquad t = 1, \ldots, T \qquad (4.15)$$

where w_t is an $r \times 1$ vector of nonstochastic, known variables, and δ is an $r \times 1$ vector of unknown parameters. The variables in w_t may, or may not be related to the elements of x_t. The parameter vectors β and δ, however, contain no elements in common. In general the first element of w_t will be unity.
    The specification (4.15) is known as *multiplicative heteroscedasticity*. It encompasses a number of special cases which are of some interest. For example, if $w_t = [1, \log x_{jt}]'$ and $\delta = [\log \sigma^2, \alpha]'$

$$\sigma_t^2 = \sigma^2 x_{jt}^\alpha \qquad (4.16)$$

Thus the variance is proportional to the jth explanatory variable raised to the power α.

The log–likelihood function is given by

$$\log L(\beta, \delta, \sigma^2) = -\frac{T}{2}\log 2\pi - \frac{1}{2}\sum w'_t \delta$$

$$-\frac{1}{2}\sum \exp(-w'_t \delta)(y_t - x'_t \beta)^2 \qquad (4.17)$$

Differentiating (4.17) with respect to β and δ yields

$$\frac{\partial \log L}{\partial \beta} = \sum x_t \exp(-w'_t \delta)(y_t - x'_t \beta) \qquad (4.18)$$

and

$$\frac{\partial \log L}{\partial \delta} = -\frac{1}{2}\sum w_t + \frac{1}{2}\sum w_t \exp(-w'_t \delta)(y_t - x'_t \beta)^2 \qquad (4.19)$$

Differentiating a second time, multiplying by minus one and taking expectations gives the information matrix

$$I(\beta, \delta) = \begin{pmatrix} \sum \sigma_t^{-2} x_t x'_t & 0 \\ 0 & \frac{1}{2}\sum w_t w'_t \end{pmatrix} \qquad (4.20)$$

The ML estimators, $\tilde{\beta}$ and $\tilde{\delta}$, are obtained by setting the first derivatives, (4.18) and (4.19), equal to zero and solving the resulting nonlinear equations; see Section 4.4 of the next Chapter. Under suitable regularity conditions it can be shown that $\tilde{\beta}$ and $\tilde{\delta}$ are asymptotically normal with a covariance matrix given by the asymptotic information matrix. In view of the block diagonality of (4.20), $\tilde{\beta}$ and $\tilde{\delta}$ are independently distributed in large samples and so

$$\tilde{\beta} \sim AN[\beta, (\sum \sigma_t^{-2} x_t x'_t)^{-1}] \qquad (4.21)$$

while

$$\tilde{\delta} \sim AN[\delta, 2(\sum w_t w'_t)^{-1}] \qquad (4.22)$$

Dependent Variable Heteroscedasticity   Regression models in which the variance of the dependent variable is proportional to the square of its expectation have been proposed in a number of instances. An example is the work of Prais and Houthakker (1955, pp. 55–56) where household consumption of tea is modelled in terms of income and family size. The variance of the disturbance term may be written

$$\sigma_t^2 = \sigma^2(x'_t \beta)^2, \qquad t = 1, \ldots, T \qquad (4.23)$$

and under the normality assumption

$$I(\beta, \sigma^2) = \begin{pmatrix} (2 + \sigma^{-2}) \sum \sigma_t^{-2} x_t x_t' & \sigma^{-2} (\sum \sigma_t^{-2} x_t x_t') x_t \\ \sigma^{-2} x_t' \sum \sigma_t^{-2} x_t x_t' & T/2\sigma^4 \end{pmatrix} \qquad (4.24)$$

Unlike (4.20), the information matrix is not block diagonal, and so the ML estimators of β and σ^2 are not independently distributed in large samples. This has some important implications. For example, suppose an estimator is constructed by first applying OLS and then computing the WLS estimator

$$\hat{\beta} = [\sum (x_t' b)^{-2} x_t x_t']^{-1} \sum (x_t' b)^{-2} x_t y_t \qquad (4.25)$$

This has the same large sample distribution as a WLS estimator in which the observations are weighted by the true variances, $\sigma_1^2, \ldots, \sigma_T^2$; see Theil (1971, Chapter 8). Thus it is asymptotically normal with mean β and variance

$$\text{Avar}(\hat{\beta}) = \sigma^2 [\sum \sigma_t^{-2} x_t x_t']^{-1} \qquad (4.26)$$

However, a comparison of (4.26) with the lower bound matrix obtained by inverting (4.24) shows $\hat{\beta}$ to be inefficient; see Amemiya (1973).

If the disturbances in a model exhibiting this type of hetero-scedasticity are normal, therefore, ML estimation is to be preferred over a feasible GLS procedure. However, if the disturbances follow a Gamma distribution, it turns out that the information matrix is block diagonal, and the WLS estimator, (4.25), is asymptotically efficient; cf. Exercise 5. In fact a Gamma distribution is more sensible for a model of this kind since it implies that the observations cannot be negative.

Seemingly Unrelated Regressions

The SURE system was introduced in Section 2.9. Writing the model in the form (2.9.15), the joint density of the N observations at time t is given by

$$p(y_t) = (2\pi)^{-N/2} \cdot |\Omega|^{-1/2} \cdot \exp\{(y_t - X_t\beta)'\Omega^{-1}(y_t - X_t\beta)\} \qquad (4.27)$$

The log–likelihood function for all T sets of observations is therefore

$$\log L = -\frac{TN}{2} \cdot \log 2\pi - \frac{T}{2} \log |\Omega|$$

$$-\frac{1}{2} \sum_{t=1}^{T} (y_t - X_t\beta)'\Omega^{-1}(y_t - X_t\beta) \qquad (4.28)$$

Differentiating with respect to β yields

$$\frac{\partial \log L}{\partial \beta} = \sum X_t' \Omega^{-1}(y_t - X_t \beta) \tag{4.29}$$

Compare this result with (A.3). Setting (4.29) equal to zero and re-arranging gives the SURE estimator (2.9.17). However, this can only be computed if Ω is known.

Differentiating $\log L$ with respect to Ω is less straightforward. However, by making use of results (A.5) and (A.6), the derivatives of $\log L$ with respect to the elements of the inverse of Ω are readily obtainable. Thus

$$\frac{\partial \log L}{\partial \Omega^{-1}} = \frac{T}{2}\Omega - \frac{1}{2}\sum_{t=1}^{T}(y_t - X_t\beta)(y_t - X_t\beta)' \tag{4.30}$$

a standard result on determinants having been used to set

$$-\frac{T}{2}\log |\Omega| = -\frac{T}{2}\log |\Omega^{-1}|^{-1} = \frac{T}{2}\log |\Omega^{-1}| \tag{4.31}$$

in the likelihood function. Because of the invariance property of ML estimators, the ML estimator of Ω is equal to the inverse of the ML estimator of Ω^{-1}. Thus setting (4.30) equal to a matrix of zeros and solving will yield the ML estimators of both Ω^{-1} and Ω. In fact it is Ω which is of primary interest, and the ML estimator is given directly as

$$\tilde{\Omega} = T^{-1}\sum_{t=1}^{T} e_t e_t' \tag{4.32}$$

where $e_t = y_t - X_t \tilde{\beta}$.

Qualitative Dependent Variables

Suppose that cross-sectional data on a sample of N households is available, and that car ownership is to be modelled in terms of a number of variables such as income, family size and geographical location. The dependent variable then takes a value of unity if a family owns a car and zero otherwise. If $f(x;\beta)$ denotes the probability of a family owning a car, as a function of the explanatory variables, x, the model may be specified as

$$Pr(y_j = 1) = f(x_j; \beta),$$
$$\qquad\qquad\qquad\qquad j = 1, \ldots, N \qquad\qquad (4.33)$$
$$Pr(y_j = 0) = 1 - f(x_j; \beta),$$

where β is a vector of unknown parameters.

Since $E(y_j) = f(x_j, \beta)$, (4.33) may be cast in familiar regression form by writing

$$y_j = f(x_j; \beta) + u_j, \qquad j = 1, \ldots, N \qquad\qquad (4.34)$$

Model (4.34) is similar to (4.6), but estimation by least squares is clearly inappropriate. The disturbance term, u_j, is far from being normal since it can take only two values. However, if the observations are independent, the log–likelihood function may be written down directly. Arranging the data so that observations for households with cars are indexed $i = 1, \ldots, M$, gives

$$\log L = \sum_{i=1}^{M} \log f(x_i; \beta) + \sum_{i=M+1}^{N} \log\{1 - f(x_i; \beta)\} \qquad (4.35)$$

An important example of a model of this kind is the *logit* specification

$$f(x_j; \beta) = 1/\{1 + \exp(-x_j'\beta)\} \qquad\qquad (4.36)$$

Although the likelihood equations are nonlinear, the solutions may be obtained relatively easily by applying one of the methods described in the next chapter. An expression for $\mathrm{Avar}(\tilde{\beta})$ is given by evaluating the information matrix in the usual way. Further details, together with applications will be found in Domencich and McFadden (1975). Although the discussion here has been solely in terms of a dichotomous dependent variable, the same techniques may be used when the variable can be classified in more than two ways.

5. Dependent Observations

In a time series model the observations are, by definition, dependent. This immediately raises the question of whether the standard results on the properties of the ML estimator are still valid. The theorems normally presented rest on the assumption that the observations are independently and identically distributed. This is needed in order to invoke the classical Central Limit Theorem. However, it was argued in the last section that the usual ML results carry through if the assumption of identically distributed observations is dropped. This is also, in general, the case with dependent observations and it will be assumed throughout this book that the ML estimator of a parameter vector, ψ, is asymptotically normal with a mean of ψ and a covariance matrix equal to the inverse of the information matrix. No proof will be given for this assertion, however, and the interested reader should refer to Hannan (1970) or Fuller (1976).

This section is concerned with the practical question of constructing the likelihood function for a set of dependent observations. The most convenient approach is usually through the *prediction error decomposition*. This result is an enormously powerful one, leading to suitable expressions for the exact likelihood function, as well as suggesting suitable approximations. Furthermore, its implications extend to multivariate series of observations, a topic which is pursued in the latter part of the section.

Prediction Error Decomposition

Consider a set of T dependent observations with a mean vector, μ and a covariance matrix $\sigma^2 V$, i.e. $y \sim N(\mu, \sigma^2 V)$. The motivation behind the introduction of the scalar quantity, σ^2, is one of convenience; cf. the treatment of GLS in Section 2.4. Its inclusion is peripheral to the main result, which would remain essentially the same if σ^2 were set equal to one.

In many time series models the elements of μ will be identical, and in some cases will be zero. However, allowing μ to be unrestricted permits greater generality than might seem to be the case at first sight. In particular, if $\mu = X\beta$, where X and β are as defined in Section 2.1, the generalised regression model of Section 2.4 is obtained.

If y is assumed to have a multivariate normal distribution, i.e. $y \sim N(\mu, \sigma^2 V)$, the joint density of the observations is

$$\log L(y) = -\frac{T}{2} \log 2\pi - \frac{T}{2} \log \sigma^2 - \frac{1}{2} \log |V|$$

$$-\frac{1}{2} \sigma^{-2} (y - \mu)' V^{-1} (y - \mu) \qquad (5.1)$$

This may be factored into two parts by writing

$$\log L(y) = \log L(y_1, \ldots, y_{T-1}) + \log l(y_T / y_{T-1}, \ldots, y_1) \quad (5.2)$$

The second term on the right-hand side of (5.2) is the distribution of the last observation, *conditional* on all the previous observations. Splitting the likelihood up in this way follows from the definition of conditional probability. In its most elementary form this states that

$$Pr(A) = Pr(A/B) \cdot Pr(B)$$

where A and B are events. Since $\log L(y)$ is, itself, a probability, namely the probability of obtaining the sample $y_1, \ldots, y_T$, equating A with y and B with the subset of y which excludes y_T gives (5.2).

Now consider the problem of estimating y_T, given that $y_{T-1}, \ldots, y_1$ are known. If $\hat{y}_{T/T-1}$ is an estimator of y_T constructed on this basis the estimation, or prediction, error may be split into two parts:

$$y_T - \hat{y}_{T/T-1} = \{y_T - E(y_T/y_{T-1}, \ldots, y_1)\}$$
$$+ \{E(y_T/y_{T-1}, \ldots, y_1) - \hat{y}_{T/T-1}\} \qquad (5.3)$$

where $E(y_T/y_{T-1}, \ldots, y_1)$ is the mean of the distribution of y_T, conditional on $y_{T-1}, \ldots, y_1$. It follows from (5.3) that

$$MSE(\hat{y}_{T/T-1}) = \text{Var}(y_T/y_{T-1}, \ldots, y_1)$$
$$+ E[\{\hat{y}_{T/T-1} - E(y_T/y_{T-1}, \ldots, y_1)\}^2] \qquad (5.4)$$

The first term on the r.h.s. of (5.4) is independent of $\hat{y}_{T/T-1}$, and so the minimum mean square estimator (MMSE) of y_T conditional on $y_{T-1}, \ldots, y_1$ is

$$\tilde{y}_{T/T-1} = E(y_T/y_{T-1}, \ldots, y_1)$$

The prediction error variance associated with $\tilde{y}_{T/T-1}$ is $E[(y_T - \tilde{y}_{T/T-1})^2]$. This is identical to $MSE(\tilde{y}_{T/T-1})$, which in view of the construction of $\tilde{y}_{T/T-1}$ is equal to $\text{Var}(y_T/y_{T-1}, \ldots, y_1)$. This quantity will be denoted by $\sigma^2 f_T$.

Since the observations are normally distributed, both components on the right hand side of (5.2) are normal. The second term may be written

$$\log l(y_T/y_{T-1}, \ldots, y_1) = -\tfrac{1}{2} \log 2\pi - \tfrac{1}{2} \log \sigma^2$$
$$- \tfrac{1}{2} \log f_T - \tfrac{1}{2}\sigma^{-2}(y_T - \tilde{y}_{T/T-1})^2/f_T$$

and interpreted as the distribution of the prediction error, $y_T - \tilde{y}_{T/T-1}$.

The decomposition of (5.2) may be repeated with respect to the likelihood of the first $T - 1$ observations, and then repeated further until we obtain

$$\log L(y) = \sum_{t=2}^{T} \log l(y_t/y_{t-1}, \ldots, y_1) + \log l(y_1) \qquad (5.5)$$

For $t = 2, \ldots, T$, the mean of y_t conditional on $y_{t-1}, \ldots, y_1$, must be equal to $\tilde{y}_{t/t-1}$, the MMSE of $\tilde{y}_t$ given the previous observations. Each of the conditional distributions is therefore the distribution of the error associated with the optimal predictor, while $l(y_1)$ is the unconditional distribution of y_1. However, if μ_1 is regarded as being the MMSE of y_1, given no previous observations, the term $y_1 - \mu_1$ is the prediction error associated with y_1. It is therefore appropriate to denote the variance of y_1 by $\sigma^2 f_1$.

Expression (5.5) allows the likelihood function to be decomposed into the joint distribution of T independent prediction errors,

$$\nu_t = y_t - \tilde{y}_{t/t-1}, \qquad t = 1, \ldots, T \tag{5.6}$$

where $\tilde{y}_{1/0} = \mu_1$. Each prediction error has mean zero and variance $\sigma^2 f_t$ and so (5.1) becomes

$$\log L(y) = -\frac{T}{2} \log 2\pi - \frac{T}{2} \log \sigma^2 - \frac{1}{2} \sum_{t=1}^{T} \log f_t - \frac{1}{2}\sigma^{-2} \sum_{t=1}^{T} \nu_t^2/f_t$$

$$\tag{5.7}$$

The prediction error decomposition may be interpreted in terms of a Cholesky decomposition of V^{-1}. If $\bar{L}$ is a lower triangular matrix with ones on the leading diagonal, V^{-1} may be factorised as $V^{-1} = \bar{L}'D\bar{L}$, where $D = \text{diag}(f_1^{-1}, \ldots, f_T^{-1})$. This factorisation is unique and the prediction errors in (5.6) are given by the transformation $\nu = \bar{L}y$. Since the Jacobian of this transformation is $|\bar{L}| = 1$, the joint distribution of the elements in the vector $\nu = (\nu_1, \ldots, \nu_T)'$ is as in (5.7). Note that $|V^{-1}| = |\bar{L}'| \cdot |D| \cdot |\bar{L}| = |D|$ and so $\log |V| = \Sigma \log f_t$.

The Cholesky decomposition can be used as the basis for factorising V^{-1} numerically and computing the prediction errors directly from the transformation $\bar{L}y$. Alternatively, the underlying model may be cast in state space form and the prediction errors computed by the Kalman filter. The theory behind state space models and the way in which the Kalman filter can be exploited in ML estimation is described in TSM.

Considerable simplification can usually be achieved in the prediction error decomposition if it is felt that an approximation to the likelihood function will be adequate in practice. As a general rule, an approximation can be justified on theoretical grounds by making certain assumptions about the initial observations. This usually results in prediction errors with a constant variance, σ^2, and so $f_t = 1$ for all t. The log–likelihood function becomes

$$\log L = -\frac{T}{2} \log 2\pi - \frac{T}{2} \log \sigma^2 - \frac{1}{2}\sigma^{-2} \sum \epsilon_t^2 \tag{5.8}$$

where the change in notation from ν_t to ϵ_t serves to indicate that each prediction error will now coincide with an underlying white noise disturbance in the model. If the likelihood function can be reduced to a form like (5.8) there are important implications for estimation. Maximising the likelihood function is equivalent to minimising the sum of squares function,

$$S(\psi) = \sum \epsilon_t^2 \qquad (5.9)$$

where ψ denotes all the parameters in the model apart from σ^2.

Example 1   The first order autoregressive model

$$y_t = \phi y_{t-1} + \epsilon_t, \qquad |\phi| < 1 \qquad (5.10)$$

where $\epsilon_t \sim NID(0, \sigma^2)$ was introduced in (1.2.10). In constructing the likelihood function it will be assumed that the process began at some time in the distant past, but is only observed at times $t = 1, \ldots, T$.

For any t the distribution of y_t, given all past observations, has a mean of ϕy_{t-1} and a variance of σ^2. Thus the last $T - 1$ prediction errors are identical to the last $T - 1$ disturbances since

$$\nu_t = y_t - \tilde{y}_{t/t-1} = y_t - \phi y_{t-1} = \epsilon_t, \qquad t = 2, \ldots, T \qquad (5.11)$$

Corresponding to (5.5), the log–likelihood function may be decomposed into the density functions of $\epsilon_2, \ldots, \epsilon_T$, plus the unconditional density of y_1. This is normally distributed with mean 0 and variance $\sigma^2/(1 - \phi^2)$. The first prediction error is therefore equal to y_1 itself rather than being equal to the disturbance ϵ_1. Setting $\nu_1 = y_1$, the log–likelihood function for all T observations, $y_1, \ldots, y_T$, may be written in the form of (5.7), i.e.

$$\log L(y) = -\frac{T}{2} \log 2\pi - \frac{T}{2} \log \sigma^2 + \frac{1}{2} \log(1 - \phi^2)$$

$$- \frac{1}{2} \sigma^{-2}(1 - \phi^2)y_1^2 - \frac{1}{2}\sigma^{-2} \sum_{t=2}^{T} (y_t - \phi y_{t-1})^2$$

$$(5.12)$$

Note that $f_t = 1$ for $t = 2, \ldots, T$, while $f_1 = (1 - \phi^2)^{-1}$.

Interpreting (5.12) in terms of a Cholesky decomposition yields a matrix $L = D^{1/2} \tilde{L}$, defined by

$$
L = \begin{bmatrix}
\sqrt{1-\phi^2} & 0 & 0 & \cdots & 0 & 0 \\
-\phi & 1 & 0 & \cdots & 0 & 0 \\
0 & -\phi & 1 & \cdots & 0 & 0 \\
\vdots & \vdots & \vdots & & \vdots & \vdots \\
0 & 0 & 0 & \cdots & 1 & 0 \\
0 & 0 & 0 & \cdots & -\phi & 1
\end{bmatrix}
\qquad (5.13)
$$

This matrix plays an important role in the construction of a GLS estimator for a regression model with first-order autoregressive disturbances.

Finally, the likelihood function simplifics to the form (5.8) if the third and fourth terms in (5.12) are removed. A formal justification for this is given by taking y_1 to be fixed. The ML estimator of ϕ is then linear, being given by a regression of y_t on y_{t-1}.

Example 2   The full likelihood function for an MA(1) model, (1.2.13), may be calculated by the Kalman filter as described in TSM. However, the usual approach is based on an approximation in which ϵ_0 is set equal to zero. Given ϵ_{t-1}, the MMSE of y_t is $\theta\epsilon_{t-1}$ and the prediction error is

$$
\epsilon_t = y_t - \theta\epsilon_{t-1}, \qquad t = 1, \ldots, T \qquad (5.14)
$$

This expression is a straightforward re-arrangement of (1.2.13) and if $\epsilon_0 = 0$, the prediction error decomposition leads to a likelihood function of the form (5.8). The ML or *conditional sum of squares* (CSS) estimator of θ is obtained by minimising (5.9) with respect to θ. This is an exercise in nonlinear least squares.

Example 3   The ARMA(1, 1) model can be handled in a similar way to the MA(1) model. Given values of ϕ and θ the CSS function (5.9) may be evaluated from the recursion

$$
\epsilon_t = y_t - \phi y_{t-1} - \theta\epsilon_{t-1}, \qquad t = 2, \ldots, T \qquad (5.15)
$$

with $\epsilon_1 = 0$. Setting $\epsilon_1 = 0$ is equivalent to letting y_1 be fixed. The summation in (5.9) runs from $t = 2$ to T, but the alternative strategy of setting the pre-sample observation, y_0, to zero and starting the recursion at $t = 1$ is not recommended as it can lead to considerable distortion.

Asymptotic Properties

The asymptotic properties of ML estimators, described in Section 3, continue to hold if the observations are dependent. Once the joint density function has been broken down by the prediction error decomposition, suitable expressions for $IA(\psi)$ can be obtained fairly easily.

Example 4   The asymptotic properties of ML estimators in dynamic models are generally unaffected by the treatment of the initial observations. As was pointed out above, the likelihood in the AR(1) model simplifies considerably if y_1 is fixed. From the second derivatives of $\log L$:

$$
IA(\phi, \sigma^2) = -T^{-1}E
\begin{bmatrix}
-\sigma^{-2}\sum y_{t-1}^2 & -\sigma^{-4}\sum \epsilon_t y_{t-1} \\
-\sigma^{-4}\sum \epsilon_t y_{t-1} & -\tfrac{1}{2}\sigma^{-6}\sum \epsilon_t^2
\end{bmatrix}
$$

$$
=
\begin{bmatrix}
1/(1-\phi^2) & 0 \\
0 & 1/2\sigma^4
\end{bmatrix}
\tag{5.16}
$$

Therefore

$$
\text{Avar}(\tilde{\phi}, \sigma^2) =
\begin{bmatrix}
(1-\phi^2)/T & 0 \\
0 & 2\sigma^4/T
\end{bmatrix}
\tag{5.17}
$$

More generally, if the likelihood function is of the form (5.8), the asymptotic information matrix is block diagonal with respect to ψ and σ^2, and

$$
\text{Avar}(\tilde{\psi}) = \sigma^2 T^{-1}\left[\text{plim } T^{-1}\sum_{t=1}^{T} z_t z_t' \right]^{-1} \simeq \sigma^2\left(\sum_{t=1}^{T} z_t z_t' \right)^{-1}
$$

$$
\tag{5.18}
$$

where the vector $z_t = -\partial \epsilon_t / \partial \psi$ is evaluated at the true value of ψ.

Decomposition of the Likelihood Function for Multivariate Models

The prediction error decomposition may be extended to the multivariate case where an $N \times 1$ vector is observed at each point in time. The argument goes through exactly as before with ν_t being an $N \times 1$ vector of prediction errors at time t, with mean zero and covariance matrix F_t. (There is less point in extracting a scalar quantity, σ^2, in the multivariate case.) The log–likelihood may be decomposed as

$$\log L(y_1, \ldots, y_T) = -\frac{TN}{2} \log 2\pi - \frac{1}{2} \sum_{t=1}^{T} \log |F_t|$$

$$-\frac{1}{2} \sum_{t=1}^{T} v_t' F_t^{-1} v_t \qquad (5.19)$$

where $y_1 \sim N(\mu_1, F_1)$ and $v_1 = y_1 - \mu$.

The Kalman filter can again be used to effect the decomposition in (5.19) provided the underlying model can be expressed in state space form. However, as in the single equation case, it is often possible to approximate $\log L$ by identifying v_t with the vector of white noise disturbances driving the process. If $\epsilon_t \sim NID(0, \Omega)$, the approximation takes the form

$$\log L = -\frac{TN}{2} \log 2\pi - \frac{T}{2} \log |\Omega| - \frac{1}{2} \sum \epsilon_t' \Omega^{-1} \epsilon_t \qquad (5.20)$$

cf. the likelihood function for the SURE system, (4.28).

Example 5   If the initial vector, y_1, in the multivariate AR(1) process, (2.5.28), is taken to be fixed, the likelihood function is of the form (5.20) with $\epsilon_t = y_t - \Phi y_{t-1}$ for $t = 2, \ldots, T$. However, because the explanatory variables, y_{t-1}, are identical for each equation, the ML estimator of Φ is given by regressing each element in y_t on the complete vector y_{t-1}. This is basically multivariate regression as described in Section 2.10.

6. Identifiability

Identifiability is a fundamental concept in model building. If a model cannot be identified, two issues are raised. Firstly, it is reasonable to ask whether the model actually makes sense in the way it has been specified. Secondly, and this is closely related to the first point, it is important to examine what meaning, if any, can be given to the estimates computed from a model which is not identified.

For any given model, therefore, it is important to specify conditions under which the model is, or is not, identifiable. In some cases these conditions may be trivial or they may be innocuously buried in the assumptions made when the model is formulated. In other situations it may be necessary to check explicitly the conditions for identifiability whenever a model is specified. As will be seen in Chapter 9, the conditions needed for identifiability play a prominent role in the construction of simultaneous equation systems. However, questions of identifiability can arise for much simpler models as some of the examples given in this section will show.

The Concept of Identifiability

In order to discuss the meaning of identifiability, it is important to be clear about the distinction between a *model* and a *structure*. A structure is a model in which all the parameters are assigned numerical values. Thus

$$y_t = \beta_1 x_{1t} + \beta_2 x_{2t} + \epsilon_t, \qquad t = 1, \dots, T \qquad (6.1)$$

with $\epsilon_t \sim NID(0, \sigma^2)$ is a model, while

$$y_t = 3x_{1t} + 2x_{2t} + \epsilon_t, \qquad t = 1, \dots, T \qquad (6.2)$$

with $\epsilon_t \sim NID(0, 4)$ is a structure of that model.

A model specifies a distribution for the endogenous variables, conditional on any exogenous variables in the system. A structure specifies the parameters of that distribution. Given this background, the following concepts may be defined.

(a) If two structures have the same joint density function they are said to be *observationally equivalent*.

(b) A *structure* is identifiable if there exists no other observationally equivalent structure. If a structure is not identifiable, it is said to be *underidentified*.

(c) A *model* is *identifiable* if all its possible structures are identifiable. If no structure is identifiable, the model is said to be *underidentified*.

Identifiability has an immediate bearing on estimation. If two structures have the same joint density function, the probability of generating a particular set of observations is the same for both structures. Thus there is no way of differentiating between them on the basis of the data. Even if one of the structures could be deduced from the observations, it might prove difficult to give a meaningful interpretation to the parameters. However, it will often be the case that attempts to estimate structures which are not identifiable will run into practical difficulties.

Multicollinearity

The consequences of trying to estimate a model which is not identifiable are best developed by starting with an example. Consider the linear regression model (6.1) and suppose that the observations on the explanatory variable are such that $2x_{1t} = x_{2t}$ for all $t = 1, \ldots, T$. This is a case of *extreme* or *perfect multicollinearity* and it means that the model is not identifiable. Since

$$E(y_t) = \beta_1 x_{1t} + \beta_2 x_{2t} = (\beta_1 + 2\beta_2)x_{1t}, \qquad t = 1, \ldots, T$$

the distribution of the observations is not dependent on the individual values of β_1 and β_2, but only on the linear combination $\beta_1 + 2\beta_2$. Thus if (6.2) is a structure of the model, any other structure in which β_1 and β_2 satisfy the constraint

$$\beta_1 + 2\beta_2 = 7 \tag{6.3}$$

will be observationally equivalent. Clearly, there are an infinite number of such structures, and so (6.2) is not identifiable. Furthermore, since any structure of (6.1) is not identifiable, the model itself is underidentified.

The lack of identifiability will become apparent if any attempt is made to estimate the model by OLS. The cross-product matrix, $X'X$, is singular and so formula (2.2.4) breaks down. If an 'inverse' of $X'X$ is obtained — which is always possible because of rounding error — the multicollinearity in the model will be signalled by enormous standard errors.

There are two possible ways of overcoming the problems raised by the lack of identifiability of (6.1). The first is to bring in *a priori* information in order to impose *restrictions* on the model. If it is assumed, for example, that $\beta_2 = 3\beta_1$, then the model is no longer underidentified. This is perhaps not surprising since the constraint effectively reduces the number of parameters in the model by one. Thus β_1 may be estimated by regressing y_t on $x_{1t} + 3x_{2t}$, while $\hat{\beta}_2$ is given by $3b_1$.

A second way of tackling the problem is to *re-parameterise* the model as

$$y_t = \beta x_{1t} + \epsilon_t, \qquad t = 1, \ldots, T \tag{6.4}$$

where $\beta = \beta_1 + 2\beta_2$. The parameter β can then be estimated by OLS. Again, the number of parameters in the model has been reduced by one. However, in contrast to the first solution, no *a priori* information has been used. Indeed, had the model originally been specified as (6.4), the question of lack of identifiability would not have arisen in the first place. Nevertheless, (6.4) may not be an attractive specification insofar as it has no natural interpretation. Although theoretical considerations may lead to the specification in (6.1) as opposed to that in (6.4), the former is not a sensible model on statistical grounds. Because x_1 and x_2 are perfectly collinear, they contain exactly the same information, at least over the sample period in question.

Common Factors

In a dynamic model, a particular structure may not be identifiable because of a common factor. Consider the ARMA(1, 1) model. Repeatedly substituting for lagged values of y_t in (1.5.21) yields

$$y_t = \sum_{j=0}^{\infty} \phi^j \epsilon_{t-j} + \theta \sum_{j=0}^{\infty} \phi^j \epsilon_{t-j-1}, \qquad t = 1, \ldots, T \qquad (6.5)$$

The first term on the right hand side of (6.5) may be split into two parts to give

$$y_t = \epsilon_t + \phi \sum_{j=0}^{\infty} \phi^j \epsilon_{t-j-1} + \theta \sum_{j=0}^{\infty} \phi^j \epsilon_{t-j-1}$$

$$= \epsilon_t + (\phi + \theta) \sum_{j=0}^{\infty} \phi^j \epsilon_{t-j-1}, \qquad t = 1, \ldots, T \qquad (6.6)$$

It can be seen from (6.6) that any structure for which $\phi = -\theta$ is not identifiable, since whenever this constraint is satisfied, y_t is white noise. The two sides of (1.5.21) are said to have a common factor, and if the model is to be regarded as identifiable this possibility must be explicitly ruled out. Indeed, this is not an unreasonable course of action, since if $\phi = -\theta$, it makes little sense to regard the data as being generated by an autoregressive-moving average process. However, from the practical point of view, it is important to be aware of the consequences of common factors for estimation. In later chapters it will become apparent that estimating an autoregressive-moving average model with common factors is not unlike estimating a regression model with perfectly collinear regressors.

Identifiability and the Information Matrix

In the classical regression model with normally distributed disturbances, the ML estimator of β is given by OLS. It has already been demonstrated that the computation of the OLS estimator will tend to break down if the explanatory variables are perfectly collinear. As it happens, this is a typical feature of ML estimators whenever a structure is not identified. The computational breakdown usually stems from the singularity of the information matrix. For a linear regression model, the information matrix is (4.5) and in the example given above the failure of the standard OLS estimation formulae was a direct result of the non-invertibility of $X'X$. In the next chapter it will be shown that many of the techniques for computing ML estimators involve the inversion of the information matrix or some approximation to it. As a rule, an iterative procedure will be involved and if the algorithm does not actually break down, it will probably fail to converge. In fact, even if an algorithm which did not involve the information matrix were used, convergence problems could well be encountered because of the 'flat' likelihood function typically associated with underidentified models.

In practice, lack of identifiability is often a matter of degree rather than kind. Thus the explanatory variables in a regression model may be highly correlated without being perfectly collinear. Similarly, in the ARMA$(1, 1)$ model, $(1.5.21)$, ϕ may be close to $-\theta$, but not exactly equal to it. In neither of these cases will the structure be underidentified. However, the information matrices will be close to singularity in both cases and the computational problems associated with structures which are truly underidentified may well be encountered. More fundamentally, the near singularity of the information matrices will be reflected in very high variances for the estimators. Because the structure is close to being underidentified, the parameters cannot be estimated with any high degree of accuracy.

Local and Global Identifiability

Although lack of identifiability is generally associated with singularity of the information matrix, this need not necessarily be so. Consider the model

$$y_t = \alpha y_{t-1} + u_t,$$
$$u_t = \phi u_{t-1} + \epsilon_t, \qquad t = 1, \ldots, T \qquad (6.7)$$

with $|\alpha| < 1, |\phi| < 1$ and $\epsilon_t \sim NID(0, \sigma^2)$. This is a rather special case of a dynamic regression model with a first order autoregressive disturbance term. Subtracting ϕy_{t-1} from both sides of (6.7) and re-arranging gives

$$y_t = (\alpha + \phi)y_{t-1} - \alpha\phi y_{t-2} + \epsilon_t, \qquad t = 1, \ldots, T \qquad (6.8)$$

Assume, for simplicity, that y_1 and y_2 are fixed. The distribution of the observations $y_3, \ldots, y_T$ then depends on the disturbances $\epsilon_3, \ldots, \epsilon_T$ and the values taken by the composite parameters $\alpha + \phi$ and $\alpha\phi$. However, the values of ϕ and α can be interchanged without affecting the values of the composite parameters. Thus suppose that $\phi = 0.7$ and $\alpha = 0.2$. The joint density function of this structure will be identical to the joint density function of a structure with $\alpha = 0.7$ and $\phi = 0.2$. Hence, neither structure is identifiable. However if, for any structure, ϕ and α are restricted to lie within the immediate neighbourhood of their true value, that structure will be identifiable. Model (6.7) is therefore *locally*, but not *globally* identifiable.

If a structure is locally identifiable it can be shown that the associated information matrix is nonsingular. This has both theoretical and practical implications. On the theoretical side, local identifiability means that the standard results on the large sample properties of the ML estimator still hold. On the practical side, problems of convergence are unlikely to arise when a ML algorithm is implemented. However, the fact that a model is not globally identifiable can have other consequences. In the case of (6.7), for example, the likelihood function will, in general, be bimodal. This follows because the likelihood function for any pair of parameter values will remain unchanged if those values are interchanged. A maximum value of the likelihood function will normally be located fairly close to the true parameter values. Unless ϕ and α are identical at this maximum, another maximum will be located elsewhere.

When there is more than one maximum to the likelihood function of a model which is locally, but not globally, identifiable, it may not matter which is located if the main object of building the model is prediction. However, if the parameters have an interpretation in terms of some underlying theory, the lack of global identifiability may suggest that the specification is unsatisfactory.

7. Robustness

Unless there are strong *a priori* reasons to suggest otherwise, it is usually assumed that the observations in an econometric model are normally distributed. This assumption is partly justified by an appeal to the central limit theorem. However, normality can rarely be taken for granted and the famous remark of Poincaré is still very apt:
'. . . everyone believes in the (Gaussian) laws of errors, the experimenters because they think it is a mathematical theorem, the mathematicians because they think it is an empirical fact.'

Although it is often recognised that normality cannot be taken for granted, the assumption is still retained for pragmatic reasons. The situation is sometimes characterised by the use of the term 'quasi maximum likelihood estimator'. The OLS estimator can be regarded in this light, although the Gauss—Markov theorem provides a quite independent justification for its adoption in linear regression.

Once the assumption of normally distributed observations is cast into doubt, robustness becomes a key issue. It becomes important to ask how well the ML estimator performs when normality is assumed, but the observations are actually drawn from some other distribution. In a linear regression model this question may be answered by considering the properties of the OLS estimator. Given the relationship between the normality assumption and least squares, it seems likely that these findings will have strong implications for the kind of results to be expected in more general models.

The attention throughout this section will be focused on the linear regression model

$$y_t = x_t'\beta + u_t, \qquad t = 1, \ldots, T \tag{7.1}$$

where u_t is a sequence of independent and identically distributed random variables with p.d.f. $p(u)$. It will be assumed that $p(u)$ is symmetric about zero, but no assumption will be made about the existence of either a first or a second moment. A good deal of the literature on robustness is concerned with the location model in which the observations are distributed around a centrality parameter, ψ. This may be regarded as a regression model in which the only independent variable is the constant term, i.e. $x_t'\beta = \psi$. The OLS estimator of ψ is then the sample mean. Most of the results on the sample mean can be extended to the OLS estimator in the general model, (7.1).

After examining the performance of OLS for different distributions, the minimum absolute deviation (MAD) estimator is introduced. This estimator is less sensitive to extreme observations than OLS. The M-estimators, which are considered next, have similar properties, but are more efficient over a much wider range of distributions.

Efficiency of OLS for Non-Normal Distributions

When $p(u)$ is symmetric about zero, the OLS estimator is symmetric about β. This follows immediately from (2.3.1), since if $p(u_t) = p(-u_t)$ for all t, the joint density at $(b - \beta)$ must be identical to the joint density at $\{-(b - \beta)\}$. If the first moment of b exists, it will therefore be an unbiased estimator of β. However, the existence of the first moment cannot always be taken for granted. Consider a location model in which the observations are drawn from the standardised Cauchy distribution, (1.11). This is similar in shape to the normal distribution, except that it has fatter tails. The chances of obtaining extreme observations are therefore relatively high, and neither the mean nor the variance of the distribution is defined. As a result, the mean and variance of $\bar{y}$ are not defined. More seriously, $\bar{y}$ is not consistent. This follows immediately from the result that the mean of a sample from a Cauchy distribution has exactly the same distribution as a single observation.

Both the Cauchy and the normal distribution are members of the family of *symmetric stable* distributions. If y has a stable distribution, then for T independent observations, there exist numbers $a_T > 0$, b_T such that $a_T \Sigma y_t + b_T$ has the same distribution as y. The normal distribution is the only member of this family with a finite variance. In all the other cases, OLS leaves something to be desired as an estimation procedure.

The normal distribution is also a special case of the 'general error distribution'. If an observation in (7.1) is drawn from such a distribution, its p.d.f. is

$$p(y_t; \beta, \phi, \theta) = [2^{1+1/\theta}\Gamma(1 + \theta^{-1})]^{-1}\phi^{-1}$$
$$\times \exp\{-\tfrac{1}{2}|(y_t - x_t'\beta)/\phi|^\theta\} \qquad (7.2)$$

where ϕ and θ are positive scalar parameters and $\Gamma(.)$ is the Gamma function. The shape of the distribution is determined by θ. Setting $\theta = 2$ yields a normal distribution while the Laplace, or double exponential distribution is given by $\theta = 1$. At the other end of scale, $\theta = \infty$ corresponds to a rectangular distribution.

Both the mean and the variance of the general error distribution are defined for all admissible values of θ. The OLS estimator is therefore unbiased, with a covariance matrix, $\sigma^2(\Sigma x_t x_t')^{-1}$ in which

$$\sigma^2 = 2^{2/\theta}\Gamma(3/\theta)[\Gamma(1/\theta)]^{-1}\phi^2 \qquad (7.3)$$

For a given value of θ, the ML estimator of β is obtained by minimising the sum of the absolute deviations raised to the power θ; i.e.

$$S(\beta; \theta) = \sum_{t=1}^{T} |y_t - x_t'\beta|^\theta \qquad (7.4)$$

This may be carried out independently of ϕ; cf. the treatment of σ^2 for a normal distribution. After some manipulation it can be shown that

$$\text{Avar}(\tilde{\beta}) = \lambda(\Sigma x_t x_t')^{-1} \qquad (7.5)$$

where

$$\lambda = \frac{\Gamma(1/\theta)2^{2/\theta}}{\theta \cdot \Gamma\{(2\theta - 1)/\theta\}}, \qquad 1 \leqslant \theta < \infty \qquad (7.6)$$

The relative efficiency of the OLS estimator is therefore characterised by the ratio λ/σ^2. Some figures are presented in table 3.1. The
low efficiency for $\theta = 20$ partly reflects the situation for the
rectangular distribution where the variance of the ML estimator is of
$0(T^{-2})$. On the other hand, the fact that the efficiency for a
Laplace distribution is only 0.5 may have more important implications in practice.

Table 3.1   Relative Efficiency of OLS and MAD Estimators for a General
Error Distribution

Estimator	$\theta = 1.0$	$\theta = 1.5$	$\theta = 2.0$	$\theta = 4.0$	$\theta = 20$
OLS	0.50	0.91	1.00	0.73	0.16
MAD	1.00	0.83	0.64	0.30	0.05
M-class ($c = 1.5$)	0.62	0.87	0.96	0.63	0.12

The MAD Estimator

The minimum absolute deviation estimator, $b^\dagger$, is obtained by
minimising (7.4) with θ set at unity. It is therefore the ML estimator
of β when the disturbances follow a Laplace distribution. Furthermore, it will be observed that in the location model, the sum of the
absolute deviations is minimised by the median. The median is well-
known to be robust to outlying observations and so there is an
obvious attraction in generalising it to regression.

Although the MAD estimator is not necessarily unique, it has a
well defined asymptotic distribution. In large samples $b^\dagger$ is normal,
with mean β and covariance matrix

$$\text{Avar}(b^\dagger) = \{2p(0)\}^{-2}(\Sigma \, x_t x_t')^{-1} \qquad (7.7)$$

cf. Bassett and Koenker (1978). The relative efficiency of MAD for
the general error distribution is indicated by the second row of
figures in table 3.1 where $4\lambda p^2(0)$ is evaluated for different values of
θ. The corresponding figure for the Cauchy distribution is $\pi/4 = 0.81$.
This can be demonstrated for the location model by noting that
$2p(0) = \sqrt{2/\pi}$ and that $\text{Avar}(\tilde{\psi}) = 2/T$; see (4.4.6).

There are a number of ways of computing the MAD estimator, the
most efficient algorithms being based on linear programming.
However, if a specific routine is not available, iterative weighted least
squares may be used since the function to be minimised can be

written as

$$S(\beta) = \sum_{t=1}^{T} w_t (y_t - x_t'\beta)^2 \tag{7.8}$$

where $w_t = |y_t - x_t'\beta|^{-1}$.

Asymptotic standard errors of a MAD estimator may be obtained by generalising known results for the median. If, in the location model, $y_{(i)}$ denotes the ith order statistic from a population with median ψ, a consistent estimator of $\{p(\psi)\}^{-1}$ is given by

$$T\{y_{(m+l)} - y_{(m-l)}\}/2l$$

where $m = (2r + 1)/2$ for r odd and l is of $0(r^d)$ with $0 < d < 1$; see Cox and Hinkley (1974, p. 470). A convenient choice of d is one-half. Carrying these results over to MAD regression suggests the estimator,

$$\text{Avar}(b^\dagger) = (T/4l)^2 \{r_{(m+l)} - r_{(m-l)}\}^2 \cdot (\Sigma\, x_t x_t')^{-1} \tag{7.9}$$

where the $r_{(i)}$'s are ordered residuals and $r_{(m)} = 0$. Note that k residuals will be identically equal to zero as a consequence of the MAD fit.

M-Estimators

Huber (1973) has defined a class of 'M-estimators', which are obtained by minimising

$$S(\beta) = \sum_{t=1}^{T} \rho(y_t - x_t'\beta) \tag{7.10}$$

where $\rho(.)$ is a suitably defined function. The idea is to choose $\rho(.)$ in such a way that the estimator is efficient over a wide range of distributions. One possibility is

$$\rho(z) = \begin{cases} z^2/2, & |z| < c \cdot \sigma^*, \\ c\,|z| - c^2/2, & |z| \geqslant c \cdot \sigma^*, \end{cases} \tag{7.11}$$

where c is a constant and σ^* is a scale parameter. This estimator combines features of both OLS and MAD. In fact, it reduces to OLS if $c = \infty$, but with c set at a typical value of around one or two, outlying observations receive relatively less weight. The estimator is therefore not sensitive to extreme observations in the same way as OLS, but it remains reasonably efficient in the cases where OLS performs well. This is illustrated by the figures in table 3.1. For $c = 1.5$ and $\sigma^* = \sigma$, the relative efficiency of the M-estimator for a normal distribution is 0.96. On the other hand, its efficiency for a standardised Cauchy distribution is 0.66 for $c = 1.5$ and $\sigma^* = 1$.

In practice the scale parameter must be estimated from the data. The usual estimator, (2.7.6), is not suitable since the variance of the underlying distribution may not be finite. A robust alternative is

$$s^* = \text{median}\{|r_{(i)}|\}, \tag{7.12}$$

where $r_{(i)}$, $i = 1, \ldots, T$, are the ordered residuals.

Robust Estimation in Econometrics

Robust estimation procedures offer a viable alternative to OLS in linear regression. Furthermore, the same techniques may be employed in more general models. The use of a MAD or M-estimator in non-linear regression is straightforward, while in the multiplicative hetero-scedasticity model, all that is required is a slight modification of the likelihood function in (4.17). Similar devices may be adopted in systems of equations.

The main concern is with long-tailed distributions because these are the cases where OLS is particularly unsatisfactory. Since extreme observations tend to be accommodated by the OLS fit, it is possible to remain unaware of their existence. There is a strong case for computing the MAD estimator and examining the residuals. If the form of the distribution is unclear, a sensible strategy would then be to construct a robust M-estimator.

If outliers are detected, however, it is important to ask whether they are to be considered as a purely random phenomenon, or whether they are really an indication of some misspecification in the systematic part of the model. At the simplest level it may be possible to 'explain' an outlying observation by the occurrence of a particular event. This may then be incorporated into the model by means of a dummy variable. Such an approach is clearly more appropriate than the application of a robust estimation technique, particularly if a similar event is likely to occur in the future.

Notes

Sections 1, 2 and 3   Silvey (1970) provides an excellent introduction. More detailed treatments will be found in Kendall and Stuart (1973, Vol. 2) and Cramér (1946).

Section 4   Heteroscedasticity models are described in Amemiya (1977), Goldfeld and Quandt (1972) and Harvey (1976).

Section 5   Schweppe (1965).

Section 6   Silvey (1970) and Rothenberg (1971).

Section 7   Robust estimators are discussed in Cox and Hinkley (1974), Huber (1973), Andrews *et al.* (1972) and Andrews (1974). Computation of the MAD estimator is treated in Schlossmacher (1973) and Armstrong and Frome (1976).

Exercises

1. Show that the sample mean, $\bar{y}$, is a sufficient statistic for the parameter θ when: (i) $y \sim N(\theta, 1)$, (ii) $p(y) = \theta^{-1} \exp(-y/\theta)$.

2. Find the ML estimator of θ given a random sample of T observations from the exponential distribution in 1(ii). Write down its asymptotic distribution.

3. If $y \sim N(0, \sigma^2)$, find the ML estimator of σ (*not* σ^2), show that it is sufficient, and calculate its large sample variance.

4. If $p(y) = \theta^2 \cdot y \cdot \exp(-\theta y)$, show that $\bar{y}$ is a sufficient statistic for θ. Find the ML estimator of θ and its large sample variance.

5. If the observations in a dependent variable heteroscedastic regression model, (4.23), follow an exponential distribution, i.e. $p(y_t) = (x_t'\beta)^{-1} \cdot \exp\{-y_t/(x_t'\beta)\}$, show that the WLS estimator is asymptotically efficient.

6. Derive (4.24).

7. Suppose the logit model, (4.26), contains a single explanatory variable, x_t. Sketch $f(x_t; \beta)$ against x_t and explain the attraction of this particular functional form. Derive an expression for the asymptotic covariance matrix of the ML estimator.

8. Consider the time trend model, $y_t = \alpha + \beta t + \epsilon_t$ in which $\epsilon_t \sim NID(0, \sigma^2)$. Is the estimator $\hat{\beta} = (y_T - y_1)/(T - 1)$: (i) unbiased, (ii) sufficient, (iii) consistent, (iv) efficient?

9. If θ in (7.2) is unknown, explain how you would estimate the complete set of parameters, (β', ϕ, θ) by ML.

10. Explain how you would construct an approximation to the likelihood function for the MA(2) model, $y_t = \epsilon_t + \theta_1\epsilon_{t-1} + \theta_2\epsilon_{t-2}$, with $\epsilon_t \sim NID(0, \sigma^2)$. How are the initial conditions handled?

4
Numerical Optimisation

1. Introduction

In general, nonlinear estimation problems are solved by *iterative* techniques. An initial estimate is obtained, and a new estimate, which (hopefully) is an 'improvement' on the original, is computed by a given rule. This process is then repeated until *convergence*. If the procedure has been successful, the final estimate should satisfy all the properties required of that particular estimation principle. The rules governing the iterative procedure provide the basis of a particular nonlinear optimisation routine.

There is a wide range of algorithms available. One important way in which they differ is in the extent to which they employ partial derivatives of the function. Some routines require second derivatives, while others require only first derivatives, and some are not based on derivatives at all. However, a number of algorithms based on first derivatives offer the option of calculating these numerically. Thus there is no need to obtain the derivatives analytically, which may be difficult, and then programme them, which may be expensive. Apart from the question of derivatives, algorithms differ in other ways, and the choice of a particular algorithm will depend, to some extent, on the type of function to be maximised or minimised. Ill-conditioned functions, sums of squares and problems involving optimisation with respect to only one parameter are all special cases.

Before high speed computing facilities became widely available, any form of nonlinear estimation was extremely time-consuming and to be avoided at all costs. This situation has now changed dramatically, and coupled with the advances in computer technology, improved methods of carrying out nonlinear optimisation have been devised. These routines are very efficient and, in addition, they avoid many of the pitfalls and difficulties inherent in nonlinear estimation.

119

This is not to imply, however, that all problems have been eliminated from nonlinear optimisation. Most nonlinear routines are subject to problems of execution, while the results are often liable to misinterpretation. A good deal of care is therefore required. Nevertheless, the fact that the best estimator of a parameter is non-linear should no longer be regarded as a deterrent to employing that estimator. Indeed, when faced with a new situation, there is a strong argument for basing estimation on one principle, that of maximum likelihood, rather than attempting to derive simpler formulae for alternative estimators which must then be shown to be asymptotically equivalent to the ML estimators.

Some of the general issues raised by numerical optimisation are discussed in Section 2. In particular, the iterative scheme underlying descent methods of optimisation is outlined, and this is contrasted with methods of 'direct search'. It is also pointed out that, in certain circumstances, nonlinear calculations as such can be avoided by the use of a step-wise optimisation procedure. The basic Newton–Raphson algorithm is described in Section 3, while Section 4 considers ways in which this scheme may be modified for ML estimation. The methods of scoring and Gauss–Newton are introduced here, although the latter scheme has somewhat wider applicability, since it is a suitable technique for minimising any sum of squares function. In certain circumstances, linearities in the likelihood equations may be exploited by 'concentrating' the likelihood function. This technique is also described in Section 4. In Section 5, the construction of asymptotically efficient two-step estimators is considered. This material is fundamental to understanding and evaluating many of the methods which have been proposed in the literature. Finally, the question of how to compute estimates of asymptotic covariance matrices is discussed.

2. Principles of Numerical Optimisation

In discussing the general problem of nonlinear estimation, it will be assumed that a *criterion function*, $f(\psi)$, is to be minimised with respect to the n parameters in the vector ψ. The $n \times 1$ vector of first derivatives, $\partial f(\psi)/\partial \psi$, will be denoted by $g(\psi)$ or simply g, while the $n \times n$ matrix of second partial derivatives, the *Hessian*, will be written as

$$\frac{\partial f(\psi)}{\partial \psi \, \partial \psi'} = G(\psi)$$

The generality of the discussion is not restricted by taking $f(\psi)$ as a function to be minimised, since a maximisation problem may be converted to one of minimisation simply by multiplying the criterion function by minus one.

If $g(\psi)$ and $G(\psi)$ exist and are continuous in the neighbourhood of a particular value, $\hat{\psi}$, sufficient conditions for $\hat{\psi}$ to be a local minimiser of $f(\psi)$ are that $g(\hat{\psi}) = 0$ and that $G(\hat{\psi})$ is p.d. However, these conditions do not guarantee that $\hat{\psi}$ is a global minimiser since there may be another local minimiser, $\tilde{\psi}$, such that $f(\tilde{\psi}) < f(\hat{\psi})$.

Univariate Search Techniques

Before considering the general problem of nonlinear optimisation, it is worth briefly considering the one-dimensional problem, in which a function is to be minimised with respect to a single parameter only. If the function is known to contain a single minimum, then *linear search* techniques may be applicable. These methods fit a low order polynomial, and calculate where the minimum lies. The process is then repeated until convergence. Algorithms are available both with and without the requirement for programming first derivatives.

If it is felt desirable to explore the function over the whole range of values of the parameter, a *grid search* may be employed. Suppose the parameter is constrained to lie in the range $0 \leqslant \psi \leqslant 1$. The function may be evaluated at the points $\psi = 0$, $\psi = 0.1, \ldots, \psi = 1.0$, i.e. at 11 equally spaced points over the range of the parameter. If the minimum is found to lie at, say, $\psi = 0.7$, a 'finer' search over the interval 0.61 to 0.79 is carried out in steps of 0.01. If desired, this may be repeated until a sufficiently accurate estimate of the minimum has been obtained.

A grid search may be carried out in two-dimensional problems, but for $n \geqslant 3$ the technique becomes unwieldy. Linear searches are restricted to univariate problems. However, their range of application is much wider than this statement might suggest, since a line search is incorporated into most descent algorithms.

Descent Methods

Most iterative schemes are of the form

$$\psi^* = \hat{\psi} + \lambda \cdot d(\hat{\psi}) \tag{2.1}$$

where $\hat{\psi}$ is the current approximation to the minimum, ψ^* is the revised estimate, λ is a positive scalar known as the *step-length* and $d(\psi)$ is the *direction vector*. The direction vector is of length n, and depends on the *gradient*, $g(\psi)$.

The simplest scheme within this class is the *method of steepest descent*. In one dimension the idea can be explained simply by noting that a small change proportional to the slope multiplied by minus one will decrease the function. More generally, if the sum of squares of the changes in the elements of ψ is constrained to be a certain length, the greatest change in the function occurs when the changes in ψ are proportional to $g(\psi)$. For these changes to lead towards a minimum rather than a maximum, the constant of proportionality must be negative. Thus the direction vector is equal to $-g(\psi)$, and the step-length must be chosen so that the function does not increase at a particular iteration.

Given a suitable rule for determining λ at each iteration, it can be shown that the method of steepest descent will eventually converge. However, its rate of convergence can be extremely slow and the algorithm is rarely of practical use. A more usual choice for the direction vector is $d(\psi) = -G(\psi) \cdot g(\psi)$. The rationale underlying this method, which is generally known as Newton—Raphson, is set out in the next section. However, whatever the choice of $d(\psi)$, convergence can almost invariably be speeded up by adopting an efficient line search procedure to determine λ. Once $d(\hat{\psi})$ has been evaluated, the problem is one-dimensional and the techniques referred to in the previous sub-section may be used to minimise the function

$$\phi(\lambda) = f\{\hat{\psi} + \lambda \cdot d(\hat{\psi})\} \tag{2.2}$$

with respect to λ.

Step-wise Optimisation

It is sometimes possible to break down the optimisation problem in such a way that complicated nonlinear calculations are avoided. Let ψ be partitioned as $\psi = (\psi_1', \psi_2')'$, and suppose that $f(\psi)$ is linear in the parameters of ψ_1 when ψ_2 is fixed, and linear in ψ_2 when ψ_1 is fixed. This suggests a *step-wise* or *zig-zag* optimisation procedure. Conditional on a value of ψ_1, $f(\psi)$ is minimised with respect to ψ_2. The value of ψ_2 obtained in this way is then held constant while $f(\psi)$ is minimised with respect to ψ_1. The function is then minimised with respect to ψ_2 conditional on this new value of ψ_1 and the process is repeated until convergence. Indeed, it can be shown that such a procedure is bound to converge; see Sargan (1964) and Oberhofer and Kmenta (1974).

Direct Search

For some objective functions it may be difficult to ascertain whether or not partial derivatives exist in the domain of interest. For other functions it may be known that partial derivatives do not exist at certain key points, or that they are ill-behaved in certain ways. When this is the case, it is imperative to seek an optimisation procedure which does not use partial derivatives and which, furthermore, does not even require the existence of partial derivatives for its theoretical justification. These requirements are satisfied by direct search methods, where the only assumption made is that $f(\psi)$ is continuous. One of the most widely used direct search algorithms is due to Powell (1964), the directions of the searches being based on the principle of '*conjugate directions*'. Another useful technique is the *simplex* method of Nelder and Mead (1965). Although the algorithm does not usually converge to the exact answer, it is relatively robust. It can be applied to functions which contain discontinuous derivatives, and it normally produces a reasonably accurate answer quite efficiently. When a 'difficult' function is encountered, the simplex method has much to recommend it, even if it is only used to provide starting values for another technique.

Convergence

An iterative scheme may be regarded as having converged if: (i) $f(\psi^*)$ is close to $f(\hat{\psi})$; (ii) ψ^* is close to $\hat{\psi}$; and (iii) $g(\psi^*)$ is close to $g(\hat{\psi})$. 'Closeness' is generally defined in terms of a small positive scalar, κ. Thus the first condition is taken to be satisfied if $|f(\psi^*) - f(\hat{\psi})| < \kappa_1$. For $n > 1$, ψ and $g(\psi)$ are vectors and there are a number of different ways of specifying the last two conditions. The usual approach is to consider the second condition as being satisfied if

$$(\psi^* - \hat{\psi})'(\psi^* - \hat{\psi}) = \| \psi^* - \hat{\psi} \| < \kappa_2$$

while the third condition is satisfied if $\| g(\psi^*) \| < \kappa_3$. Although different algorithms differ in the way in which these conditions are implemented, it is generally advisable that they be applied in one way or another. Of course, if a direct search method which does not involve first derivatives is being used, only the first two conditions are appropriate.

Achieving convergence is not always straightforward. One problem which bedevils econometric work is that the objective function is often relatively 'flat' around the optimum. This is illustrated in figure 4.1. Once at $\hat{\psi}_1$ or $\hat{\psi}_2$, the algorithm either stops or moves towards the minimum very slowly. In the latter case, a typical algorithm may well terminate before $\tilde{\psi}$, due to an upper limit on the number of iterations. The practical consequences are, firstly, that a good deal of computer time is likely to be burnt up to very little

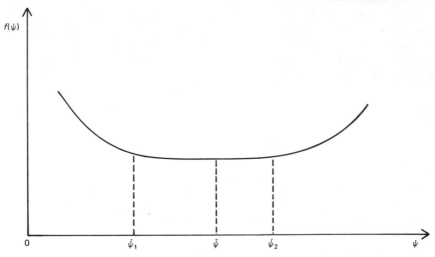

Figure 4.1   *A 'Flat' Objective Function*

effect and, secondly, that different results may be obtained
depending on the position of the starting values.

   Although figure 4.1 provides a good illustration of the consequences of a flat objective function, such problems are normally
encountered only when $f(\psi)$ depends on two or more parameters.
From the statistical point of view, a flat objective function arises
when a model is close to being underidentified. Thus consider the
ARMA(1, 1) process, (1.5.21). It was pointed out in Section 3.6 that
the model is not identifiable if $\phi = -\theta$. If, on the other hand, ϕ is
merely close to $-\theta$, the model is identifiable, but any pair of
estimates, $\hat{\phi}$ and $\hat{\theta}$, which satisfy the constraint $\hat{\phi} + \hat{\theta} = 0$, or are near
to satisfying it, will yield a value of $S(\phi, \theta)$ which is close to the
minimum. Similarly, if $2x_{1t} \simeq x_{2t}$ in (3.6.1), any estimates of β_1 and
β_2 which are close to satisfying (3.6.3) will be close to minimising the
residual sum of squares. Since the model is linear in the parameters,
the convergence problems which might be encountered in the
ARMA(1, 1) model will not arise here. However, the ill-conditioning
of the cross-product matrix and the numerical instability of the OLS
estimates are a reflection of the same underlying phenomenon. Given
the multicollinearity of economic time series, and the small samples,
it is perhaps not surprising that flat objective functions are the rule,
rather than the exception, in nonlinear estimation.

   An even more serious problem arises when the objective function
has more than one minimum. This situation is illustrated in figure
4.2. If the starting point is at $\hat{\psi}$, it is more than likely that a
'shortsighted' iterative procedure will converge to a *local* minimum,

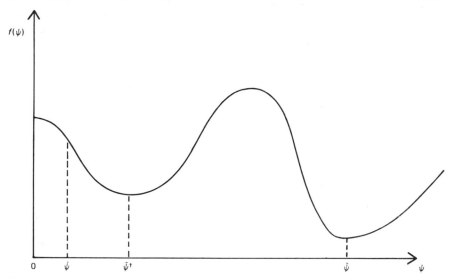

Figure 4.2 *Local and Global Minima*

$\tilde{\psi}^\dagger$, rather than to the *global* minimum, $\tilde{\psi}$. One attraction of a grid search is that it will usually locate the global minimum, although if the function exhibits the type of behaviour shown in figure 4.3 even a grid search might fail. Fortunately, this kind of behaviour is not common in statistical applications. As a result, a relatively coarse grid search can sometimes be used to provide a suitable starting value for a more refined iterative procedure. If this is not done, it is at least worth applying the optimisation procedure from a number of different starting points. This provides a check on whether a global solution has actually been obtained.

The introduction of an exogenous variable into (3.6.7) provides a simple illustration of how multiple optima can arise in econometrics. The model becomes

$$y_t = \alpha y_{t-1} + \beta x_t + u_t \tag{2.3}$$

where $u_t \sim AR(1)$, and it is now globally, as well as locally, identifiable. However, by a continuity argument, the sum of squares function will still be bimodal if the variance of x_t is sufficiently small, although the two minima will not, in general, be equal.

Numerical Evaluation of Derivatives

The first partial derivative of $f(\psi)$ with respect to the jth element of ψ, ψ_j, may be estimated from the formula

$$\{f(\hat{\psi} + h_j e_j) - f(\hat{\psi})\}/h_j, \qquad j = 1, \ldots, n \tag{2.4}$$

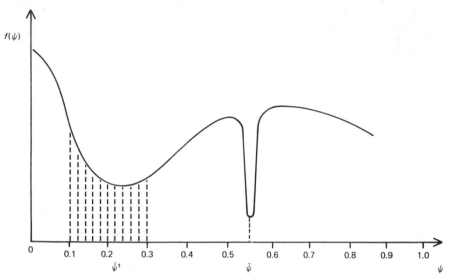

Figure 4.3   *Pitfalls with a Grid Search*

where the jth element of the $n \times 1$ vector e_j is unity, while all the other elements are zero. The scalar, h_j, is sometimes referred to as the step-length, although this should not be confused with the step-length, λ, defined in the iterative scheme (2.1). Each h_j must be chosen so as to be small enough for the difference estimates to be close to the true derivatives, but not so small that the calculated differences are dominated by rounding errors. Determining suitable values for the h_j's often requires some experimentation, and a good deal of care is needed in making a judicious choice.

The calculation of a vector of numerical derivatives from (2.4) involves n evaluations of the function. This can be quite time-consuming. Because numerical derivatives are also subject to inaccuracies, it is usually preferable to use analytic derivatives if these can be obtained reasonably easily. In many instances, however, it is not easy to obtain analytic derivatives, and hence the use of numerical derivatives is common. Indeed, it could be argued that since differentiation and computer programming are tasks which may both be time-consuming and very much open to human error, the use of numerical derivatives is, in some instances, safer as well as being more efficient.

Second derivatives may also be calculated numerically. An estimate of the ijth element in $G(\hat{\psi})$ is given by

$$\{g_j(\hat{\psi} - h_i e_i) - g_j(\hat{\psi})\}/h_i, \qquad i, j = 1, \ldots, n \qquad (2.5)$$

Constrained Optimisation: The Use of Transformations

In some problems there are restrictions on the values which the parameters can take. Algorithms have been designed to handle constraints of various types and details of some of these methods will be found in Box *et al.* (1969, Chapter 5) and Dixon (1972, Chapter 6). However, simple constraints can sometimes be handled by means of transformations.

In econometrics and time series analysis, it is often desirable to impose constraints of the form:

(a) $\psi \geq k$, or
(b) $k_1 \leq \psi \leq k_2$

where k, k_1 and k_2 are known constants. The first type of constraint arises when ψ is a variance and in this case k is zero. Negative values of ψ may, however, be avoided very easily, simply by defining a new variable, θ, such that $\theta^2 = \psi$. Unconstrained optimisation may then be applied with respect to θ.

Constraining a parameter to lie within a particular range, i.e. $k_1 \leq \psi \leq k_2$, can also be handled relatively easily. One possible transformation is $\psi = k_1 + (k_2 - k_1)\sin^2 \theta$, with θ taking any value in the range $-\infty$ to ∞. If the inequality is a strict one, the transformation,

$$\psi = \frac{k_2 + k_1}{2} \frac{\theta}{1 + |\theta|} + \frac{k_2 - k_1}{2} \tag{2.6}$$

might be appropriate. In the important special case where ψ is constrained to lie within the unit circle, (2.6) simplifies to $\psi = \theta/(1 + |\theta|)$.

As a final point, note that when a grid search is to be carried out, transformations may often be applied in such a way that the role they play is exactly opposite to the one considered above. Suppose, for example, that a parameter is unconstrained, i.e. $-\infty \leq \psi \leq \infty$. It will generally be convenient to carry out a grid search over the range $(-1, 1)$, for which the transformation

$$\theta = \psi/(1 + |\psi|) \tag{2.7}$$

is appropriate. Once the function has been minimised with respect to θ the required value of ψ is given by $\tilde{\psi} = \tilde{\theta}/\{1 - |\tilde{\theta}|\}$.

3. Newton–Raphson

The problems inherent in the method of steepest descent led to the consideration of new methods of optimisation. It was soon realised that approximating the function by a quadratic could form the basis of relatively efficient computational schemes. Such schemes employ second, as well as first, derivatives and the basic procedure obtained from this approach is known as *Newton—Raphson* or simply *Newton's method.*

A Taylor series expansion of $f(\psi)$ around the minimum $\tilde{\psi}$ gives

$$f(\psi) = f(\tilde{\psi}) + (\psi - \tilde{\psi})g(\tilde{\psi}) + \tfrac{1}{2}(\psi - \tilde{\psi})^2 G(\tilde{\psi}) \qquad (3.1)$$

Differentiating expression (3.1) with respect to ψ yields

$$g(\psi) = g(\tilde{\psi}) + (\psi - \tilde{\psi})G(\tilde{\psi}) \qquad (3.2)$$

However, since $g(\tilde{\psi}) = 0$, (3.2) may be rearranged to give

$$\tilde{\psi} \simeq \psi - G^{-1}(\tilde{\psi})g(\psi) \qquad (3.3)$$

If $f(\psi)$ were a quadratic, ψ could be set equal to any initial estimate of $\tilde{\psi}$, and $\tilde{\psi}$ would be given exactly by the right-hand side of expression (3.3). (It is instructive to verify this by taking a simple quadratic, say $f(\psi) = \psi^2 + 4\psi + 3$, arbitrarily choosing a starting value, and then using (3.3) to calculate the minimum.) For more general functions, this will not be the case, but expression (3.3) does suggest the iterative scheme

$$\psi^* = \hat{\psi} - G^{-1}(\tilde{\psi})g(\hat{\psi}) \qquad (3.4)$$

where the revised estimate, ψ^*, is expected to be closer to the minimum than the initial estimate, $\hat{\psi}$.

The evaluation of the Hessian matrix poses no problem for a quadratic function since its elements do not depend on ψ. More generally, however, an iterative scheme of the form (3.4) raises a difficulty, since it presupposes that the Hessian is evaluated at the minimum $\tilde{\psi}$, which is, of course, unknown. The solution adopted in Newton—Raphson is to evaluate the Hessian at the current estimate, $\hat{\psi}$, on the grounds that this will yield an acceptable approximation to $G(\tilde{\psi})$ if $\hat{\psi}$ is 'reasonably close' to $\tilde{\psi}$. The iterative scheme therefore becomes

$$\psi^* = \hat{\psi} - G^{-1}(\hat{\psi})g(\hat{\psi}) \qquad (3.5)$$

This is sometimes written more concisely as

$$\psi^* = \hat{\psi} - G^{-1}g \qquad (3.6)$$

where it is understood that G and g are evaluated at the current estimate.

Example   Consider the function

$$f(\psi) = \psi^3 - 3\psi^2 + 5, \qquad 0 \leqslant \psi < \infty \qquad (3.7)$$

The first and second derivatives are

$$g(\psi) = 3\psi(\psi - 2) \tag{3.8}$$

and

$$G(\psi) = 6(\psi - 1) \tag{3.9}$$

respectively. The function is sketched in figure 4.4.

Suppose that we decide to find the minimum by the Newton–Raphson iteration. If the starting value, $\hat{\psi}$, is chosen to be 1.500, then $g = -2.25$ and $G = 3$. Thus

$$\psi^* = 1.5000 - (-2.25)/3 = 2.2500$$

For the second iteration $\hat{\psi}$ is set equal to the previous ψ^*, 2.2500, and the revised estimate is

$$\psi^* = 2.25 - (0.16875)/7.5 = 2.0250$$

The third iteration yields $\psi^* = 2.0003$. This is very close to the actual minimum, $\tilde{\psi}$, which is exactly equal to 2.

The above sequence seems very straightforward. However, consider what would have happened if the initial starting value had been $\hat{\psi} = 0.5$. Since $G(0.5)$ is *negative*, the effect of the first iteration is to send the estimate in the wrong direction, i.e.

$$\psi^* = 0.5 - (-2.25)/(-3) = -0.2500$$

If this process is left unchecked, the algorithm converges to the maximum at $\psi = 0$.

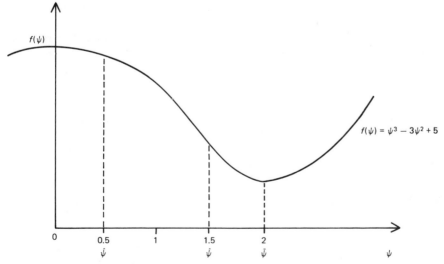

Figure 4.4   *Function in Example 3.7*

Safeguards and Modifications

The Newton–Raphson scheme will only progress towards a minimum if $G(\hat{\psi})$ is p.d. While it is true that the Hessian is always p.d. for a concave function, it is not true in general. If $G(\hat{\psi})$ is singular, the direction vector is not defined and the method breaks down. If G^{-1} exists, but is not p.d., the direction vector may be wrongly aligned and the value of the objective function may increase. This is precisely what happened in the example above, when the starting value was taken to be $\hat{\psi} = 0.5$.

Even if $G(\hat{\psi})$ is p.d., it is still possible for $f(\psi^*)$ to exceed $f(\hat{\psi})$. However, this represents a case of 'overshooting', and such an occurrence can always be avoided by introducing a variable step-length, λ, into the scheme. Thus (3.6) becomes

$$\psi^* = \hat{\psi} - \lambda G^{-1}g \qquad (3.10)$$

and it can be shown that if G^{-1} is p.d. and $\hat{\psi}$ is not a minimum, there exists a positive value of λ such that $f(\psi^*) < f(\hat{\psi})$. As was observed in Section 2, the most efficient way of fixing the value of λ is by a line search. Procedures of this kind are a standard feature in most descent algorithms and Newton–Raphson is no exception.

A number of techniques have been devised for modifying the basic Newton–Raphson method so as to ensure that the gradient is always premultiplied by a p.d. matrix. The iterative scheme for such methods may be written

$$\psi^* = \hat{\psi} - \lambda Hg \qquad (3.11)$$

where $H = H(\hat{\psi})$ is an $n \times n$ p.d. matrix. In the method of *quadratic hill-climbing*, proposed by Goldfeld, Quandt and Trotter (1966), H is set equal to $(G + \mu I)^{-1}$. The positive scalar μ varies as the iterations proceed, its value always being chosen so as to ensure that H is p.d. When G is n.d., this may be guaranteed by setting μ to a value greater than the modulus of the largest negative eigenvalue of G. When G is singular, any positive value of μ ensures that H is p.d. If μ is large, the iteration will be similar to the method of steepest descent. However, as the minimum is approached, G will tend to become p.d., and in this case μ may be set equal to zero, so that the method collapses to Newton–Raphson. Although quadratic hill-climbing is a useful technique in many problems, its efficiency is likely to fall markedly as n increases, since the matrix inversions and eigenvalue evaluations will impose a heavy computational burden.

Another way of carrying out an iterative scheme of the form (3.11) is to adopt an updating procedure, whereby a 'rank-one' or 'rank-two' matrix is added to a p.d. matrix at the current iteration to produce a p.d. matrix at the next iteration. If the initial Hessian matrix is not p.d., it is amended so that it is p.d. As in quadratic hill-climbing, the matrix H will be equal to the Hessian at the minimum. A description of several of these updating methods will be found in Gill and Murray (1972). In the method they themselves propose, H is equal to $G + D$ where D is a diagonal matrix. Neither G nor D is actually computed, the updating being carried out in terms of a lower triangular matrix, L, such that $L'L = H$. This method of updating is both computationally efficient, and numerically stable, in that it avoids the rounding errors which are always liable to be present when a set of simultaneous equations is solved by matrix inversion.

Quasi-Newton Methods

The main advantage of *quasi-Newton*, or *variable metric*, methods is that they do not require the Hessian to be explicitly evaluated. The iterative scheme is of the form (3.11), but the initial H matrix may be any p.d. matrix. At each iteration H is updated in such a way as to yield a series of p.d. matrices which eventually converge to the inverse of the Hessian. A common choice for the starting matrix is the identity matrix. The first iteration is therefore carried out by the method of steepest descent, but the procedure gradually tends towards Newton–Raphson as the minimum is approached.

Examples of quasi-Newton algorithms include Davidon–Fletcher–Powell and the more recent programme designed by Gill, Murray and Pitfield (1972). As in the updating procedure described in the previous sub-section, the key to efficiency and numerical stability in Gill–Murray–Pitfield lies in storing H as triangular factors.

4. Maximisation of a Likelihood Function

Maximum likelihood estimation is usually carried out by maximising the logarithm of the likelihood function, $\log L(\psi)$, although it is often more convenient to implement a particular algorithm by considering the problem as one of minimising $\{-\log L\}$. However, whether the problem is viewed as one of maximisation or minimisation makes little difference to the discussion below.

The Newton–Raphson iteration, (3.5), for the log–likelihood function may be written

$$\psi^* = \hat{\psi} - [D^2 \log L(\hat{\psi})]^{-1} D \log L(\hat{\psi}) \tag{4.1}$$

where both the first and second derivatives are evaluated at current estimate, $\hat{\psi}$. A number of variants of (4.1) are available for ML estimation. These can sometimes exploit particular features of the problem, thereby producing a more efficient algorithm. The Gauss–Newton procedure, for example, is particularly appropriate for cases where maximising the likelihood function is equivalent to minimising a sum of squares.

The Method of Scoring

Rather than work with the matrix of second derivatives, it is sometimes more efficient to use its expectation. This yields the information matrix when multiplied by minus one and the iterative scheme in (4.1) becomes

$$\psi^* = \hat{\psi} + I^{-1}(\hat{\psi})D \log L(\hat{\psi}) \tag{4.2}$$

The procedure is known as the *method of scoring*.

The method of scoring is likely to have a slower rate of convergence than Newton–Raphson, since the information matrix is only an approximation to the Hessian. However, in many applications the information matrix has a simple form and is much easier to compute. Each iteration will therefore be performed more quickly and this may compensate for a slower rate of convergence. Furthermore, provided the model is identifiable, the information matrix will always be p.d. Some of the difficulties which may be encountered with Newton–Raphson are therefore avoided.

As with most descent methods, it will usually be advisable to introduce a variable step-length, λ, into the iterative scheme; cf. (2.1). This point should always be borne in mind as it will not be mentioned explicitly in the applications cited.

Example 1   The standardised Cauchy distribution was introduced in Example 3.1.2 and it was observed that the likelihood equation was nonlinear. Differentiating $\log L$ twice with respect to ψ gives

$$\frac{\partial^2 \log L}{\partial \psi^2} = \sum_{t=1}^{T} \frac{2(y_t - \psi)^2 - 2}{\{1 + (y_t - \psi)^2\}^2} \tag{4.3}$$

and this term may be used as the basis of a Newton–Raphson iterative scheme. However, the scoring algorithm takes a much simpler form. Taking the expectation of (4.3) and using a standard result on integrals yields

$$\frac{T}{\pi} \int_{-\infty}^{\infty} \frac{2(y_t - \psi) - 2}{\{1 + (y_t - \psi)^2\}^3} d\psi = -\frac{T}{2} \tag{4.4}$$

Thus $I(\psi) = T/2$ and the scoring algorithm is

$$\psi^* = \hat\psi + \frac{4}{T} \sum_{t=1}^{T} \frac{(y_t - \hat\psi)}{1 + (y_t - \hat\psi)^2} \tag{4.5}$$

Example 2   The multiplicative heteroscedasticity regression model is defined by (3.4.15). Because the information matrix, (3.4.20), is block-diagonal with a much simpler form than the Hessian, the method of scoring yields a more attractive algorithm than Newton–Raphson, i.e.

$$\beta^* = \hat\beta + [\Sigma \exp(-w_t'\hat\delta) x_t x_t']^{-1} \Sigma x_t \exp(-w_t'\hat\delta)(y_t - x_t'\hat\beta) \tag{4.6a}$$

and

$$\delta^* = \hat\delta + [\Sigma w_t w_t']^{-1} \Sigma w_t \{\exp(-w_t'\hat\delta)(y_t - x_t'\hat\beta)^2 - 1\} \tag{4.6b}$$

The first iterative equation, (4.6a), reduces to WLS, while the second equation has the attraction that the matrix $\Sigma w_t w_t'$ does not depend on any unknown parameters, and so it need be inverted only once.

Gauss–Newton

The computation of estimators in econometrics and time series analysis frequently entails the minimisation of a sum of squares function of the form

$$f(\psi) = S(\psi) = \sum_{t} \epsilon_t^2 \tag{4.7}$$

where $\epsilon_t = \epsilon_t(\psi)$ is a 'residual' which depends on the value taken by ψ. The vector of first derivatives is

$$g(\psi) = \frac{\partial S(\psi)}{\partial \psi} = 2\Sigma \frac{\partial \epsilon_t}{\partial \psi} \cdot \epsilon_t \tag{4.8}$$

while the Hessian is

$$G(\psi) = \frac{\partial^2 S(\psi)}{\partial \psi \, \partial \psi'} = 2\Sigma \left\{ \frac{\partial \epsilon_t}{\partial \psi} \cdot \frac{\partial \epsilon_t}{\partial \psi'} - \frac{\partial^2 \epsilon_t}{\partial \psi \, \partial \psi'} \cdot \epsilon_t \right\} \tag{4.9}$$

The Newton–Raphson scheme is therefore

$$\psi^* = \hat\psi - \left[\Sigma \left\{ \frac{\partial \epsilon_t}{\partial \psi} \cdot \frac{\partial \epsilon_t}{\partial \psi'} - \frac{\partial^2 \epsilon_t}{\partial \psi \, \partial \psi'} \cdot \epsilon_t \right\} \right]^{-1} \Sigma \frac{\partial \epsilon_t}{\partial \psi} \cdot \epsilon_t \tag{4.10}$$

where all the quantities on the right-hand side of (4.10) are evaluated at the current estimate, $\hat{\psi}$. The terms in the expression for the Hessian matrix, (4.9), which involve second derivatives will, however, be small relative to the terms which involve first derivatives only. This suggests the use of the modified iterative scheme

$$\psi^* = \hat{\psi} - \left[\Sigma \frac{\partial \epsilon_t}{\partial \psi} \cdot \frac{\partial \epsilon_t}{\partial \psi'} \right]^{-1} \Sigma \frac{\partial \epsilon_t}{\partial \psi} \cdot \epsilon_t \qquad (4.11)$$

which is generally known as *Gauss–Newton*.

The above procedure may be written more concisely by defining

$$z_t = -\partial \epsilon_t / \partial \psi \qquad (4.12)$$

where z_t, like ϵ_t, is understood to depend on ψ. Substituting (4.12) into (4.11) yields

$$\psi^* = \hat{\psi} + (\Sigma z_t z_t')^{-1} \Sigma z_t \epsilon_t \qquad (4.13)$$

Writing the iterations in the form (4.13) stresses even more clearly the interpretation of Gauss–Newton as a series of OLS regressions. At each step z_t and ϵ_t are evaluated at the point $\psi = \hat{\psi}$, and $\hat{\psi}$ is then updated by means of a regression of ϵ_t on z_t.

As in Newton–Raphson, it is generally desirable to incorporate a line search into the algorithm, and a variable step length may be introduced in the manner suggested by (2.1). A slightly different scheme, due to Marquardt (1963), is based on the same principle as the quadratic hill-climbing method described in Section 3. The iterations are of the form

$$\psi^* = \hat{\psi} + (\Sigma z_t z_t' + \mu I)^{-1} \Sigma z_t \epsilon_t \qquad (4.14)$$

where μ is a non-negative scalar which may be varied with each iteration. For a large value of μ, this scheme imparts a bias in the direction of steepest descent which may be advantageous if the starting point is some distance from the minimum. Combining (4.14) with a line search produces a powerful algorithm for which strong results are available on global convergence; see Wolfe (1978, Chapter 7).

A slightly different rationalisation may be given for the Gauss–Newton iteration for the nonlinear regression model, (3.4.6). A first-order Taylor series expansion of $f(x_t; \psi)$ about $\hat{\psi}$ yields

$$f(x_t; \psi) \simeq f(x_t; \hat{\psi}) + (\psi - \hat{\psi}) z_t \qquad (4.15)$$

since $z_t = \partial f(x_t; \psi)/\partial \psi$. Substituting in (3.4.6) then gives

$$y_t - f(x_t; \hat{\psi}) = (\psi - \hat{\psi}) z_t + \epsilon_t, \qquad t = 1, \ldots, T \qquad (4.16)$$

This again suggests the iterative scheme (4.13), in which successive estimates of ψ are obtained by regressing the residuals from the current estimate on the partial derivatives of $f(x_t; \psi)$ and subtracting from the current estimate.

Example 3   As a specific example of a nonlinear regression model, consider

$$y_t = \alpha \exp(\beta x_t) + \epsilon_t, \qquad t = 1, \ldots, T \tag{4.17}$$

The vector of partial derivatives is given by

$$z_t = \begin{pmatrix} \exp(\beta x_t) \\ \alpha x_t \exp(\beta x_t) \end{pmatrix} \tag{4.18}$$

Regressing the residuals, $\epsilon_t = y_t - \hat{\alpha} \exp(\hat{\beta} x_t)$ on z_t, produces 'corrections', $\Delta \hat{\alpha}$ and $\Delta \hat{\beta}$, to the original estimates. The new estimates $\alpha^* = \hat{\alpha} + \Delta \hat{\alpha}$ and $\beta^* = \hat{\beta} + \Delta \hat{\beta}$ are then used to construct residuals and their derivatives, and the process is repeated until convergence.

Example 4   In example 3.5.2 it was shown that maximising the likelihood function for an MA(1) model is equivalent to minimising a sum of squares function if ϵ_0 is taken to be fixed and equal to zero. The residuals, ϵ_t, are built up recursively from (3.5.14). The first derivatives may be constructed in a similar way since

$$\frac{\partial \epsilon_t}{\partial \theta} = -\theta \frac{\partial \epsilon_{t-1}}{\partial \theta} - \epsilon_{t-1}, \qquad t = 1, \ldots, T \tag{4.19}$$

with $\partial \epsilon_0 / \partial \theta = 0$. Thus, given an initial estimate, $\hat{\theta}$, ϵ_t and $\partial \epsilon_t / \partial \theta$ may be evaluated and the estimate updated by (4.11).

Berndt, Hall, Hall and Hausman (BHHH)

Gauss–Newton has two main advantages over Newton–Raphson. Firstly it requires only first, and not second, derivatives. Secondly, the matrix used to approximate the Hessian is either positive definite or singular, and a singular matrix can always be avoided by a modification such as (4.14).

Berndt, Hall, Hall and Hausman (1974) have proposed an algorithm which retains the advantages of Gauss–Newton while being applicable to problems more general than the minimisation of a sum of squares. They suggest approximating the Hessian by

$$\sum_{t=1}^{T} \frac{\partial \log p(y_t; \psi)}{\partial \psi} \frac{\partial \log p(y_t; \psi)}{\partial \psi'}$$

The iterative scheme for $n = 1$ is therefore

$$\psi^* = \hat{\psi} + \lambda \left[\sum_{t=1}^{T} \left\{ \frac{\partial \log p(y_t; \psi)}{\partial \psi} \right\}^2 \right]^{-1} \sum_{t=1}^{T} \frac{\partial \log p(y_t; \psi)}{\partial \psi} \quad (4.20)$$

where λ is the variable step length.

Example 5   For the estimation of the location parameter for a Cauchy distribution, the BHHH scheme is

$$\psi^* = \hat{\psi} + \frac{1}{2} \left[\sum_{t=1}^{T} \frac{(y_t - \psi)^2}{\{1 + (y_t - \psi)^2\}^2} \right]^{-1} \sum_{t=1}^{T} \frac{(y_t - \psi)}{1 + (y_t - \psi)^2}$$

$$(4.21)$$

cf. (4.5).

Systems of Equations

Consider the SURE model (2.9.15). The joint density function depends on two sets of parameters, those in β and those in Ω. Each set of parameters may be estimated very easily conditional on the other set. Thus the ML estimator of Ω conditional on β is given by

$$\tilde{\Omega}(\beta) = T^{-1} \sum_{t=1}^{T} \epsilon_t \epsilon_t' \quad (4.22)$$

where $\epsilon_t = y_t - X_t\beta$; cf. (3.4.32). The ML estimator of β conditional on Ω is given by GLS as in (2.9.17). The step-wise optimisation procedure described in Section 2 therefore provides a natural means of computing the unconditional ML estimates. The process starts off by applying OLS to each equation in turn and using the residuals to form an estimate of Ω. A feasible GLS estimate of β is then computed and the whole process carried on until convergence.

Maximising the likelihood function for the SURE model is equivalent to minimising

$$S = T \log |\Omega| + \sum_{t=1}^{T} \epsilon_t' \Omega^{-1} \epsilon_t \quad (4.23)$$

A criterion function of this form often emerges in more general situations, as in the prediction error decomposition of (3.5.20). However, even with Ω known, such models tend to be nonlinear. Hence an optimisation technique of the type described in previous sub-sections will usually be appropriate. When Ω is unknown, the optimisation procedure adopted for the parameters other than Ω may be embedded within a step-wise algorithm of the kind described in the previous paragraph.

For a criterion function of the form (4.23), in which ϵ_t is an $N \times 1$ vector, a generalisation of the Gauss–Newton algorithm is appropriate. Conditional on Ω, the iterative scheme is

$$\psi^* = \hat{\psi} + \left[\sum_{t=1}^{T} Z_t \Omega^{-1} Z_t' \right]^{-1} \sum_{t=1}^{T} Z_t \Omega^{-1} \epsilon_t \tag{4.24}$$

where Z_t is an $n \times N$ matrix defined by $Z_t = -\partial \epsilon_t' / \partial \psi$. The relationship of (4.24) to (4.13) on the one hand, and to (2.9.17) on the other should be apparent. The scheme may be embedded within a step-wise procedure by updating the estimate of Ω after each iteration. Such a procedure will be referred to as *multivariate Gauss–Newton*.

Concentrating the Likelihood Function

The discussion so far has assumed that the likelihood function is to be maximised with respect to all n parameters in the vector ψ using a nonlinear optimisation technique. However, since there may be a heavy penalty attached to having a large number of parameters in a nonlinear optimisation routine, it often pays to explore ways in which this number may be reduced. In general, this is done by recognising that the model is *linear* in a subset of parameters. This means that if the remaining parameters are held constant, ML estimates of the first set of parameters may be obtained directly, without resorting to an iterative procedure. These estimates may be substituted into the likelihood function, and the resulting *concentrated* likelihood function maximised with respect to the second set of parameters only.

Example 6   The CES production function (2.1.12), is linear with respect to two parameters, β and α, if the disturbances are multiplicative. If $\epsilon_t \sim NID(0, \sigma^2)$, ML estimation is equivalent to nonlinear least squares. For given values of γ and δ, the ML estimators of β and α are given by regressing $\log Q_t$ on $-\gamma^{-1} \log\{(1 - \delta)L_t^{-\gamma} + \delta^{-\gamma} K_t\}$. Maximising the concentrated likelihood function is then equivalent to minimising the residual sum of squares with respect to γ and δ. Griliches and Ringstad (1971), estimate a series of CES functions in this way using a grid search.

Example 7   In the SURE model, Ω may be concentrated out of the likelihood function using (4.22). Since

$$\Sigma \, \epsilon_t' \Omega^{-1} \epsilon_t = tr(\Omega^{-1} \Sigma \, \epsilon_t \, \epsilon_t') = TN, \tag{4.25}$$

it follows from (4.23) that maximising the concentrated likelihood function is equivalent to minimising the generalised sum of squares function, $S = |\Omega(\beta)|$.

An important practical point to note about working with a concentrated likelihood function is that analytic first derivatives may be obtained from the original likelihood. Suppose that the parameter vector is partitioned as $\psi = (\psi_1', \psi_2')'$, and that ψ_2 is concentrated out of the likelihood function by solving

$$\frac{\partial \log L(\psi_1, \psi_2)}{\partial \psi_2} = 0 \tag{4.26}$$

to yield $\tilde{\psi}_2(\psi_1)$. Maximising the concentrated log–likelihood function, $\log L_c \{\psi_1, \psi_2(\psi_1)\} = \log L_c(\psi_1)$, will usually be accomplished more effectively if first derivatives can be employed. Using a standard result on implicit functions,

$$\frac{\partial \log L_c(\psi_1)}{\partial \psi_1} = \frac{\partial \log L(\psi_1, \psi_2)}{\partial \psi_1} + \frac{\partial \log L(\psi_1, \psi_2)}{\partial \psi_2} \cdot \frac{\partial \psi_2}{\partial \psi_1} \tag{4.27}$$

When evaluated at $\tilde{\psi}_2(\psi_1)$, the last term in (4.27) disappears because of (4.26). Hence the first derivatives of $\log L_c(\psi_1)$ may be obtained by setting $\psi_2 = \tilde{\psi}_2(\psi_1)$ in the expression for $\partial \log L(\psi_1, \psi_2)/\partial \psi_1$.

5. Two-step Estimators

The iterative procedures described in Section 4 are of the form

$$\psi^* = \hat{\psi} + [I^*(\hat{\psi})]^{-1} D \log L(\hat{\psi}) \tag{5.1}$$

where $I^*(\hat{\psi})$ is the information matrix, or an approximation to it, evaluated at $\psi = \hat{\psi}$. In the Newton–Raphson procedure $I^*(\psi)$ is defined as the Hessian, multiplied by minus one, while in the method of scoring it is actually equal to the information matrix. However, (5.1) defines a fairly general iterative scheme in which a variety of expressions may be used for $I^*(\psi)$. The essential point is that under standard regularity conditions, $I^*(\psi)$ is asymptotically equivalent to the information matrix in the sense that

$$\lim T^{-1} I^*(\psi) = IA(\psi) \tag{5.2}$$

A two-step estimator is constructed by evaluating the r.h.s. of (5.1) at a *consistent* estimator of ψ. In other words only one iteration is carried out, with $\hat{\psi}$ being the initial consistent estimator of ψ^* the two-step estimator. The attraction of this procedure is that the two-step estimator has the same asymptotic distribution as the ML estimator. Thus, under the usual regularity conditions, it is asymptotically efficient. As such, it will be denoted by $\tilde{\psi}^*$.

The initial estimator must be consistent in order to ensure that

$$\text{plim } T^{-1}I^*(\hat{\psi}) = IA(\psi) \tag{5.3}$$

The proof given below also requires that $\hat{\psi}$ should have a well-defined asymptotic distribution, although this is not generally necessary for the result to be true.

Example 1   Consider the problem of estimating the location parameter for a Cauchy distribution. As shown in the previous section, the scoring algorithm provides a convenient iterative scheme for finding the ML estimator. It is therefore the natural approach to adopt for finding a two-step estimator. Taking the median to be the initial consistent estimator, an asymptotically efficient estimator of ψ is given directly by (4.5).

Example 2   When maximising the likelihood function is equivalent to minimising a sum of squares function, the two-step estimator may be computed by linear regression. In the MA(1) model a consistent estimator of θ is given by

$$\hat{\theta}_c = [1 - (1 - 4r_1^2)^{1/2}]/2r_1 \tag{5.4}$$

A single iteration of Gauss–Newton, i.e. a regression of ϵ_t on $\partial \epsilon_t / \partial \theta$, then yields an asymptotically efficient estimator of θ.

Estimators formed by one iteration of (5.1) are sometimes referred to as *linearised maximum likelihood*. This conveys their method of derivation, though 'two-step' gives a better insight into their construction, and is the term adopted here. Historically, two-step estimators in econometrics have not usually been derived through (5.1), although in a single equation context, at least, most asymptotically efficient two-step estimators may be obtained as special cases of this approach. This is of some importance, since for any model, (5.1) provides the basis for constructing a two-step estimation procedure. In most cases efficiency can be established formally, but even when this is difficult to do, it will still be generally safe to assume that (5.1) is a reasonably sensible guide to estimation.

Iterating (5.1) further will usually result in an estimate which is closer to the ML solution, although this certainly cannot be guaranteed. However, iterating beyond the first round will not change the asymptotic distribution of the estimator even if continuing the process eventually produces the ML estimator.[1] Small sample properties may be improved by iterating, but this can usually be determined only by Monte Carlo experiments. It is not possible to make general statements about the small sample efficiency of two-step estimators. In some cases a two-step estimator gives essentially the same performance as the ML estimator, while in others carrying out the full ML procedure shows a clear gain.

Statistical Basis of the Two-Step Estimator *

The proof that one iteration of (5.1) yields an asymptotically efficient estimator will be given for the case of a single parameter only. A general proof may be constructed along similar lines.

A Taylor series expansion of $D \log L(\hat{\psi})$ around ψ gives

$$D \log L(\hat{\psi}) \simeq D \log L(\psi) + (\hat{\psi} - \psi)D^2 \log L(\psi) \qquad (5.5)$$

Compare the result of (3.3.4). Subtracting ψ from both sides of (5.1), multiplying by $\sqrt{T}$ and substituting for $D \log L(\hat{\psi})$ from (5.5) gives

$$\sqrt{T}(\tilde{\psi}^* - \psi) \simeq \sqrt{T}(\hat{\psi} - \psi) + \sqrt{T}\frac{D \log L(\psi)}{I^*(\hat{\psi})}$$
$$+ \frac{\sqrt{T}(\hat{\psi} - \psi)D^2 \log L(\psi)}{I^*(\hat{\psi})}$$

Since $\hat{\psi}$ is a consistent estimator of ψ, (5.3) is satisfied. Furthermore $\lim\{-T^{-1}D^2 \log L(\psi)\} = IA(\psi)$, and so if $\hat{\psi}$ has a well defined asymptotic distribution, it follows from Cramér's theorem that

$$-\sqrt{T}(\hat{\psi} - \psi)D^2 \log L(\psi)/I^*(\hat{\psi})$$

has the same limiting distribution as $\sqrt{T}(\hat{\psi} - \psi)$. The limiting distribution of $\sqrt{T}(\tilde{\psi}^* - \psi)$ may therefore be determined through

$$\frac{\sqrt{T}\, D \log L(\psi)}{I^*(\hat{\psi})} \qquad (5.6)$$

Multiplying the numerator and denominator of (5.6) by $D^2 \log L(\psi)$ and rearranging gives

---

[1]ML estimators, however, have a property known as second-order efficiency, which two-step estimators do not, in general, possess; see for example, Rothenberg (1979). Nevertheless, if *two* iterations of (5.1) are carried out from a consistent starting point, it can be shown that second order efficiency is guaranteed also. This has implications for certain three-step procedures.

$$\sqrt{T}(\tilde{\psi}^* - \psi) \simeq \frac{\sqrt{T} \, D \log L(\psi)}{D^2 \log I(\psi)} \left(\frac{-D^2 \log L(\psi)}{I^*(\psi)} \right) \tag{5.7}$$

In view of (5.3), the term in parentheses has a probability limit of unity, and on comparing (5.7) with (3.3.5) it will be seen that the limiting distribution of $\sqrt{T}(\tilde{\psi}^* - \psi)$ is the same as that of $\sqrt{T}(\tilde{\psi} - \psi)$, where $\tilde{\psi}$ is the ML estimator. This is the desired result.

Regression

The generalised regression model of Section 2.4 plays a prominent role in econometrics. Suppose that the elements in the matrix V depend on a finite number of parameters in a vector, δ, but are independent of the regression parameters in β. If the disturbances are normally distributed, the information matrix is block diagonal with respect to β and δ. Furthermore

$$\frac{\partial \log L(\beta, \delta)}{\partial \beta} = X'V^{-1}(y - X\beta) \tag{5.8}$$

and so one iteration of the method of scoring for β yields

$$\beta^* = \hat{\beta} + (X'\hat{V}^{-1}X)^{-1}X'\hat{V}^{-1}(y - X\hat{\beta})$$
$$= (X'\hat{V}^{-1}X)^{-1}X'\hat{V}^{-1}y \tag{5.9}$$

where $\hat{V}$ is formed from an initial estimator of δ, $\hat{\delta}$. If $\hat{\beta}$ and $\hat{\delta}$ are consistent, (5.9) is an asymptotically efficient two-step estimator. However, it is also the expression for the feasible GLS estimator.

Example 3   In the multiplicative heteroscedasticity model, (3.4.15), a consistent estimator of δ may be obtained by regressing $\log \hat{u}_t^2$ on w_t, where $\hat{u}_t$ is the tth OLS residual. An efficient two-step estimator of β is then given by a feasible GLS regression of the form (4.6a). Less obviously, an efficient two-step estimator of δ is obtained from (4.6b).

A *three-step* estimator of β may be constructed from the two-step estimator of δ. While basing the GLS estimator of β on an efficient estimator of δ has no effect on its asymptotic distribution, it seems reasonable to suppose that its small sample properties will be improved.

6. Test Statistics and Confidence Intervals

All the methods described in Sections 3, 4 and 5 are based on the information matrix or an approximation to it. An estimator of the covariance matrix of $\hat{\psi}$ will therefore be automatically available. If, on the other hand, a direct search is used, a numerical approximation based on (2.5) will usually be appropriate.

From the point of view of constructing asymptotically valid test statistics, all that is required is a consistent estimator of $IA(\psi)$. Thus, if $\text{avar}(\tilde{\psi})$ denotes an estimator of the asymptotic covariance matrix, (3.3.2), it must satisfy the condition that

$$\text{plim } T \cdot \text{avar}(\tilde{\psi}) = IA^{-1}(\psi) \tag{6.1}$$

Asymptotic Standard Errors

An *asymptotic standard error* of any estimator, $\hat{\psi}$, is defined as a consistent estimator of the standard deviation of the limiting distribution of $\sqrt{T}(\hat{\psi} - \psi)$, divided by $\sqrt{T}$. When $\hat{\psi}$ is an ML estimator, or an estimator with the same asymptotic distribution, any approximation to $\text{Avar}(\tilde{\psi})$ which satisfies (6.1) yields a set of valid asymptotic standard errors.

Once valid asymptotic standard errors have been computed, the asymptotic 't-ratios',

$$\hat{\psi}_i/\sqrt{\text{avar}(\hat{\psi}_i)}, \qquad i = 1, \ldots, n \tag{6.2}$$

may be constructed. It follows directly from the argument used in (1.4.19) that these are distributed as $N(0, 1)$ variables in large samples. Similarly large sample confidence intervals may also be formed.

In small samples, different methods of computing asymptotic standard errors can give conflicting results. This can happen even where an estimate can be obtained without an iterative procedure.

Example 1   For the AR(1) model it was shown in (3.5.17) that

$$\text{Avar}(\tilde{\phi}) = (1 - \phi^2)/T \tag{6.3}$$

The square root of

$$\text{avar}(\tilde{\phi}) = (1 - \tilde{\phi}^2)/T \tag{6.4}$$

therefore provides an obvious asymptotic standard error. However, if $\tilde{\phi}$ is obtained by regressing y_t on y_{t-1}, it may be more convenient to compute the asymptotic standard error as

$$s \bigg/ \left(\sum_{t=2}^{T} y_{t-1}^2 \right)^{1/2} \tag{6.5}$$

This is the expression suggested by (2.3.17). It is perfectly valid in this context as

$$\text{plim}\left\{Ts^2\bigg/\sum_{t=2}^{T} y_{t-1}^2\right\} = \sigma^2/\text{plim } T^{-1}\sum_{t=2}^{T} y_{t-1}^2 = 1 - \phi^2$$

The advantage over (6.4) is that hypotheses on ϕ can be tested within the standard OLS framework.

Two-Step Estimators

In a two-step procedure, an estimate of the asymptotic covariance matrix of $\tilde{\psi}^*$ is given by the inverse of $I^*(\hat{\psi})$. This must satisfy (6.1) if $\tilde{\psi}^*$ is to be asymptotically efficient; cf. (5.3). A more accurate estimate may be obtained if $I^*(\psi)$ is evaluated at $\psi = \tilde{\psi}^*$, but this will usually be less convenient.

Example 2   In Example 5.2 it was shown that an efficient estimator of θ in an MA(1) model can be constructed by regressing (3.5.14) on (4.19) with $\theta = \hat{\theta}_c$. An asymptotic standard error for $\hat{\theta}^*$ may be obtained by treating the derivative of ϵ_t as though it were an explanatory variable in a standard regression model.

Notes

General treatments of numerical optimisation will be found in Wolfe (1978), Dixon (1972) and Box *et al.* (1969). In the UK the NAG collection of programmes is widely available and the manual contains a discussion of the various algorithms for numerical optimisation.

Exercises

1. Suppose that the disturbances in a linear regression model have a standardised Cauchy distribution. Show that the Newton—Raphson scheme may be interpreted in terms of a series of regressions.

2. Why might the transformation $\psi = \{1 - \exp(-\theta)\}/\{1 + \exp(\theta)\}$ be useful in constrained optimisation?

3. Show that Gauss—Newton is identical to the method of scoring in a non-linear regression model.

4. What would happen to the Newton—Raphson algorithm in Example 3.1 if the starting value was $\hat{\psi} = 1.0$? Explain briefly how a quasi-Newton algorithm would work in these circumstances.

5. For the ARMA(1, 1) model, (1.5.21), show that Gauss—Newton breaks down if $\hat{\phi} = \hat{\theta} = 0$. Show that the modified scheme (4.14) avoids this problem if μ is strictly positive. Explain how you would construct an asymptotic covariance matrix for the ML estimates in this model. How would you test the hypothesis that $\phi = 0.5$?

6. Derive expressions for the first two theoretical autocorrelations, ρ_1 and ρ_2, in an ARMA(1, 1) model. Hence obtain expressions for calculating consistent estimators of ϕ and θ from r_1 and r_2. Show how these estimates may be used to construct an asymptotically efficient two-step estimator of (ϕ, θ).

5
Test Procedures and
Model Selection

1. Introduction

The discussion in the previous chapters has centred on the estimation of a correctly specified model. In the context of maximum likelihood, this means that the distribution of the observations on the sample space is known apart from a finite number of real parameters. The previous chapters discussed how these parameters could be estimated efficiently, but nothing was said regarding the way in which one might arrive at a suitable specification in the first place.

Before proceeding further, it is worth going back to Section 1.1 and recalling exactly what is meant by a 'correctly specified' or 'true' model in a practical context. In the real world the process by which observations are generated will almost invariably be complex. The motivation behind model construction is not to provide a description of this process, but rather to extract its main features.

Bearing this in mind, what is required is a model which is relatively simple, but has high predictive power. It is against this yardstick that a model must be assessed. The concept of a 'true' model is therefore a purely theoretical one, but it provides a basis upon which criteria for detecting misspecification may be developed. In addition it enables the consequences of inappropriate modelling to be explored and discussed in a concrete form.

Consequences of Misspecification

The consequences of adopting an inappropriate model will depend on its relationship to the true model. Three cases may be distinguished. In the first, the adopted model is *underparameterised* in the sense that it imposes invalid constraints on the true model. In the second, it is *overparameterised*, being more general than is necessary. In the third, the assumed and true models are *non-nested*: neither can be obtained as a special case of the other.

Overparameterisation tends to have rather different consequences from underparameterisation or the adoption of a model which is not nested within the true model. This has important implications for model selection strategies. Although no general conclusions can be drawn from the example below, the trade-off it reveals between bias and efficiency is not untypical.

Example   Suppose that the true model is

$$y_t = \alpha + \beta x_t + \epsilon_t, \qquad \epsilon_t \sim NID(0, \sigma^2) \tag{1.1}$$

while in the assumed model y_t is taken to be a linear function of x_t and another variable, z_t. The consequence of regressing y_t on both z_t and x_t is to yield an estimator of β, $\hat{b}$, which is unbiased, but has a variance,

$$\text{Var}(\hat{b}) = \sigma^2 / \{ \Sigma x_{*t}^2 (1 - r_{xz}^2) \} \tag{1.2}$$

where $x_{*t} = x_t - \bar{x}$ and r_{xz} is the correlation coefficient between x_t and z_t. Hence, $\hat{b}$ is an inefficient estimator of β since

$$\text{Eff}(\hat{b}) = \text{Var}(b)/\text{Var}(\hat{b}) = 1 - r_{xz}^2 \leqslant 1 \tag{1.3}$$

where b is the OLS estimator in (1.1). The estimator of the coefficient of z_t will have an expectation of zero and so any predictions made from the overparameterisated model will also be unbiased. However, they will have a higher MSE than predictions from the true model. Note that overfitting can be disastrous in certain circumstances. If x_t and z_t are perfectly collinear, the model ceases to be identified (cf. the discussion in Section 3.6).

   If the true model is

$$y_t = \alpha + \beta x_t + \gamma z_t + \epsilon_t, \qquad \epsilon_t \sim NID(0, \sigma^2) \tag{1.4}$$

but it is incorrectly assumed that $\gamma = 0$, the estimator of β will generally be biased. If $b^\dagger$ is the estimator obtained from regressing y_t on x_t, then

$$E(b^\dagger) = \beta + \gamma \Sigma x_{*t} z_{*t} / \Sigma x_{*t}^2 = \beta + \gamma r_{xz} \sqrt{\Sigma x_{*t}^2 / \Sigma z_{*t}^2} \tag{1.5}$$

Thus $b^\dagger$ will be biased unless x_t and z_t are orthogonal, and this bias will persist as the sample size increases. However, if r_{xz}^2 is close to unity, the omission of z_t will have little effect on predictions.

Test Procedures

There is an important distinction to be drawn between tests of specification and tests of misspecification. This can be illustrated by model (1.4). Suppose that the only two formulations under consideration are (1.4) itself and the same model with z_t omitted.

A natural way to proceed would be to estimate (1.4) and carry out a t-test of the hypothesis that $\gamma = 0$. This is a test of *specification* and in Section 3 it is shown that a t-test of this type would, in a certain sense, be an optimal test procedure. There would be little virtue in estimating the restricted model and then using the residuals to test whether or not z_t should have been included in the regression. On the other hand, including z_t in the model might not have been an obvious possibility at the outset. Therefore y_t could have been regressed on x_t with no clear alternative in view. In such a case, it would be prudent to try and assess whether the model was adequate. If it were not, there should be some indication in the residuals. Unless x_t and z_t are orthogonal, the residuals will have non-zero expectation, and so it should be possible to detect some pattern either visually or by means of an appropriate test statistic. This would be a test of *misspecification*.

Unlike a test of specification, therefore, a test of misspecification is constructed with no clear alternative hypothesis in mind. As such, it is a procedure designed for assessing the goodness of fit of the model implied by a particular maintained hypothesis. However, although the alternative does remain rather vague, the choice of a particular test statistic is usually motivated by a suspected departure from the maintained hypothesis in some particular direction. Thus, for example, general tests against 'serial correlation' or 'heteroscedasticity' may be constructed, even though these phenomena may be modelled in a number of different ways.

Both types of test have a role to play in model building. It will be argued later that it is generally advisable to consider all possible forms of the model before any estimation is carried out, and then to plan to test the various hypotheses in a systematic fashion. Nevertheless it is not usually possible to be certain that all the alternatives have been considered, and so such a process should be supplemented by tests of misspecification.

Not all test procedures may be neatly classified as tests of specification or misspecification. Suppose that the models

$$H_1 : y_t = \beta x_t + \epsilon_t, \qquad \epsilon_t \sim NID(0, \sigma^2) \tag{1.6}$$

and

$$H_2 : y_t = \gamma z_t + \epsilon_t, \qquad \epsilon_t \sim NID(0, \sigma^2) \tag{1.7}$$

are under consideration as alternative explanations of the behaviour of y. These are non-nested in the sense that neither one can be obtained as a special case of the other.

There are basically two ways of approaching a non-nested problem. The first is to view it as a question of discrimination: *either H_1 or H_2* must be chosen and the choice must be based on some suitable criterion indicating goodness of fit. Alternatively the investigator may entertain two further possibilities, namely that both hypotheses may be accepted or both may be rejected. The first of these conclusions implies that there is insufficient data to separate the two hypotheses, while the second implies that neither model is satisfactory. Thus although two well-defined hypotheses are under consideration, the fact that both may be rejected suggests that the adopted procedure is, in part, a test of misspecification.

The test procedures described so far are all carried out within the sample. Hence the same observations being used to select a model are also being employed to estimate it. In general, the model which finally emerges must be regarded as being the product of a certain amount of 'data mining'. Hence its validity can really only be assessed with respect to a completely fresh set of data. This leads to the idea of *post-sample predictive testing.* The observations used in this exercise are often made available by the researcher deliberately omitting a set of points from the data used to construct the model.

Four different aspects of specification testing have therefore been distinguished, and these are discussed in the remainder of the chapter. Tests of misspecification are introduced in Section 2, with examples being drawn from linear regression. Most of these tests are based on residuals. However, because the residuals often have a natural interpretation under certain types of misspecification, graphical procedures can also play a useful role in the general process of 'diagnostic checking'.

Section 3 deals with the classical theory of hypothesis testing, and the construction of tests based on the likelihood ratio principle. Two related test procedures are discussed in Sections 4 and 5. These lead to what are commonly known as Wald and Lagrange multiplier tests, and they supplement the likelihood ratio principle in forming the basis for tests of specification. All three tests have certain desirable properties, although the justification for their use relies, in general, on large sample arguments. The choice between the three approaches rests mainly on computational considerations. The likelihood ratio approach requires that the model be estimated under both the null and alternative hypotheses, i.e. it requires estimation of both the *restricted* and *unrestricted* models. A Wald test, on the other hand, may be based on estimates from the unrestricted model, while a Lagrange multiplier statistic requires only that the model be estimated under the null. The question of which statistic is easiest to compute depends on the structure of the particular problem.

Because the Lagrange multiplier approach is based on estimation under the null hypothesis, many of the tests constructed in this way may be associated with tests of misspecification. The recognition that certain diagnostics are basically Lagrange multiplier tests provides a good deal of insight into their potential effectiveness, as well as suggesting viable generalisations. However, it is always important to distinguish the sense in which such tests are used. If a test of misspecification rejects the null hypothesis, this implies that the model is inadequate, without necessarily pointing to the more general model underlying the Lagrange multiplier interpretation.

Criteria for discriminating between models are introduced in Section 6. Although discussed primarily in the context of non-nested hypotheses, there is no reason why they should not be used in nested situations, and a number of examples of such applications are given later in the book. The remaining part of Section 6 describes test procedures applicable to the non-nested case. Section 7 deals with post-sample predictive testing, while Section 8 considers an overall strategy for model selection.

2. Tests of Misspecification

The purpose of constructing a model is to systematically account for as much of the variation in the observations as possible. The movements not captured by the fitted model are termed *residuals*, and if the model is reasonably adequate these residuals should be approximately random. Departures from randomness are an indication that the model is failing to pick up a systematic component in the observations, and an attempt should therefore be made to find a better model.

Residuals play an important role in procedures for detecting misspecification. Such procedures may embody a number of test statistics, since lack of randomness may manifest itself in a number of different ways. However, although test statistics provide a concise summary of the behaviour of residuals, it is generally advisable to plot the residuals. This may be done in a number of different ways, and an examination of the various plots can often prove extremely useful.

The discussion of tests of misspecification will be carried out mainly within the context of a classical linear regression model estimated by ordinary least squares. As well as being of importance in itself, this serves as a convenient vehicle for developing points of much wider applicability. The assumed model is therefore of the form (2.1.1), with the disturbances independently distributed with mean zero and variance σ^2. Since the construction of exact tests generally requires that the disturbances be normally distributed, this will be taken to hold throughout.

Residuals and Misspecification

The properties of the OLS residuals in a correctly specified model were derived in Section 2.7, where it was shown that the vector e has a multivariate normal distribution with mean vector 0 and covariance matrix $\sigma^2 M$. Thus even though the disturbances may be homoscedastic and uncorrelated, this will not, in general, be true for the OLS residuals. On the other hand, it can be shown that e converges in distribution to u as $T \to \infty$. This result holds even if the disturbance term is misspecified. In that case $E(e) = 0$ while $E(ee') = \sigma^2 MVM'$, where $\sigma^2 V$ is the covariance matrix of u. However, in large samples e is normally distributed with zero mean and a covariance matrix given approximately by $\sigma^2 V$. Thus the OLS residuals will tend to mirror the properties of the true disturbances, even though their respective distributions are not, strictly speaking, identical.

The behaviour of the OLS residuals under structural misspecification is less obvious, but more interesting. Suppose that the assumed model is of the form (2.1.5), while the true model is

$$y = X_r \gamma_r + f + \epsilon, \qquad \epsilon \sim N(0, \sigma^2 I) \tag{2.1}$$

where X_r is a $T \times r$ matrix whose columns are a subset of the columns of X corresponding to correctly specified variables, and γ_r is an $r \times 1$ vector of parameters. This set-up covers both functional misspecification and omitted variables, depending on the composition of the elements of the $T \times 1$ vector f.

Since

$$e - My = Mf + M\epsilon \tag{2.2}$$

it follows that e will have a mean vector Mf and a covariance matrix $\sigma^2 M$. Thus the expectation of the residual vector is equal to the residual vector obtained by regressing f on X. As an example, suppose that the true model is

$$y_t = \alpha + \beta x_t + \gamma x_t^2 + \epsilon_t \tag{2.3}$$

but that the term in x_t^2 is omitted. The expectation of the residual vector will be equal to γ multiplied by the vector of OLS residuals obtained by regressing x_t^2 on x_t.

The properties of other residuals may be analysed in a similar way. The LUS residuals, (2.7.5), have $E(e^*) = Cf$ and $\text{Var}(e^*) = \sigma^2 I$ under

misspecification of the type analysed in (2.1). The expectation of a
LUS residual is therefore given by the corresponding vector of
residuals obtained by regressing f on X.

The expectations of the residuals determine their overall pattern.
This is important insofar as it gives us some insight into the likely
patterns of residuals under various types of misspecification. One
attraction of the recursive residuals is that their behaviour in a
misspecified model is very different from that of the OLS residuals.
This means that test procedures based on recursive residuals may be
complementary to those based on OLS residuals. As an example,
consider the functional misspecification which arises when x_t^2 is
omitted from (2.3). Figure 5.1 shows the expectations of OLS,
BLUS and recursive residuals when the data are arranged with the
x_t's in ascending order and $\gamma = 3$. While the BLUS and OLS residuals
show a similar pattern, that of the recursive residuals is entirely
different, with all the expectations having the same sign. A heuristic
argument as to why this is the case can be given with the aid of
figure 5.2. Since the underlying relationship between y and x is
quadratic rather than linear, the prediction errors, $\tilde{v}_t = y_t - x_t' b_{t-1}$,
$t = k + 1, \ldots, T$, all tend to have the same sign.

The distinctive pattern of the recursive residuals may be exploited
to yield useful test statistics as well as graphical procedures. By way
of contrast, BLUS residuals tend to follow the OLS residuals more
closely and hence yield little additional information. In describing
procedures for detecting misspecification, therefore, the emphasis
will be primarily on OLS and recursive residuals.

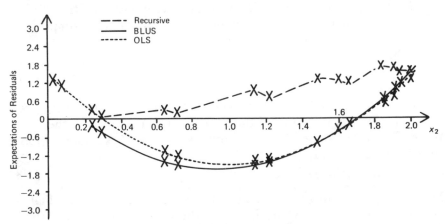

Figure 5.1   *Pattern of Expectations of Residuals for Model (2.3)*

Source: This figure is reproduced from Harvey and Collier (1977) with the kind
permission of the *Journal of Econometrics* and the authors.

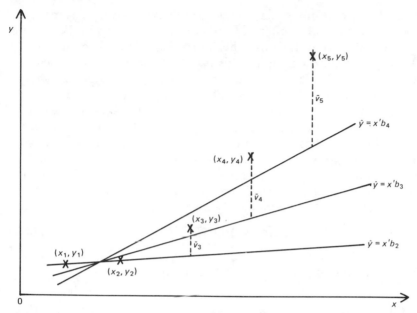

Figure 5.2 *Derivation of Recursive Residuals for a Misspecified Bivariate Regression Model*

Source: See Figure 5.1.

Graphical Techniques

A good deal of information may be obtained simply by plotting residuals and examining them for some distinctive pattern. The plots may be against time, or against one of the explanatory variables. This last type of plot may be particularly valuable if it is suspected that the variable in question is subject to functional misspecification.

The information contained in the residuals may also be assessed by plotting the cumulative sum (CUSUM) and cumulative sum of squares (CUSUMSQ). The first of these procedures is used primarily with recursive residuals, a plot based on OLS residuals being difficult to interpret because of the constraint that the residuals finally sum to zero. The CUSUM of recursive residuals is defined by

$$W_t = \hat{\sigma}^{-1} \sum_{j=k+1}^{t} v_j, \qquad t = k+1, \ldots, T \tag{2.4}$$

where

$$\hat{\sigma}^2 = \Sigma (v_t - \bar{v})^2 / (T - k - 1)$$

$\bar{v}$ is the arithmetic mean of the residuals.

If the model is incorrectly specified, there may be a tendency for a disproportionate number of recursive residuals to have the same sign. The cumulative effect of this will then tend to show up in W_t moving away from the horizontal axis. The functional misspecification for (2.3) provides a good example. This type of behaviour also arises when a structural break occurs, the point at which it happens being indicated by the beginning of a secular increase or decrease in W_t.

Two lines drawn symmetrically above and below $W_t = 0$ provide a means of assessing the significance of departures from the horizontal axis. The lines are constructed in such a way that the probability of crossing either one or both of them is equal to a given significance level. The underlying theory, which is based on results in Brownian motion, will not be dealt with here. However, the equation of the lines is given quite straightforwardly by

$$W = \pm\{a\sqrt{T-k} + 2a(t-k)/(T-k)^{1/2}\} \qquad (2.5)$$

where $a = 0.948$ for a significance level of 5% and 0.850 for 10%.

The CUSUMSQ is based on a plot of the quantities

$$WW_t = \sum_{t=k+1}^{t} v_t^2 \bigg/ \sum_{t=k+1}^{T} v_t^2 \qquad (2.6)$$

When the model is correctly specified, WW_t has a Beta distribution with a mean of $(t-k)/(T-k)$. This suggests drawing a pair of lines

$$WW = \pm c_0 + (t-k)/(T-k) \qquad (2.7)$$

parallel to the mean value line and rejecting H_0, the null hypothesis of correct specification, if either of the lines is crossed. The value of c_0 is determined such that the probability that either line is crossed under H_0 is equal to the desired significance level. Table C provides the required values of c_0 (see page 364).

The CUSUMSQ plot provides a useful complement to the CUSUM plot. It may be applied with OLS as well as recursive residuals, although the test procedure just described assumes that the residuals are $NID(0, \sigma^2)$. When used with recursive residuals it should provide a means of detecting a structural break. As regards other types of misspecification, both OLS and recursive residual plots will be sensitive to variances increasing over time. However, although both the CUSUM and CUSUMSQ techniques can be used as test procedures against particular types of misspecification, they are best regarded as 'data analytic' techniques; i.e. the value of the plots lies in the information to be gained simply by inspecting them. The significance lines constructed are, to paraphrase Brown *et al.* (1975, p. 155), best regarded as *yardsticks* against which to assess the observed plots rather than as formal tests of significance.

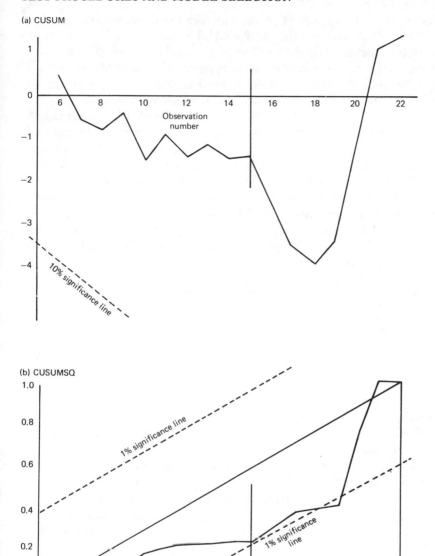

Figure 5.3 *CUSUM and CUSUM of Squares of Recursive Residuals for Local Telephone Calls*

Source: This figure is reproduced from Brown, Durbin and Evans (1975) with the kind permission of the *Journal of the Royal Statistical Society* and the authors.

Example   Figure 5.3a reproduced from Brown *et al.* (1975, p. 159), illustrates the use of the CUSUM technique for detecting a structural break. The data relate to a model in which the growth in local telephone calls is explained in terms of four other variables and a constant term. The model shows no sign of instability up to 1964/5, but after that point the CUSUM shows a definite downward movement. This accorded with the limited prior knowledge of the data which suggested that the estimates of local telephone calls might be subject to some uncertainty. Thus the 'structural break' detected in this case really represents a break in the data series on the dependent variable. Figure 5.3b which is also reproduced from Brown *et al.* shows the plot of the CUSUM of squares, (2.6). Like the CUSUM this indicates some instability, although it is less easy to pinpoint exactly where the break occurs.

Although the CUSUM plot in figure 5.3a clearly indicates some instability beyond 1964/5, it is not formally significant. This is quite likely to be the case, even if there is a genuine structural break in the data. Thus figure 5.3a highlights the point that the CUSUM and CUSUM of squares plots contain more information that can be summarised in a single test statistic.

Test Statistics

The departures from randomness exhibited by the residuals in a misspecified model may be captured by a limited number of test statistics. Such statistics provide a compact summary of the information contained in the residuals. However, they should not be regarded as an alternative to visual checks, but rather as complementary to them.

Misspecification test statistics should have a number of desirable properties. The first is that they should be relatively easy to compute. The second is that their distribution under the null hypothesis should be known, and that significance points for this distribution should be readily available. If the distribution is known exactly, the critical region may be defined so as to yield a test of the desired size. However, for many statistics the distribution may only be derived asymptotically, in which case the actual size of the test will not be equal to the nominal size. Such tests are often termed *large sample* tests. This terminology indicates that the choice of critical region can only be justified on an asymptotic basis, in contrast to the situation for *exact* tests. However, it must be stressed that this consideration by no means precludes the use of large sample tests when the sample size is small. Choosing a test simply because its size is known exactly may not be good practice if its power characteristics are poor.

Power, of course, is a crucial aspect of any test procedure, but in diagnostic checking the issue is complicated by the fact that the researcher may have several different types of misspecification in mind. Furthermore, the alternative hypotheses implied by these misspecifications are bound to be vague. Nevertheless, a test statistic should, as a minimum requirement, exhibit a 'reasonably' high power against at least one well-defined alternative. If a test statistic exhibits high power against several types of alternative hypothesis, it is obviously useful as a general test of misspecification. Nevertheless, test statistics which are *robust* to certain classes of specification error, can play a valuable complementary role, since they help to provide some indication of the direction of misspecification.

The *Durbin—Watson d-statistic* is defined by

$$d = \frac{\sum_{t=2}^{T} (e_t - e_{t-1})^2}{\sum_{t=1}^{T} e_t^2} \qquad (2.8)$$

and this forms the basis for what is arguably the most important test of misspecification in linear regression. The d-statistic was originally constructed for use in a test against a first-order autoregressive process in the disturbances, and it will be fully described in that context in Chapter 6. For the moment, it is important to note simply that the distribution of d is centred around two when the model is correctly specified, but tends towards zero if the residuals exhibit positive serial correlation. Now, while serial correlation in the disturbances implies serial correlation in the residuals, the converse is not true. Thus although a low value of d could indicate a serially correlated disturbance term, it could also be symptomatic of some other type of misspecification. In particular, it should be clear from Figure 5.1 that positive serial correlation can be expected when the functional form is inappropriate. A similar pattern may also emerge when a variable has been omitted.

Unfortunately the distribution of d depends on X, the matrix of regressors and so exact significance points cannot be tabulated. However, d is asymptotically normal with a mean of 2 and a variance of $4/T$, and this result provides the basis for a large sample test procedure.

There is no difficulty in basing an exact test on a set of LUS residuals. In the context of serial correlation, the von Neumann Ratio or the modified von Neumann Ratio is the appropriate statistic. Using recursive residuals, the *modified von Neumann Ratio* takes the form

$$\text{MVNR} = \frac{T-k}{T-k-1} \cdot \frac{\sum\limits_{t=k+2}^{T} (v_t - v_{t-1})^2}{\sum\limits_{t=k+1}^{T} v_t^2} \tag{2.9}$$

Significance points may be obtained from table B, although if the sample size is large, a satisfactory approximation is obtained from a normal distribution with mean 2 and variance $4/(T-k)$. As with the d-statistic, a one-tailed test against positive autocorrelation provides the most appropriate basis for a general test of misspecification.

In time series regression, it is natural to take the first k observations as the 'base' and to calculate recursive residuals corresponding to observations $y_{k+1}, \ldots, y_T$. However, other orderings of the data may sometimes be more appropriate. For example, if functional misspecification is suspected, it may be preferable to have the observations on a particular explanatory variable in increasing order of magnitude.

A second test statistic based on recursive residuals is given by

$$\psi = \left\{ (T-k-1)^{-1} \sum_{t=k+1}^{T} (v_t - \bar{v})^2 \right\}^{-1/2} (T-k)^{-1/2} \sum_{t=k+1}^{T} v_t \tag{2.10}$$

When the model is correctly specified, ψ has a t-distribution with $T-k-1$ degrees of freedom, a result which follows immediately from the properties of the recursive residuals. A two-sided '*recursive t-test*' may therefore be carried out. This is likely to be particularly effective in situations where the recursive residuals tend to have the same sign, as in figure 5.1. A second example concerns structural change. This is precisely the kind of misspecification against which the CUSUM of recursive residuals is effective. In fact the two procedures are closely related since $\psi = (T-k)^{-1/2} W_T$.

A Sign test provides a simple alternative to (2.10). Under the null hypothesis, the expected number of positive recursive residuals is equal to $(T-k)/2$. A critical region may therefore be constructed from the binomial distribution. However, the evidence presented in Harvey and Collier (1977) suggests that the Sign test tends to lack power compared with the t-test. Nevertheless, it is very simple and it may be more robust to non-normality.

There is no OLS counterpart to (2.10) since $\Sigma e_t = 0$. BLUS and other types of LUS residual may be used to form analogous t-statistics, but there is no obvious reason as to why they should be effective in the kind of situation just described. The great advantage of the recursive residuals is that they have a natural interpretation, and this can be exploited in deriving suitable test procedures.

A number of test statistics may be used to detect systematic patterns in the variance of the residuals. These may be termed *heteroscedasticity* tests, although heteroscedasticity in the residuals may be symptomatic of a good deal of other things besides heteroscedasticity in the disturbances.

Heteroscedasticity tests often involve some grouping of observations. Residuals which are relatively homogeneous with regard to variance are then contrasted with other groups of residuals. Thus if it is felt that there may be a tendency for the variances of the residuals to increase over time, a natural test statistic to adopt is

$$H = \sum_{t=T-m+1}^{T} e_t^2 \Big/ \sum_{t=1}^{m} e_t^2 \qquad (2.11)$$

where $m \simeq T/3$. The purpose in dropping some observations is to accentuate any difference between the numerator and denominator of H.

High values of (2.11) indicate a rejection of the null hypothesis of correct specification, but the problem with using this statistic is that its distribution depends on the matrix of regressors, X. A large sample test may, however, be obtained by noting that mH is asymptotically χ_m^2 under the null hypothesis. This follows because the OLS residuals converge in distribution to the true disturbances and

$$\text{plim } m^{-1} \sum_{t=1}^{m} e_t^2 = \sigma^2$$

An exact heteroscedasticity test may be based on any set of LUS residuals. For recursive residuals, the statistic

$$H^\dagger = \sum_{t=T-m+1}^{T} v_t^2 \Big/ \sum_{t=k+1}^{k+m} v_t^2 \qquad (2.12)$$

has an $F_{m,m}$ distribution under the null hypothesis, a suitable choice of m being $(T-k)/3$.

A final method of testing for heteroscedastic residuals is provided by the CUSUMSQ plot. This may have wider applicability than (2.12), since it can still be effective in situations where the variances are not monotonically increasing.

All the tests considered so far in this subsection have been based directly on residuals. However, tests of misspecification may easily be constructed in other ways. An example is the RESET test developed by Ramsey and Schmidt (1976) in which powers of the explanatory variables are added to the regression equation and tested for significance. Thus if the maintained hypothesis were (1.1), x_t^2, x_t^3 and x_t^4 might be added to the regression and the joint significance of their coefficients tested by the usual F-statistic. The rationale for such a procedure in the context of suspected functional misspecification is obvious. However, RESET has also been proposed as a general test for specification error in regression.

The form of the test statistics described so far has been justified mainly on an intuitive basis. Given the kind of misspecification one has in mind, a reasonably sensible test statistic can often be constructed on the basis of the expected behaviour of the residuals. However, the appropriate form for such a test statistic may not always be apparent. In these circumstances, a more formal approach based on the Lagrange multiplier test principle can often prove useful. The Lagrange multiplier procedure is introduced in Section 5, and some of the examples given there show how this approach can lead to suitable tests of misspecification.

Concluding Remarks

A number of tests for checking the specification of a regression model have been suggested. As with the accompanying graphical procedures, these tests are based primarily on OLS and recursive residuals. Other LUS residuals could also be used, but there is no evidence, as yet, to indicate that they might yield additional information.

Because a whole battery of tests is being applied to the model, the results must be treated with some care. In particular, the probability of rejecting the model when it is true may be relatively high. A further conceptual problem arises when a test is carried out after a visual inspection of the residuals has suggested that such a test might be appropriate. For reasons such as these, misspecification test procedures are, strictly speaking, invalid when viewed in a classical framework, and the term 'diagnostic' is often more appropriate.

It has already been noted that a 'significant' value of a particular test statistic may be indicative of more than one type of specification error. Thus the maintained hypothesis may be rejected, but without a clear indication of where the fault lies. However, a combination of the results from several test statistics can be rather more informative. For example, the Durbin—Watson d-statistic is, amongst other things, sensitive to structural change and functional misspecification, as well as serial correlation in the disturbances. On the other hand, the recursive t-statistic has moderately high power against certain types of structural change and functional misspecification, yet is relatively robust to serial correlation in the disturbances.

All the discussion so far has assumed a normally distributed disturbance term. If this assumption is not reasonable the test procedures described may still, in general, be used, although they will have only asymptotic validity. However, difficulties may be encountered in cases where certain moments of the distribution do not exist. If the OLS estimator is inconsistent, as would, for example, be the case with a Cauchy disturbance term, none of the tests will have even asymptotic validity. When there is some doubt about the distribution of the disturbance term, the most prudent course of action is first to examine the residuals from a robust estimator such as MAD.

3. Classical Test Procedures: The Likelihood Ratio Test

The tests of misspecification discussed in the previous section provide a means of assessing 'goodness of fit'. No specific alternative hypothesis is entertained at the outset, although a knowledge of the power of the tests against particular alternatives is obviously useful insofar as it suggests ways of modifying the model. Tests of specification, on the other hand, employ a specific alternative hypothesis, H_1, as well as a null, H_0.

The foundations of the classical approach to hypothesis testing were laid by Neyman and Pearson. The sample space is partitioned into two regions. One is a region of acceptance, in which the data are consistent with the null hypothesis, and the other a region of rejection, known as the *critical region*. The essence of a 'good' test procedure is the choice of a critical region which is in some sense optimal.

A test procedure of the above form implies two types of error. The first, a Type I error, arises when the null hypothesis is rejected when it is true, while a Type II error is made when H_0 is incorrectly accepted. Generally speaking, the probability of a Type II error can be determined only with respect to a specific alternative hypothesis, and given such an alternative, a 'good' test should, other things being equal, make this probability small. In order that other things should be equal, the (maximum) probability of making a Type I error is fixed. This probability is termed the *significance level* or *size* of the test, and a test can then be considered optimal if it minimises the probability of making a Type II error over the complete set of parameter values associated with the alternative. Such a test is termed uniformly most powerful (UMP), since the term power is defined as $\{1 - Pr(\text{Type II Error})\}$.

The above criterion is not the only one which could be used in judging the desirability of a particular test. A natural question to ask, therefore, is whether it suggests a form for a suitable test statistic, together with an appropriate critical region. Consider first the simplest possible situation, in which the observations are drawn at random from a distribution which depends on a single parameter ψ. Furthermore, suppose that the null and alternative hypotheses each consist only of a single value of ψ, i.e. $H_0 : \psi = \psi_0$ is to be tested against $H_1 : \psi = \psi_1$. Let $L(\psi_0)$ and $L(\psi_1)$ denote the likelihood function of the sample under H_0 and H_1 respectively. The form of the most powerful test — the qualification 'uniformly' is superfluous when ψ_1 is a single point — is then determined by a fundamental result known as the *Neyman—Pearson lemma*. This states that if there exists a critical region, C, of size α and a constant κ such that

$$\frac{L(\psi_0)}{L(\psi_1)} \leqslant \kappa \qquad \text{inside } C$$

and

$$\frac{L(\psi_0)}{L(\psi_1)} > \kappa \qquad \text{outside } C$$

then C is a most powerful critical region of size α for testing $\psi = \psi_0$ against $\psi = \psi_1$.

A proof of the Neyman—Pearson lemma may be found in almost any text on mathematical statistics, and will not be given here. However, it is worth noting the plausibility of the test suggested by the lemma: if H_0 is true, we would expect $L(\psi_0)$ to be large relative to $L(\psi_1)$ and so the ratio of the likelihoods would tend to fall outside the critical region. Conversely, if H_1 were true, $L(\psi_0)/L(\psi_1)$ would tend to be small.

Although the Neyman—Pearson lemma cases can be extended to slightly more general situations, its direct applicability to many of the situations of practical interest is limited. In fact, it is often the exception rather than the rule that a UMP test exists. Nevertheless, even in circumstances where a test satisfying the requirements for optimality cannot be constructed, the Neyman—Pearson lemma at least suggest a general method for constructing tests based on the likelihood ratio.

The Likelihood Ratio Test

The likelihood ratio (LR) takes the form

$$\lambda = \lambda(y_1, \ldots, y_T) = L(\tilde{\psi}_0)/L(\tilde{\psi}_1), \tag{3.1}$$

where $L(\tilde{\psi}_0)$ and $L(\tilde{\psi}_1)$ denote the maximised likelihood functions obtained under the null and alternative hypotheses respectively. When the alternative hypothesis holds, λ will tend to be small, which suggests setting a critical region according to the criterion $\lambda < \kappa$, where κ is determined by the desired size of the test, α. Thus when the null hypothesis is true, κ is determined by the condition that

$$Pr\{\lambda < \kappa\} = \alpha \tag{3.2}$$

What we have therefore is a sample statistic, λ, together with an indication of the form of the critical region. However, for a given size of test, a value of κ which results in (3.2) being satisfied *exactly* can only be found if the small sample distribution of λ, or a monotonic function of it, is known. This is rarely the case in more complex models, with the result that testing must be carried out on the basis of large-sample approximations. Before moving on to discuss such approximations, a case where the LR principle does lead to an exact test is considered.

Example 1 Consider the problem of testing for the inclusion of a subset of $m < k$ explanatory variables in a classical linear regression model. If the disturbances are normally distributed, the maximised likelihood for the general model is

$$L(\tilde{\psi}_1) = (2\pi\tilde{\sigma}_1^2)^{-T/2} \exp(-T/2) \tag{3.3}$$

where $\tilde{\sigma}_1^2 = SSE_1/T$. The expression for $L(\tilde{\psi}_0)$ takes a similar form except that $\tilde{\sigma}_1^2$ is replaced by an estimator, $\tilde{\sigma}_0^2 = SSE_0/T$, constructed from the OLS residuals in the regression with the m explanatory variables omitted. The LR is therefore

$$\lambda = \left(\frac{\tilde{\sigma}_0^2}{\tilde{\sigma}_1^2}\right)^{-T/2} = \left(\frac{SSE_0}{SSE_1}\right)^{-T/2} \tag{3.4}$$

A critical value, κ, which satisfies (3.2) cannot be obtained directly since the distribution of (3.4) is unknown. However, the distribution of

$$\lambda* = \frac{(SSE_0 - SSE_1)/m}{SSE_1/(T-k)} \tag{3.5}$$

is known. Under H_0 it has an F-distribution with $(m, T-k)$ degrees of freedom; cf. (2.8.10). On comparing (3.5) with (3.4) it will be found that one is a monotonic function of the other, since

$$\lambda^* = (\lambda^{-2/T} - 1)(T - k)/m \tag{3.6}$$

A critical region of the form $\lambda < \kappa$ for the LR statistic corresponds to a critical region of the form $\lambda^* > \kappa^*$ for (3.5) and κ^* may be determined such that

$$Pr\{\lambda^* > \kappa^*\} = \alpha \tag{3.7}$$

is satisfied exactly. An exact LR test could therefore be carried out by comparing λ with a critical value, k, obtained indirectly from the F-distribution via (3.6). However, since an F-test based on λ^* will give exactly the same result, it is this statistic which is normally used.

When $m = 1$, the LR test is equivalent to the classical 't-test' of significance. This test may be shown to be UMP against a one-sided alternative, and so in this case there is a direct justification for the LR test in terms of Neyman–Pearson theory. However, when $m > 1$, a UMP test no longer exists. Nevertheless, the results for $m = 1$ at least support the notion that the LR test and the equivalent F-test are intuitively reasonable.

Example 2 A regression model is said to exhibit a structural break if the parameter vector changes at some point in time. Under such a hypothesis the model may be written as

$$y_t = x_t'\beta_1 + \epsilon_t, \qquad t = 1, \ldots, T_1$$
$$y_t = x_t'\beta_2 + \epsilon_t, \qquad t = T_1 + 1, \ldots, T$$

The null hypothesis is that $\beta_1 = \beta_2$ and so if $\epsilon_t \sim NID(0, \sigma^2)$ under both 'regimes', the LR statistic is

$$\lambda = \left(\frac{SSE_0}{SSE_1 + SSE_2}\right)^{-T/2}$$

where SSE_1, SSE_2 and SSE_0 are the residual sums of squares from the regression on the first T_1 observations, the second $T - T_1$ observations and the complete set of observations respectively.

The statistic λ is, however, a monotonic transformation of

$$\lambda^* = \frac{\{SSE_0 - (SSE_1 + SSE_2)\}/k}{(SSE_1 + SSE_2)/(T - 2k)}$$

which has an F-distribution with $(k, T - 2k)$ degrees of freedom. Hence the LR test may be based on λ^* with the upper tail of the F-distribution used to construct the critical region.

Large Sample Theory

Although the LR principle led to the construction of exact tests in the examples above, it is not generally possible to determine a suitable value for κ by linking the likelihood ratio to a statistic whose distribution is already known. However, provided the null hypothesis is nested within the alternative, and the conditions for the asymptotic normality and efficiency of the ML estimators are satisfied, large sample theory provides the basis for a viable approach to LR testing.

In Example 1 above, H_0 was nested within H_1, since setting the coefficients of the m explanatory variables under test equal to zero gave the model under the null hypothesis. The restricted model is obtained from the general model with k explanatory variables by imposing a set of m *restrictions* or *constraints* on it. Such constraints take the general form

$$R\psi = r, \tag{3.8}$$

where R is an $m \times n$ matrix of full rank, and r is an $m \times 1$ vector. The values taken by the elements of R and r are specified under the null hypothesis. In Example 1 $n = k + 1$ and $\psi = (\beta, \sigma^2)$. The vector r is null, while if the independent variables are arranged such that it is the last m which are being tested for inclusion in the model then $R = [0_{k-m} \ I_m \ 0]$. The restrictions imposed by the null hypothesis need not be linear. In general, the m constraints may be written

$$r_j(\psi) = 0, \qquad j = 1, \ldots, m \tag{3.9}$$

where $r_j(\psi)$ denotes some function of ψ for which the first derivatives exist.

When the models are nested, the likelihood ratio statistic (3.1), must be less than, or equal to unity, since the value of the likelihood for the restricted model cannot exceed the likelihood for the unrestricted model.

A test of the validity of the restrictions imposed by the null hypothesis may be carried out by invoking the following result. *Subject to regularity conditions, the statistic*

$$\text{LR} = -2 \log \lambda = 2 \log L(\tilde{\psi}) - 2 \log L(\tilde{\psi}_0) \tag{3.10}$$

is asymptotically distributed as χ^2_m *under* H_0, *where m is the number of restrictions needed to define the null hypothesis.*

The proof of this result relies on a Taylor series expansion of $\log L(\tilde{\psi}_0)$ about $\tilde{\psi}$. This gives

$$\log L(\tilde{\psi}_0) \simeq \log L(\tilde{\psi}) + (\tilde{\psi} - \tilde{\psi}_0)D \log L(\tilde{\psi})$$
$$+ \tfrac{1}{2}(\tilde{\psi} - \tilde{\psi}_0)'D^2 \log L(\tilde{\psi})(\tilde{\psi} - \tilde{\psi}_0)$$

The second term in the expansion is zero since $\tilde{\psi}$ is the unrestricted ML estimator, and so

$$-2 \log \lambda \simeq (\tilde{\psi} - \tilde{\psi}_0)' \{-D^2 \log L(\tilde{\psi})\} (\tilde{\psi} - \tilde{\psi}_0) \qquad (3.11)$$

Silvey (1970, p. 114) sketches out the remaining details of the proof.

Given the above result, the LR test is carried out by computing $-2 \log \lambda$, and rejecting H_0 if $-2 \log \lambda > \kappa'$. The value of κ' is chosen such that

$$Pr(\chi_m^2 > \kappa') = \alpha, \qquad (3.12)$$

where α is the *nominal* size of the test. Unless the small sample distribution of $-2 \log \lambda$ can be determined, the true size of the test will be unknown, and it is said to be *asymptotic* rather than exact. The concept of power is subject to some ambiguity for an asymptotic test, although in any discussion which follows power will be taken to mean the probability of rejecting H_0 for a given nominal test size.

Although the LR principle can lead to UMP tests in special cases, this is not generally so. Nevertheless, it can be shown that a likelihood ratio test is *consistent*, in the sense that as the sample size increases, the probability of rejecting the null hypothesis when any particular member of the alternative holds, tends to unity. The consistency of the LR test follows from the consistency of the ML estimator. If H_0 is true, then $\lambda \to 1$ as $T \to \infty$, and so the critical region will have its boundary κ' approaching one. When H_0 does not hold, the limiting value of λ will be some constant, $\bar{\lambda}$, satisfying

$$0 \leqslant \bar{\lambda} < 1$$

Therefore $Pr\{\lambda \leqslant \kappa'\}$ will tend to one as $T \to \infty$.

Example 3   A natural hypothesis to test in the multiplicative heteroscedasticity model, (3.4.15), is that the variances are constant, i.e. $H_0 : \delta_2 = \cdots = \delta_r = 0$. The LR statistic is

$$\text{LR} = 2 \log L(\tilde{\beta}, \tilde{\delta}) - 2 \log L(\bar{\beta}_0, \bar{\delta}_0)$$

$$= T \log \tilde{\sigma}_0^2 + T - \Sigma w_t' \tilde{\delta} - \Sigma \exp(-w_t' \tilde{\delta})(y_t - x_t' \tilde{\beta})^2 \quad (3.13)$$

However, when the first element in (3.4.19), the vector $\partial \log L / \partial \delta$, is evaluated at $\delta = \tilde{\delta}$ and $\beta = \tilde{\beta}$, the second and fourth terms in (3.13) cancel. The test statistic becomes

$$\text{LR} = T \log \tilde{\sigma}_0^2 - \Sigma w_t' \tilde{\delta} \qquad (3.14)$$

and this is asymptotically distributed as χ_{r-1}^2 under the null hypothesis of homoscedasticity.

Example 4 The maximised log–likelihood function for a SURE model is equal to a constant plus $(-T/2)\log|\Omega|$; see (3.4.28), (3.4.32) and the result in (4.4.25). Any hypothesis concerning restrictions on the covariance matrix, Ω, may therefore be tested by a statistic of the form:

$$\text{LR} = T\log\{|\tilde{\Omega}_0|/|\tilde{\Omega}|\} \tag{3.15}$$

Thus if Ω_0 is diagonal, LR is treated as a χ^2 variate with $N(N-1)/2$ degrees of freedom.

In the investment example given in Section 2.9, it follows from (2.9.24) that

$$|\tilde{\Omega}_0| = 660.83 \times 88.67 = 58596$$

The unrestricted estimate of Ω should be calculated from the residuals from ML estimation. However, using the residuals from the feasible SURE estimator (2.9.25) makes little difference in large samples. On this basis Pagan and Byron (1977) report $|\Omega| = 25768$. The sample size is $T = 20$ and so LR = 16.43. Since the LR statistic is asymptotically χ_1^2 under the null hypothesis, this is rejected quite convincingly.

4. Wald Tests

The form of the LR statistic implies that the model must, in general, be estimated both under the null and alternative hypotheses. Wald tests, on the other hand, may be carried out on the basis of the unrestricted model only, and so they are particularly appealing when the restricted model is difficult to estimate.

If the constraints are linear as in (3.8), the elements of the vector $R\tilde{\psi} - r$ should be 'close' to zero when the null hypothesis is true. The problem is to decide on some criterion for determining when $R\tilde{\psi} - r$ is close enough to zero to be regarded as consistent with H_0. Since

$$\sqrt{T}(\tilde{\psi} - \psi) \xrightarrow{L} N[0, IA^{-1}(\psi)]$$

it follows that

$$\sqrt{T}R(\tilde{\psi} - \psi) = \sqrt{T}(R\tilde{\psi} - r) \xrightarrow{L} N[0, R \cdot IA^{-1}(\psi) \cdot R']$$

Hence the statistic

$$T \cdot (R\tilde{\psi} - r)'[R \cdot IA^{-1}(\psi) \cdot R']^{-1}(R\tilde{\psi} - r) \tag{4.1}$$

has a limiting χ_m^2 distribution. This distribution remains unchanged if the matrix $T \cdot IA(\psi)$ is replaced by any matrix, $I^*(\hat{\psi})$, with the property (4.5.3). Equating $\hat{\psi}$ with the unrestricted ML estimator, $\tilde{\psi}$, gives the usual definition for the Wald statistic:

$$W = (R\tilde{\psi} - r)'[R \cdot I^{*-1}(\tilde{\psi}) \cdot R']^{-1}(R\tilde{\psi} - r) \qquad (4.2)$$

When the null hypothesis is false, W will tend to be large and so a natural choice of critical region is $W > k$. If the distribution of W cannot be determined exactly, a suitable value of κ may be obtained from χ^2 tables, and the test is an asymptotic one.

As a very simple example of a Wald test, consider an hypothesis which specifies that a single parameter, ψ, takes a particular value, ψ_0. The test statistic is then

$$W = (\tilde{\psi} - \psi_0)^2/\text{avar}(\tilde{\psi}) \qquad (4.3)$$

However, it may be more convenient to carry out an equivalent two-sided test on the statistic

$$W = (\tilde{\psi} - \psi_0)/\sqrt{\text{avar}(\tilde{\psi})} \qquad (4.4)$$

since this will have a standardised normal distribution in large samples.

Wald Tests in Regression Models

The Wald principle provides a convenient approach to the testing of restrictions in both linear and nonlinear regression models. As a specific example, suppose that we wish to test the hypothesis that one of the regression coefficients in a classical linear model is equal to zero. If the coefficient is taken to be β_k, then in terms of (3.8), $R = [0 \ldots 0, 1]$ and $r = 0$. The information matrix is $\sigma^{-2}(X'X)$, but if σ^2 is replaced by the unbiased estimator s^2, W is the square of (2.8.3), the familiar t-statistic. The likelihood ratio approach also leads to the t-test in this case, but only after a good deal of algebraic manipulation.

More generally, the Wald statistic for testing a set of m linear constraints, $R\psi = r$, may be modified to give

$$(Rb - r)'[R(X'X)^{-1}R']^{-1}(Rb - r)/ms^2 \qquad (4.5)$$

Under the null hypothesis, (4.5) is distributed as the F-ratio with $(m, T - k)$ degrees of freedom; see Theil (1971, p. 144). After some manipulation, (4.5) becomes

$$(SSE_0 - SSE)/ms^2 \qquad (4.6)$$

which is identical to the test statistic obtained from likelihood ratio considerations; cf. (3.5).

For a nonlinear model, the Wald approach suggests the use of asymptotic t-ratios for testing individual coefficients. However, the justification for basing such tests on the t-distribution rests solely on the analogy with the linear case; in general, all that can be proved about the distribution of such t-statistics is that they are asymptotically normal. Asymptotic F-ratios may be employed in the same spirit.

Nonlinear Constraints

If the constraints are nonlinear, a Wald test may still be constructed. A Taylor series expansion of $r_j(\tilde{\psi})$ about the true parameter, ψ, gives

$$r_j(\tilde{\psi}) \simeq r_j(\psi) + (\tilde{\psi} - \psi)' \frac{\partial r_j(\psi)}{\partial \psi} \tag{4.7}$$

If the restrictions are satisfied, then $r_j(\psi) = 0$ for $j = 1, \ldots, m$, and the Wald statistic is defined by

$$W = \{r(\tilde{\psi})\}'[\tilde{R}'I^{*-1}(\tilde{\psi}) \cdot \tilde{R}]^{-1}\{r(\tilde{\psi})\} \tag{4.8}$$

where $r(\tilde{\psi})$ is an $m \times 1$ vector in which the jth element is $r_j(\tilde{\psi})$, while $\tilde{R}$ is now a matrix of derivatives in which the jth column is $\partial r_j(\psi)/\partial \psi$ evaluated at $\psi = \tilde{\psi}$. Subject to regularity conditions, W is χ^2_m under H_0.

Example Consider the model

$$y_t = \beta_1 x_{1t} + \beta_2 x_{2t} + \beta_3 x_{3t} + \epsilon_t \tag{4.9}$$

The hypothesis to be tested is that $\beta_3 = \beta_1\beta_2$, and so an LR test would be relatively unattractive since the restricted model is nonlinear. However, the unrestricted model may be estimated by OLS. The Wald statistic is of the form (4.8) with $r_1(\tilde{\psi}) = b_1 b_2 - b_3$, $I^{*-1}(\tilde{\psi}) = s^2(X'X)^{-1}$, and $\tilde{R} = (b_2, b_1, -1)'$. The use of s^2 rather than the ML estimator, $\tilde{\sigma}^2$, in the information matrix is arbitrary. Any consistent estimator of σ^2 will leave W with a χ^2_1 distribution in large samples.

5. The Lagrange Multiplier Test

As with the Wald test, the Lagrange multiplier (LM) procedure is applicable to testing nested hypotheses. It also leads to a test statistic with the same asymptotic distribution as the LR statistic under the null hypothesis, and it is consistent. The distinguishing feature of the LM test is that it only entails the estimation of the restricted model.

Let $\tilde{\psi}_0$ denote the ML estimator of ψ in the restricted model. If the restrictions are valid, $\tilde{\psi}_0$ will be 'close' to the unrestricted estimator, $\tilde{\psi}$, and so the partial derivatives in the vector $D \log L(\tilde{\psi}_0)$ will also be 'close' to zero. On the other hand if H_0 is not true, there is no reason to believe that this will be the case.

A procedure for testing whether $D \log L(\tilde{\psi}_0)$ is close to zero may be developed by first taking a Taylor series expansion around $\tilde{\psi}$, i.e.

$$D \log L(\tilde{\psi}_0) \simeq D \log L(\tilde{\psi}) + (\tilde{\psi}_0 - \tilde{\psi})' D^2 \log L(\tilde{\psi}) \tag{5.1}$$

Since $\tilde{\psi}$ is the unrestricted ML estimator, $D \log L(\tilde{\psi}) = 0$, and so (5.1) reduces to

$$D \log L(\tilde{\psi}_0) \simeq (\tilde{\psi}_0 - \tilde{\psi})' D^2 \log L(\tilde{\psi}) \tag{5.2}$$

Now consider the test statistic

$$\text{LM} = \{D \log L(\tilde{\psi}_0)\}' I^{-1}(\tilde{\psi}_0)\{D \log L(\tilde{\psi}_0)\} \tag{5.3}$$

Substituting from (5.2) yields

$$\text{LM} = (\tilde{\psi}_0 - \tilde{\psi})' D^2 \log L(\tilde{\psi}) I^{-1}(\tilde{\psi}_0) D^2 \log L(\tilde{\psi})(\tilde{\psi}_0 - \tilde{\psi})$$

However, under H_0, $T^{-1} I(\tilde{\psi}_0)$ has the same probability limit as $T^{-1} D^2 \log L(\tilde{\psi})$. For a large sample, therefore, the distribution of the LM statistic, (5.3), will be the same as the distribution of

$$(\tilde{\psi}_0 - \tilde{\psi})' D^2 \log L(\tilde{\psi})(\tilde{\psi}_0 - \tilde{\psi})$$

when the null hypothesis is true. This expression is exactly the same as that given in (3.11) for the LR statistic, and so LM has a χ_m^2 distribution under H_0.

Expression (5.3) is the usual definition of the LM statistic. However, $I(\tilde{\psi}_0)$ may be replaced by any matrix, $I*(\tilde{\psi}_0)$, which is asymptotically equivalent in the sense implied by (4.5.3), under the null hypothesis. This will not affect the distribution in large samples, and any statistic of the form

$$\{D \log L(\tilde{\psi}_0)\}' I*^{-1}(\tilde{\psi}_0)\{D \log L(\tilde{\psi}_0)\} \tag{5.4}$$

will be referred to as a *Lagrange multiplier statistic*. A critical region for the test is then found with respect to the χ_m^2 distribution in exactly the same way as for the LR statistic.

The use of the term 'Lagrange multiplier' in this context is not obvious from the above argument. However, given a set of restrictions of the form (3.9), a restricted likelihood function may be defined by

$$\log L* = \log L + \sum_{j=1}^{m} \lambda_j r_j(\psi) \tag{5.5}$$

where the λ_j's are Lagrange multipliers. Maximising $\log L^*$ with respect to ψ implies

$$D \log L(\psi_0) = \Sigma \lambda_j \frac{\partial r_j(\psi)}{\partial \psi} = \tilde{R} \tilde{\lambda} \qquad (5.6)$$

where $\tilde{R}$ is the $n \times m$ matrix of derivatives and $\tilde{\lambda}$ is the $m \times 1$ vector of Lagrange multipliers, evaluated at $\psi = \tilde{\psi}_0$. Substituting for $D \log L(\tilde{\psi}_0)$ in (5.3) yields the alternative form of the LM statistic,

$$\text{LM} = \tilde{\lambda}'\tilde{R}' \cdot I^{-1}(\tilde{\psi}_0) \cdot \tilde{R} \tilde{\lambda} \qquad (5.7)$$

A third form for the LM statistic has recently been suggested by Breusch and Pagan (1980). If $\tilde{\psi}_0$ is updated by a single iteration of the method of scoring as defined by (4.4.2), then

$$D \log L(\tilde{\psi}_0) = I(\tilde{\psi}_0) \cdot (\psi^* - \tilde{\psi}_0)$$

Substituting this expression in (5.3) then gives

$$\text{LM} = (\psi^* - \tilde{\psi}_0)'I(\tilde{\psi}_0)(\psi^* - \tilde{\psi}_0) \qquad (5.8)$$

which may be interpreted as a test of the 'hypothesis' $\psi^* = \tilde{\psi}_0$.

This form of the statistic may be easier to evaluate than (5.3) in certain circumstances. If it is the case that an alternative procedure such as Gauss—Newton or Newton—Raphson is more readily applied, then as with (5.4) it is appropriate to write (5.8) with $I(\tilde{\psi}_0)$ replaced by $I^*(\tilde{\psi}_0)$.

Testing Subsets of Parameters

In many problems the restrictions under test involve only a subset of the parameters. Thus if ψ is partitioned as

$$\psi = \begin{pmatrix} \psi^{(1)} \\ \psi^{(2)} \end{pmatrix}$$

only the $m \times 1$ vector $\psi^{(2)}$ is subject to constraints imposed by the null hypothesis. An important special case is

$$H_0 : \psi^{(2)} = 0$$

Partitioning the information matrix conformably with $\psi^{(1)}$ and $\psi^{(2)}$ yields

$$I(\psi) = \begin{pmatrix} I_{11} & I_{12} \\ I_{21} & I_{22} \end{pmatrix} \qquad (5.9)$$

Let $D_j \log L(\psi)$ denote the vector of partial derivatives of $\log L(\psi)$

with respect to $\psi^{(j)}$ for $j = 1, 2$. From (5.6) it follows that
$D_1 \log L(\tilde{\psi}_0) = 0$ since the parameters in $\psi^{(1)}$ do not enter into the
restrictions. The LM is therefore

$$\text{LM} = [0 \quad D_2 \log L(\tilde{\psi}_0)]' \begin{pmatrix} I_{11} & I_{12} \\ I_{21} & I_{22} \end{pmatrix}^{-1} \begin{pmatrix} 0 \\ D_2 \log L(\tilde{\psi}_0) \end{pmatrix}$$

and using the formula for a partitioned inverse this becomes

$$\text{LM} = \{D_2 \log L(\tilde{\psi}_0)\}' [I_{22}^{-1} - I_{21}' I_{11}^{-1} I_{12}] \{D_2 \log L(\tilde{\psi}_0)\} \quad (5.10)$$

Note that all the components of the information matrix are evaluated
at $\psi = \tilde{\psi}_0$.

If $I_{12} = I_{21} = 0$ under H_0, the information matrix $I(\tilde{\psi}_0)$ is block
diagonal and expression (5.10) reduces to

$$\text{LM} = D_2 \log L(\tilde{\psi}_0)' \cdot I_{22}^{-1}(\tilde{\psi}_0) \cdot D_2 \log L(\tilde{\psi}_0) \quad (5.11)$$

This formulation of the LM statistic is particularly convenient in
generalised regression models where the parameters in $\psi^{(2)}$ determine
the form of the disturbance covariance matrix. Provided the expectation of the dependent variable is independent of the elements of
$\psi^{(2)}$, the information matrix will be block diagonal and (5.11) may
be applied.

Example   The LR test for constant variance in the multiplicative
heteroscedascity model was obtained in Section 3. For the derivation of the LM statistic, it is convenient to partition the parameter
vector as $\psi^{(1)} = \beta$ and $\psi^{(2)} = \delta$ even though the first parameter in
δ, $\log \sigma^2$, is not restricted by the null hypothesis. This takes
advantage of the block diagonality of the information matrix and
the LM statistic is given by

$$\text{LM} = \left(\frac{\partial \log L}{\partial \delta}\right)' \left(-E\left(\frac{\partial^2 \log L}{\partial \delta \, \partial \delta'}\right)\right)^{-1} \left(\frac{\partial \log L}{\partial \delta}\right)$$

This expression is evaluated at $\tilde{\delta}_0 = (\log \tilde{\sigma}_0^2 \quad 0 \; \ldots \; 0)'$, where $\tilde{\sigma}_0^2$ is
the ML estimator of σ^2 in the OLS regression. From (3.4.19)

$$\frac{\partial \log L}{\partial \delta} = \tfrac{1}{2} \Sigma w_t v_t$$

where $v_t = \{(y_t - x_t' b)^2 / \tilde{\sigma}_0^2\} - 1$. Thus

$$\text{LM} = (\Sigma w_t v_t)' (\Sigma w_t w_t')^{-1} (\Sigma w_t v_t)/2 \quad (5.12)$$

and this will be asymptotically χ_{r-1}^2 under H_0.

An alternative form of the LM statistic is given by observing
that in the classical linear regression model the formula for R^2,

(2.2.13), may be expressed as:

$$R^2 = \frac{\Sigma y_t^2 - T\bar{y}^2}{\Sigma y_t^2 - T\bar{y}^2} = \frac{b'X'Xb - T\bar{y}^2}{y'y - T\bar{y}^2} = \frac{y'X(X'X)^{-1}X'y - T\bar{y}^2}{y'y - T\bar{y}^2}$$

(5.13)

The last two steps follow from (2.2.6) and (2.2.4) respectively. Bearing this in mind, and noting that $\Sigma v_t = 0$, the statistic

$$\text{LM*} = \frac{T(\Sigma w_t v_t)'(\Sigma w_t w_t')^{-1}(\Sigma w_t v_t)}{\Sigma v_t^2}$$

(5.14)

is equal to TR^2 where R^2 is the coefficient of multiple correlation obtained from regressing v_t on w_t. Since plim $T^{-1}\Sigma v_t^2 = 2$, LM* has the same asymptotic distribution as LM.

One attraction of the LM statistic in this context is that it is appropriate for a wide range of heteroscedastic formulations. It can be shown that (5.12) is the LM statistic for any model in which σ_t^2 is a function of $w_t'\delta$. Thus, for example, in the model

$$\sigma_t^2 = \delta_1 + \delta_2 x_{2t}^2 + \cdots + \delta_k x_{kt}^2$$

an LM test of $H_0 : \delta_2 = \delta_3 = \cdots = \delta_k$ is carried out via (5.12) or (5.14) with $w_t = (1 \ x_{2t}^2 \ \ldots \ x_{kt}^2)'$. This suggests that an LM test based on (5.12) or (5.14) may be useful as a general test for heteroscedasticity.

The LM Test in Least Squares Problems

In Section 3.5 it was observed that for many models, maximising the likelihood is equivalent to minimising a sum of squares function. From the definition of the likelihood function in (3.5.8)

$$\partial \log L(\psi, \sigma^2)/\partial \psi = -\sigma^{-2}\Sigma z_t \epsilon_t$$

where $z_t = z_t(\psi) = -\partial \epsilon_t/\partial \psi$, the vector ψ denoting all the parameters apart from σ^2. Since σ^2 will never be subject to constraints in this type of model, the block diagonality of the information matrix may be exploited via (5.11). Replacing

$$E\left(\frac{\partial \log L}{\partial \psi}\right)\left(\frac{\partial \log L}{\partial \psi'}\right) = \sigma^{-4}E(\Sigma \epsilon_t^2 z_t z_t')$$

by the asymptotically equivalent expression $\sigma^{-2}\Sigma z_t z_t'$ gives

$$\tilde{\sigma}_0^{-2}(\Sigma z_t \epsilon_t)(\Sigma z_t z_t')^{-1}(\Sigma z_t \epsilon_t)$$

(5.15)

where both ϵ_t and z_t are evaluated at $\psi = \tilde{\psi}_0$ and $\tilde{\sigma}_0^2 = T^{-1}\Sigma \epsilon_t^2(\tilde{\psi}_0)$.

When the model contains a constant term, the sum of the residuals will be identically equal to zero. In view of the definition of the coefficient of multiple correlation in (5.13), the statistic (5.15) may be written

$$LM^* = TR^2 \tag{5.16}$$

This result will still be approximately true even if the model does not contain a constant term. In all cases a valid LM test may be carried out simply by regressing $\epsilon_t(\tilde{\psi}_0)$ on $z_t(\tilde{\psi}_0)$ and testing TR^2 as a χ^2 variate with m degrees of freedom.

Example   The demand for money in an economy depends on real national income and the rate of interest. Konstas and Khouja (1969) adopt the functional form

$$y_t = \beta_1 x_{1t} + \beta_2 (x_{2t} - \gamma)^{-1} + \epsilon_t \tag{5.17}$$

where x_{1t} is real national income, x_{2t} is the rate of interest, and y_t is real money demand. A positive value of γ implies the existence of a liquidity trap: once the rate of interest has fallen to $x_2 = \gamma$, further increases in the money supply cannot lower it further because the additional money is all held as speculative balances.

Konstas and Khouja attempted to carry out an LR test of the hypothesis that $\gamma = 0$. Unfortunately, as Spitzer (1976) observed, the test was misleading as their algorithm failed globally to maximise the likelihood function. The LM approach avoids this difficulty because when $\gamma = 0$ the model is *linear*. Regressing y on x_1 and x_2^{-1} yields a set of residuals which may then be regressed against x_1, x_2^{-1} and x_2^{-2}. If R^2 is the coefficient of determination from the second regression, an LM test of the liquidity trap may be based on the statistic TR^2 which is taken to be χ_1^2 under H_0. In carrying out this test, Breusch and Pagan (1980) find a value of $LM^* = 11.47$. This indicates a rejection of the null hypothesis that there is no liquidity trap, a result which agrees with Spitzer's LR test.

In terms of the notation of (5.15), $\psi = (\beta_1, \beta_2, \gamma)'$

$$z_{1t} = -\partial \epsilon_t / \partial \beta_1 = x_1, \tag{5.18a}$$

$$z_{2t} = -\partial \epsilon_t / \partial \beta_2 = (x_{2t} - \gamma)^{-1} \tag{5.18b}$$

and

$$z_{3t} = -\partial \epsilon_t / \partial \gamma = \beta_2 (x_{2t} - \gamma)^{-2} \tag{5.18c}$$

These derivatives are all evaluated at $\gamma = 0$ in constructing the LM test statistic. Although the expression for z_{3t} implies the use of the regressor $b_2 x_{2t}^{-2}$, the omission of b_2 makes no difference to the value of R^2.

The Modified LM Test

Suppose we wish to test the hypothesis that a subset of m parameters in a classical linear regression model are zero. Both the LR and Wald principles lead to the F-statistic defined by (3.5). The LM statistic, (5.16), however, is based on the R^2 obtained by taking the residuals from the OLS regression on the restricted model and regressing them on the complete set of independent variables. Suppose the vector of residuals from this second regression is denoted by e^*, while the original OLS residuals are e_0. The explained sum of squares in the second regression is then $e_0' e_0 - e^{*'} e^*$ and so

$$R^2 = (e_0' e_0 - e^{*'} e^*)/e_0' e_0$$

However

$$e^* = M e_0 = M\left\{ y - X \binom{b_0}{0} \right\} = My = e \tag{5.19}$$

since $M = I - X(X'X)^{-1}X'$ and $MX = 0$. Thus the residuals obtained from regressing e^* on e_0 are identical to those obtained from regressing y on X. Hence $e^{*'} e^* = e'e$ and so the LM test statistic is given by

$$LM = TR^2 = T(e_0' e_0 - e'e)/e_0' e_0 = T(SSE_0 - SSE)/SSE_0 \tag{5.20}$$

This is a monotonic function of the classical F-statistic, (2.8.18), and so the LM principle leads to the standard F- and t-tests just as the LR and Wald approaches do.

The way in which the LM statistic may be modified to produce an exact test for the linear model suggests that similar modifications may be useful for least squares problems in general. In terms of the Gauss–Newton scheme, (4.4.13), $\epsilon_t(\bar{\psi}_0)$ is regressed on $z_t(\bar{\psi}_0)$ and hypotheses regarding the validity of any restrictions are tested using the standard F- and t-statistics. Such tests will be termed *modified LM tests*. The F-tests are based not on TR^2 but on

$$LM\ F = \frac{T-n}{m} \cdot \frac{R^2}{1 - R^2} \tag{5.21a}$$

The critical region is based on an F-distribution with $(m, T-n)$ degrees of freedom.

Example   A test of $\gamma = 0$ in (5.17) may be carried out by regressing the OLS residuals from the restricted model on x_1, $x_{\bar{2}}^{-1}$ and $x_{\bar{2}}^{-2}$, and examining the t-statistic associated with $x_{\bar{2}}^{-2}$. A one-sided test may be carried out very easily using this approach. This would seem to be appropriate, since the liquidity trap hypothesis implies $\gamma > 0$.

Because (5.17) is linear under the null hypothesis, the modified LM test can be carried out on the basis of only one regression. Since the OLS residuals in the restricted model are given by $y_t - z_t' \bar{\psi}_0$, where $\bar{\psi}_0 = (b_1, b_2, 0)'$, the first iteration of Gauss–Newton yields

$$\psi^* = (\Sigma z_t z_t')^{-1} \Sigma z_t y_t$$

Thus the test may be carried out directly by regressing y_t on x_{1t}, x_{2t}^{-1} and x_{2t}^{-2} and examining the t-statistic associated with x_2^{-2}.

Apart from being convenient to carry out, the modified LM test is attractive for the following reasons. Firstly, the results for the linear regression model suggest that the true significance levels for modified LM tests might be reasonably close to their nominal values when an exact test is not available. A second point relates to the question of power, which in this context means the probability of rejecting the null hypothesis for a given *nominal* test size. In terms of (5.15) the modification to the LM statistic amounts to replacing $\tilde{\sigma}_0^2$ by $\tilde{\sigma}_1^2$. This last statistic is constructed from the residuals obtained from the revised estimates given by one iteration of Gauss–Newton. Thus the modified LM statistic is given by

$$\text{LM}^\dagger = TR^2/(1 - R^2), \tag{5.21b}$$

and this must be greater than, or equal to, (5.20). Hence the 'power' of the standard LM test cannot exceed the power of a similar test based on the modified statistic, $\text{LM}^\dagger$. While this statement is no longer strictly true when the modified LM test is based on the F-distribution as in (5.21a), any reduction in power is likely to be small. On the other hand, dividing by $1 - R^2$ could make an enormous difference.

Comparison between the Wald, LR and LM Procedures

In comparing the three classical test procedures, there are essentially two issues at stake. The first is computational convenience, and the second is power. As has already been noted, this last issue cannot be resolved unequivocally since in general all three tests are asymptotic. This must be borne in mind when assessing the results of Monte

Carlo experiments carried out for special cases. Although no general results exist regarding the properties of the three tests, it can be shown that under certain circumstances the test statistics obey the inequality

$$W \geqslant LR \geqslant LM \qquad (5.23)$$

see Berndt and Savin (1977) and Breusch (1979). Thus, for example, the Wald test will tend to have a higher power than the LM test, but at the expense of a greater probability of making a Type I error.

When the sample size is large, the ambiguity over the size of the tests is resolved since the test statistics all have the same distribution under the null hypothesis. Furthermore, if the parameters are 'near' to the values taken under H_0, it can still be shown that the test statistics have the same asymptotic distribution. However, this result by no means supports the notion that the three tests will have similar power in small samples.

As regards computational convenience, the LR statistic is very simple to construct but suffers from the disadvantage that it requires ML estimates for both the restricted and unrestricted models. This suggests that as far as computer time is concerned, the LM and Wald tests are both preferable, the choice being dependent on whether the restricted or unrestricted model is easiest to estimate.

As a final point it is perhaps worth stressing what should by now be obvious, namely that the computation of the test statistics given by all three approaches is, in principle, straightforward, if testing, like estimation, is approached in an ML framework. The main point in examining the form of a test statistic in a particular case is to determine whether it can be transformed into a statistic whose exact distribution under H_0 is known.

6. Non-Nested Models

So far the discussion has been confined to testing nested hypotheses. In such cases the restricted model is a special case of the unrestricted model. However, this need not always be so. A problem which frequently arises is when the models specified under two hypotheses are non-nested, in that one cannot be obtained as a special case of the other. In these circumstances the conventional approaches are no longer applicable.

There are two ways of tackling this problem. The first is to choose between the two models on the basis of some appropriate criterion. This is termed *discrimination,* and a simple example of such a procedure would be choosing between linear regression models on the basis of R^2. Alternatively, the problem may be approached within a framework which relies on considerations similar to those in classical hypothesis testing. One possibility is to embed the competing models within a more general model. Each hypothesis may then be tested separately as a set of restrictions in the general case. On the other hand, Cox has suggested another method in which a test of significance is carried out on one of the models using a statistic similar to the LR statistic. If the roles of the two models are reversed, however, a different test statistic is usually obtained.

While only two outcomes are possible if the problem is taken as one of discrimination, the hypothesis testing approach leads to four. This is because both models can be accepted or rejected.

Discrimination

Suppose H_1 and H_2 denote two hypotheses concerning the distribution of a variable, y. One possible way of discriminating between the models would be to compare the maximised values of their respective likelihood functions. However, it would seem desirable to make some allowance for n, the number of parameters estimated. The *Akaike Information Criterion* (AIC) attempts to make such a correction, the decision rule being to select the model for which

$$\text{AIC} = -2 \log L(\tilde{\psi}) + 2n \tag{6.1}$$

is a minimum.

The AIC may be used for testing nested as well as non-nested hypotheses. In doing so it breaks away from conventional procedures which test whether a parameter is 'significant' using a test the size of which is essentially arbitrary. Thus the emphasis is on comparing the 'goodness of fit' of various models with an appropriate allowance made for parsimony. This has a good deal of appeal in problems where the specification is based primarily on pragmatic considerations.

In the case of regression models with white noise disturbances, a comparison of the likelihood functions reduces to a comparison of the residual sum of squares. When the competing models contain the same number of parameters, therefore, the AIC decision rule leads to the selection of the model for which R^2 is greatest. However, when the number of parameters differs, an adjustment must be made. A number of criteria which are different from those already considered have been suggested. The best known is probably $\bar{R}^2$ which was defined in (2.2.14).

Example 1   In comparing the linear and log–linear production functions, (2.1.9) and (2.1.10), we are faced with the immediate problem that while one of the specifications concerns y_t itself, the second is an hypothesis relating to the distribution of $\log y_t$. Hence the maximised likelihoods cannot be compared directly. However, the joint distribution of $y_1, \ldots, y_T$, in (2.1.10) may be obtained by a change of variables. Thus

$$p(y_1, \ldots, y_T; \alpha, \beta) = |J| \cdot p(\log y_1, \ldots, \log y_T; \alpha, \beta) \quad (6.2)$$

where the Jacobian, J, is a $T \times T$ diagonal matrix in which the tth diagonal element is given by $\partial \log y_t / \partial y_t = 1/y_t$. The maximised log–likelihood function is therefore

$$\log L = -\tfrac{1}{2}T(\log 2\pi + 1) - \tfrac{1}{2}T \log \tilde{\sigma}_M^2 - \Sigma \log y_t \quad (6.3)$$

where $\tilde{\sigma}_M^2$ is the ML estimator of σ^2 based on regressing $\log Q_t$ on $\log K_t$ and $\log L_t$. Since both the additive and multiplicative models contain the same number of parameters, the AIC decision rule is to accept the additive model if $T \log \sigma_A^2$ is less than $T \log \sigma_M^2 \pm 2\Sigma \log y_t$.

Embedding within a General Model

If the competing models are embedded within a general model, then H_1 and H_2 may each be tested separately. A Wald or LR test will usually be appropriate for this purpose.

A general procedure for carrying out such tests was suggested by Cox (1961, 1962) and further developed by Atkinson (1970). If $p_1(y_1, \ldots, y_T; \psi_1)$ and $p_2(y_1, \ldots, y_T; \psi_2)$ are the p.d.f.'s under H_1 and H_2 respectively, the p.d.f. of the general model can be made proportional to

$$\{p_1(y_1, \ldots, y_T; \psi_1)\}^\lambda \{p_2(y_1, \ldots, y_T; \psi_2)\}^{1-\lambda}$$

Tests of the two hypotheses may then be carried out as tests on λ. However, although this procedure is straightforward in theory, the calculations involved may be relatively heavy.

In many cases there is no need to resort to the general procedure outlined above. For example, suppose the two hypotheses concern linear regression models of the form:

$$H_1 : y_t = x_t' \beta + \epsilon_t \quad (6.4)$$

$$H_2 : y_t = z_t' \gamma + \epsilon_t \quad (6.5)$$

where x_t and β are $n_1 \times 1$ vectors and x_t and γ are $n_2 \times 1$ vectors. These may be embedded within the more general model

$$y_t = x_t'\beta + z_t'\gamma + \epsilon_t \tag{6.6}$$

A significant F-statistic in a test of the hypothesis that $\gamma = 0$ leads to a rejection of H_1, while H_2 is rejected if the estimate of β is significant. Of course, if z_t and x_t contain some elements in common, these are only included once in the general model, and are not involved in the test procedures.

Example 2   Consider again the problem of multiplicative versus additive disturbances. Both (2.1.9) and (2.1.10) may be embedded within a more general model by means of the Box–Cox transform,

$$y^{(\lambda)} = \begin{cases} (y^\lambda - 1)/\lambda, & 0 < \lambda \leqslant 1 \\ \log y, & \lambda = 0 \end{cases} \tag{6.7}$$

The general model is

$$y_t^{(\lambda)} = \{AK_tL_t\}^{(\lambda)} + \epsilon_t \tag{6.8}$$

Setting $\lambda = 1$ yields (2.1.10) while (2.1.9) is obtained with $\lambda = 0$.
Since $\epsilon_t \sim NID(0, \sigma^2)$, the joint density function is

$$\log L(y_1, \ldots, y_T; \alpha, \beta, \sigma^2, \lambda) = -\tfrac{1}{2}T \log 2\pi - \tfrac{1}{2}T \log \sigma^2$$
$$-\tfrac{1}{2}\sigma^{-2}\Sigma[y_t^{(\lambda)} - \{AK_tL_t\}^{(\lambda)}]^2 + (\lambda - 1)\Sigma \log y_t \tag{6.9}$$

The last term in (6.9) is the determinant of the Jacobian which results from transforming $y_t^{(\lambda)}$ to y_t; cf. (6.3).

The parameter σ^2 may be concentrated out of the likelihood function, leaving nonlinear optimisation to be carried out with respect to α, β and λ. The hypotheses H_1 and H_2 can then be conveniently tested by setting up a confidence interval around $\tilde{\lambda}$.

It is quite possible that the above procedure will fail to reject both models. This simply means that the sample is insufficiently large to allow a definite conclusion to be drawn. On the other hand, if both models are rejected, this should not necessarily imply acceptance of the more general model, particularly if $\lambda \neq 0, 1$ yields a model for which the theoretical interpretation is unclear.

The Cox Procedure

The method developed by Cox is based on a modification of the Neyman–Pearson likelihood ratio. Let $L_1(\psi_1)$ and $L_2(\psi_2)$ denote the likelihood functions under H_1 and H_2 respectively. A test of the hypothesis H_1 is then based on

$$T_1 = \{\log L_1(\hat{\psi}_1) - \log L_2(\hat{\psi}_2)\}$$
$$- E_{\tilde{\psi}_1}\{\log L_1(\hat{\psi}_1) - \log L_2(\hat{\psi}_2)\} \tag{6.10}$$

This statistic compares the observed difference of log–likelihoods with an estimate of the difference to be expected when H_1 is true. If H_1 is true, T_1 should be close to zero, whereas under H_2 it will tend to be negative. A large sample test may be based on the normal distribution. The test is one-sided, a general formula for deriving the asymptotic variance of T_1 being given by Cox (1962). This must be estimated consistently from the data.

In carrying out a test based on T_1 only, the hypotheses are being considered asymmetrically with H_2 serving only to indicate the type of alternative for which high power is required. However, the roles of the hypotheses may be interchanged, and a second test statistic, T_2, calculated. This takes the form

$$T_2 = \{\log L_2(\tilde{\psi}_2) - \log L_1(\tilde{\psi}_1)\}$$
$$- E_{\tilde{\psi}_2}\{\log L_2(\tilde{\psi}_2) - \log L_1(\tilde{\psi}_1)\} \tag{6.11}$$

In general, T_1 and T_2 will be different functions of the observations. As with the embedding procedure, the application of T_1 and T_2 raises the possibility that both or neither of the hypotheses will be rejected.

Example 3 When H_1 and H_2 relate to non-nested linear models of the form (6.4) and (6.5), Pesaran (1974) has shown that the Cox procedure is relatively straightforward to apply. Let $\tilde{\sigma}_1^2$ and $\tilde{\sigma}_2^2$ be the estimated variances from H_1 and H_2 respectively; let e_{21} be the vector of OLS residuals in the regression of $x_t'b$ on z_t and let e_{121} be the vector of OLS residuals in the regression of e_{21} on x. Finally let $\tilde{\sigma}_{21}^2 = \tilde{\sigma}_1^2 + T^{-1}e_{21}'e_{21}$. The statistic (6.10) is then given by $T_1 = (T/2)\log(\tilde{\sigma}_2^2/\tilde{\sigma}_1^2)$ while a consistent estimate of its asymptotic variance is given by $\mathrm{avar}(T_1) = (\tilde{\sigma}_1^2/\tilde{\sigma}_{21}^4)e_{121}'e_{121}$. Thus the statistic

$$N_1 = T_1/\sqrt{\mathrm{avar}(T_1)}$$

is asymptotically $N(0,1)$ when H_1 is true, and a significant negative value of N_1 implies a rejection of H_1 in favour of H_2. A similar statistic, N_2 may be constructed for testing H_2 against H_1.

7. Post-Sample Predictive Testing

The aim of the test procedures described in the preceding sections is to select an appropriate model from the data. Once a suitable model has been identified, its parameters are then estimated using the same set of data. The only way in which the adequacy of the chosen model may then be assessed is by its ability to make accurate predictions *outside* the sample period. It is therefore important to retain a subset of the available observations for post-sample predictive testing.

Suppose that a dynamic model of the form (2.3.18) has been estimated. Given values of the exogenous variables, x_t, predictions may be generated for the dependent variable over the whole of the post sample period, $t = T + 1$ to $T + l$. Various plots of the prediction errors, $e_{T+j/T}$, $j = 1, \ldots, l$ can be constructed, one possibility being the CUSUM

$$W_\tau^\dagger = s^{-2} \sum_{j=1}^{\tau} e_{T+j/T}, \qquad \tau = 1, \ldots, l \qquad (7.1)$$

Formal tests may also be carried out, the most important being based on a comparison of the sum of squares of the post-sample prediction errors with the residual sum of squares from the fitted model.

The Post-Sample Goodness of Fit Test

The basic post-sample predictive test may be applied to any static or dynamic model including those with nonlinearities and moving average disturbances. However, attention will be focused on the dynamic model, (2.3.18). Let $\psi = (\alpha_1, \ldots, \alpha_r, \beta')'$ and suppose that $\epsilon_t \sim NID(0, \sigma^2)$. If ψ is known, the prediction errors

$$\epsilon_{T+j/T} = y_t - \alpha_1 y_{t-1} - \cdots - \alpha_r y_{t-r} - x_t'\beta, \qquad j = 1, \ldots, l$$

$$(7.2)$$

will be $NID(0, \sigma^2)$ and if σ^2 is known the statistic

$$\xi^\dagger(l) = \sigma^{-2} \sum_{j=1}^{l} \epsilon_{T+j/T}^2 \qquad (7.3)$$

will have a χ_l^2 distribution.

When ψ is estimated, the variance of each prediction error, $\epsilon_{T+j/T}(\hat{\psi}) = e_{T+j/T}$, consists of two parts. The first part is σ^2, the variance of the random term in the model, while the second part arises from the estimation of ψ. However, if this second term is of $0(T^{-1})$, which is almost invariably the case, it becomes small relative to σ^2 when the sample size is large. Therefore, if $\hat{\sigma}^2$ denotes a consistent estimator of σ^2, the statistic

$$\xi(l) = \hat{\sigma}^{-2} \sum_{j=1}^{l} e_{T+j/T}^2 \qquad (7.4)$$

is asymptotically χ_l^2. A significant value of $\xi(l)$ may indicate one of two things: *either* that the observations in the sample and post-sample periods are generated by different models, *or* that the model fitted to the sample observations is misspecified. The first of these conclusions implies a structural break, but if the division into sample and post-sample observations were essentially arbitrary this would seem unlikely. In general, the conclusion to be drawn from a significant value of $\xi(l)$ is that the chosen model is in some sense inadequate.

Example Davidson *et al.* (1978, p. 675) report the following equation for the UK consumption function:

$$\hat{y}_t = 509 + 0.75\,c_{t-1} + 0.18x_t + 68D_t^0 - 812Q_{1t}$$
$$\phantom{\hat{y}_t = 509 +} (0.11) \phantom{c_{t-1}} (0.08) (49) (63)$$

$$+ 32Q_{2t} - 169Q_{3t}$$
$$ (46) \phantom{Q_{2t}} (26)$$

$$R^2 = 0.990, \quad d = 2.1, \quad \xi(20) = 190,$$

where y is consumers' expenditure, x is personal disposable income, D^0 is a dummy variable allowing for the effect of the 1968 budget and the introduction of VAT in 1973, and Q_1, Q_2 and Q_3 are seasonal dummies. The high R^2 indicates a close fit to the data, while the Durbin–Watson statistic, although not strictly speaking valid because of the lagged dependent variable, nevertheless gives no hint of misspecification. However, twenty observations were retained for past sample predictive testing, and the value of $\xi(20)$ is such that the model is convincingly rejected.

Post-Sample Predictive Tests in the Classical Regression Model

Since the variance of each prediction error, $e_{T+j/T}$, will be greater than σ^2, the size of a test based on $\xi(l)$ will tend to exceed the nominal significance level. However, in the classical linear regression model, an exact test is available. If $X_T^{(l)} = (x_{T+1}, \ldots, x_{T+l})'$ and $y_T^{(l)} = (y_{T+1}, \ldots, y_{T+l})'$, the $l \times 1$ vector of post sample predictions is given by $\tilde{y}_T^{(l)} = X_T^{(l)}{}'b$, where b is the OLS estimator of β in (2.1.5). For normally distributed disturbance, it follows from (1.3.13) that

$$(y_T^{(l)} - \tilde{y}_T^{(l)})'\{I + X_T^{(l)}(X'X)^{-1}X_T^{(l)}{}'\}^{-1}(y_T^{(l)} - \tilde{y}_T^{(l)})/\sigma^2$$

has a χ_l^2 distribution. Therefore, if s^2 is the unbiased estimator of σ^2 from the first T observations,

$$\xi^*(l) = (y_T^{(l)} - \tilde{y}_T^{(l)})'\{I + X_T^{(l)}(X'X)^{-1}X_T^{(l)}{}'\}^{-1}(y_T^{(l)} - \tilde{y}_T^{(l)})/ls^2$$

$$(7.5)$$

follows an F-distribution with $(l, T - k)$ degrees of freedom when the model is correctly specified and the parameters, including σ^2, remain constant throughout time. It will be observed that the crucial difference between $\xi(l)$ and $\xi^*(l)$ is that the latter allows for the estimation of β through the matrix term $X_T^{(l)}(X'X)^{-1}X_T^{(l)'}$.

The test statistic (7.5) may be written in two alternative forms. The first is

$$\xi^*(l) = (SSE^* - SSE)/ls^2 \tag{7.6}$$

where SSE and SSE^* are the residual sums of squares from regressions on the first T and $T + l$ observations respectively. Expression (7.6) forms the basis for the 'Chow test'. This is the usual test for a structural change in the vector of regression coefficients when the number of observations occurring after the hypothesised break is less than, or equal to, k. A proof of the equivalence of (7.5) and (7.6) will be found in Chow (1960).

Further insight into the test statistic may be obtained by noting that the numerator of (7.6) is equal to the sum of squares of the recursive residuals computed at $t = T + 1, T + 2, \ldots, T + l$; This follows immediately from the residual sum of squares recursion, (2.6.10). Hence

$$\xi^*(l) = \frac{\displaystyle\sum_{j=1}^{l} v_{T+j}^2/l}{\displaystyle\sum_{t=k+1}^{T} v_t^2/(T-k)} \tag{7.7}$$

The fact that (7.7) has an F-distribution for a correctly specified model follows from the properties of recursive residuals.

The original formulation of the post sample predictive test statistic in (7.5) is based on the errors arising from predictions made with an estimator constructed from the first T observations. That the quadratic form involving these quantities may be re-expressed in terms of a series of one-step ahead prediction errors based on updating that estimator is certainly not obvious. Nevertheless it is extremely useful, since the independence of the one-step ahead prediction errors suggests that they may be used to construct other tests. Thus if σ^2 is estimated from the sample observations, the CUSUM and CUSUMSQ procedures may be carried out using the recursive residuals from $t = T + 1$ to $t = T + l$. The same set of post-sample recursive residuals may be subjected to a Sign test, or a t-test based on the statistic,

$$\sum_{j=1}^{l} v_{T+j}/s\sqrt{l} \tag{7.8}$$

With the exception of the Sign test, the powers of the various post-sample prediction tests hinge on the estimation of σ^2. If σ^2 were known for the true model, then the χ^2 test, (7.3), could be carried out and there would be few qualifications regarding its effectiveness. However, σ^2 will not generally be known. It must be estimated, and if the model is inappropriate, s^2 will tend to over-estimate σ^2. Hence the denominator as well as the numerator of (7.5) will be inflated with a consequent loss in power. The t-test based on (7.8) suffers from the same drawback. A Sign test or, alternatively, a Wilcoxon test, may therefore be more attractive in certain circumstances.

8. A Strategy for Model Selection

A number of alternative approaches for testing the specification of a model have been discussed. In this section an attempt is made to bring these various components together and develop an overall strategy for model selection. All the different test procedures have a role to play in such an approach, and if used correctly they are complementary rather than competitive. However, before proceeding a word of caution is in order. The issues raised by the problem of model selection are extremely complex, and it would be foolish to pretend that there is a generally acceptable solution which can be applied in most cases. What follows should therefore be regarded as no more than an outline of what seems to me to be a reasonably sensible and systematic approach.

In many areas, particularly in economics, models are constructed on a somewhat haphazard basis. A typical approach is to begin by formulating the simplest model consistent with the theory of the subject in question. This model is then estimated. If it is a regression model, the researcher may well be tempted to proceed no further if he obtains a 'high' R^2, say over 0.9. However, a high R^2 may be obtained very easily with time series data, even when the variables in question are totally unrelated. Assuming that the researcher is sophisticated enough to appreciate this point, he will supplement his consideration of R^2 with some tests of misspecification. Should one or more of these tests be 'significant' he will reject his model, but the direction in which the model is subsequently revised will be essentially arbitrary.

The *ad hoc* approach to specification testing described above is avoided by imposing some structure on the problem at the outset. The various alternative formulations suggested by theory are first delineated and a systematic procedure for testing the hypotheses implicit in them is determined. All the available *a priori* knowledge is used to set up a decision framework *before* any estimation is carried out. Specification testing may then proceed, the starting point being the most *general* model, or models, under consideration.

The theory underlying this systematic approach is dealt with in the next sub-section. However, it is extremely rare, in practice, for a researcher to be able to enumerate confidently all the possible models he is prepared to consider. For this reason tests of mis-specification still play an important part in any overall specification strategy. Nevertheless, their role is to complement the tests of specification, rather than to provide the basis for an alternative strategy.

Testing Ordered Hypotheses

Given a set of possible model specifications, the most obvious structure to look for is one of *nesting*. Consider the three models

(0) $y_t = \beta_1 x_{1t} + \epsilon_t$

(1) $y_t = \beta_1 x_{1t} + \beta_2 x_{2t}^{-1} + \epsilon_t$ (8.1)

(2) $y_t = \beta_1 x_{1t} + \beta_2 (x_{2t} - \gamma)^{-1} + \epsilon_t$

The last of these is the demand for money equation, (5.17). However, model (1) is nested within model (2), while (0) is, in turn, nested within (1). The hypotheses implicit in this nest form a uniquely ordered set and this enables a *sequential testing procedure* to be carried out. The hypotheses are tested in increasing order of restrictiveness, with the sequence indicated by

$H_1 : \gamma = 0$

$H_0 : \gamma = 0, \beta_2 = 0$ (8.2)

The first test is of the null hypothesis $\gamma = 0$ against the alternative $\gamma \neq 0$. The second test sets $\beta_2 = 0$ against $\beta_2 \neq 0$. This is a test for the inclusion of x_{2t}^{-1} in the regression model, but it is only carried out if the first test does not reject the null hypothesis that $\gamma = 0$. If $H_1 : \gamma = 0$ is rejected, all succeeding hypotheses (i.e. $H_0 : \gamma = 0$, $\beta_2 = 0$) are rejected as well.

The tests employed in a sequential procedure may be based on the Wald, LR or LM principles. If it were felt desirable to estimate the most general model in (8.1), then a test of $\gamma = 0$ could, following the Wald principle, be based on the relevant asymptotic t-ratio. It would, however, be equally valid to employ the LR statistic, although there is no guarantee that it would give the same result. If the hypothesis were not rejected, then a test of $\beta_2 = 0$ would be carried out straight-forwardly using a t-test from a linear regression based on model (1). The sequential test procedures used in the study by Mizon (1977a) are all based on tests derived from the Wald or LR principles. However, in example (8.1), an LM test of $\gamma = 0$ has considerable appeal since it renders estimation of the nonlinear model, (2), unnecessary.

The sequential testing procedure starts from the most general, or *maintained*, hypothesis and tests hypotheses in increasing order of restrictiveness. Approaching the problem in this way yields a procedure with certain optimal properties with regard to power; see Anderson (1971, pp. 34–43, 116–134, 270–276). However, an alternative strategy, which is often used in practice, is to start from the most restricted model. In the case of (8.1), this would mean starting at (0) and treating the hypotheses in reverse order. Such an approach has obvious computational advantages if the more general models in the nest are highly nonlinear. However, even if the full testing framework is specified at the beginning, the test statistics will not, in general, be independent, and this makes the analysis of statistical power difficult. Indeed, one of the main reasons for testing the hypotheses in increasing order of restrictiveness is that the incremental test statistics are all independent in large samples.

Suppose the hypotheses in (8.2) are tested using Wald statistics, W_1 for H_1 and W_0 for H_0. W_1 and W_0 will be asymptotically χ^2_1 and χ^2_2 respectively, but the important result is that the incremental statistic, $W_0 - W_1$, is asymptotically χ^2_1 and is distributed independently of W_1 in large samples. Thus if the two hypotheses in (8.2) are tested at the 5% level of significance, the nominal significance level of the most restricted hypothesis, (0), against the maintained, (2), will be $1 - (0.95)^2 = 0.0975$.

Non-Nested Hypotheses

The sequential testing procedure is obviously inapplicable when some of the models under consideration are non-nested. However, problems of non-nestedness also arise when there is no natural ordering of hypotheses. Consider model (1) in (8.1). There are two possible sequences for testing the 'significance' of the estimates of β_1 and β_2. In the first, a t-test is carried out on the estimate of β_2 in a regression based on (1). This is followed by a t-test on the estimate of β_1 in a regression of y_t on x_{1t} only. The second sequence, which may be formalised as $H_1 : \beta_1 = 0$, $H_0 : \beta_1 = 0$, $\beta_2 = 0$, treats the parameters the other way round. If x_{1t} and x_{2t}^{-1} are strongly

correlated, the chances of the two testing sequences yielding
conflicting inferences are quite high. One approach may suggest that
y_t depends only on x_{1t}, while in the other it depends only on x_{2t}.
These two models are non-nested.

The above considerations lead to what Mizon (1977b) has called an
exhaustive test procedure. Optimal sequential testing procedures are
applied to all possible orderings of the complete set of hypotheses.
An appropriate method is then used to discriminate between the
preferred hypotheses from all the orderings.

Tests of Misspecification

The scheme set out so far has avoided the use of tests of misspecification. By first considering the various options available before any
estimation or testing is carried out, an overall structure is imposed on
the problem. A series of tests of specification are then carried out
within this framework. However, although this approach is very
attractive, there has never been any suggestion that it rules out the
use of tests of misspecification. It is rarely the case that the researcher
can be certain at the outset that his most general models are in fact
adequate. He may not become aware of certain formulations until
recognition of the shortcomings of the proposed models forces him
to revise his initial preconceptions. Tests of misspecification and
other diagnostic checking procedures clearly play a key role in this
process. However, they are a *supplement* to the formal procedure,
rather than an alternative.

Post-Sample Predictive Testing and Model Evaluation

The strategy outlined so far employs a number of procedures all of
which are based on the sample. They may therefore be regarded as a
means of arriving at a specification which is *internally valid*, in the
sense that it provides a good fit to the data in the sample. However,
it is now important to determine whether the model is *externally
valid*, i.e. does it perform well outside the sample period. This leads
to the tests based on prediction, and it is particularly important that
they be applied when tests of misspecification have led to some
revision of the hypotheses entertained. That being the case, the preferred model will, to some extent, be the product of data mining.
The 'good fit' obtained is merely the final result of a series of
experimental specifications suggested by the data themselves.

The aim of post-sample predictive testing is to provide an independent check on the specification suggested by the sample.
However, if the post-sample test rejects the null hypothesis of correct

specification, it is not altogether clear what the next step should be. The route most likely to be taken is to return to the sample data, and attempt to find a model which will satisfy the post-sample predictive test. This cycle could be repeated several times, and while the end result may be a perfectly satisfactory model, the procedure as a whole could be open to a charge of data mining. The post-sample predictive test is effectively being used in the same way as a conventional test of misspecification, and it no longer serves the purpose of providing an independent assessment of the adequacy of the fitted model.

A failure to reject the preferred model on the basis of its performance over the prediction period will often mean that the model is accepted. However, in doing this it is important to be aware of the shortcomings of such criteria. In particular, if the fit in the sample is relatively poor, the estimate of the error variance, σ^2, will be inflated. This will tend to weaken the standard post-sample predictive tests such as (7.6). It was suggested in Section 7, that the Wilcoxon and Sign tests might be more effective in certain circumstances, but even they would fail if the prediction errors were approximately random. Thus suppose a variable is omitted from a regression equation, and that this variable is uncorrelated with the variables included. Neither within-sample nor post-sample tests are likely to detect the omission. Thus, unless there is some absolute standard for determining whether the size of the prediction errors is acceptable, there are no statistical grounds on which the model may be deemed unsatisfactory. The model is therefore likely to be retained until it is challenged by an alternative specification. This leads to the adoption of a criterion such as

$$s_p^2(l) = \sum_{j=1}^{l} e_{T+j/T}^2 / l \qquad (8.3)$$

as a measure of post-sample goodness of fit. Comparing different models on the basis of (8.3) is more meaningful than a within sample comparison based on s^2 or R^2.

Notes

Section 2   Ramsey (1969) discusses the properties of residuals generally. Tests involving recursive residuals are described in Harvey and Phillips (1974), Brown, Durbin and Evans (1975) and Harvey and Collier (1977). Harrison and McCabe (1979) give exact and bounds tests for heteroscedasticity based on OLS residuals.

Sections 3, 4 and 5   Silvey (1970), Seber (1966), Breusch and Pagan (1979, 1980), Harvey and Phillips (1980).

Section 6   Sargan (1964), Akaike (1973), Seber (1978, pp. 359–370), Box and Cox (1964).

Sections 7 and 8   Gadd and Wold (1964), Leamer (1978), Mizon (1977a, 1977b).

Exercises

1. Prove (1.2) and (1.3), and comment on the interpretation of (1.3).

2. The CES production function was defined in (2.1.12). Given a multiplicative disturbance term, dividing through by L_t and taking logarithms yields,

$$\log q_t = \alpha + (\beta - 1)\log L_t - \beta\gamma^{-1}\log(1 - \delta + \delta k_t^{-\gamma}) + \epsilon_t$$

where $q_t = Q_t/L_t$ and $k_t = K_t/L_t$. When $\gamma = 0$, the Cobb–Douglas form is obtained. If $\epsilon_t \sim NID(0, \sigma^2)$ show that the modified LM test of $\gamma = 0$ may be based on the t-statistic obtained by regressing $\log q_t$ on $\log L_t$, $\log k_t$ and $(\log k_t)^2$.

3. The VES function is

$$Q = e^\alpha K^{\beta(1-\gamma\delta)}[L + (\gamma - 1)K]^{\beta\gamma\delta}$$

where α, β, γ and δ are parameters; see Revankar (1971). If the disturbances are multiplicative, construct a test of $\gamma = 1$ using a similar approach to that employed in Question 2 above. If the test rejected the null hypothesis, explain how you would proceed to estimate the model.

4. How would you construct a Wald test of $H_0: \delta_2 = \cdots = \delta_r$ in the multiplicative heteroscedasticity model (3.4.15); cf. (3.14) and (5.12)?

5. Construct an LR test for the inclusion of an equation in a SURE system.

6. How would you carry out a Wald test of $H_0: \beta_2 = \beta_4$ in (2.9.18)?

7. Construct a modified LM test of the hypothesis $\gamma = 1$ in (8.1).

8. How would you carry out an LM test of $\theta = 0$ in the ARMA $(1, 1)$ model, (1.5.21)? Compare this test with the Wald and LR tests.

6

Regression Models with Serially Correlated Disturbances

1. First Order Autoregressive Disturbances

The classical linear regression model was introduced in Chapter 2. Written in vector notation it has the form

$$y_t = x_t'\beta + u_t, \qquad t = 1, \ldots, T \tag{1.1}$$

with the symbols defined as in (2.1.3). A key assumption in the classical model is that the disturbances are uncorrelated. However, in dealing with time series regression this may not always be reasonable. If the disturbance term is regarded as being made up of a number of omitted variables, then in a time series context it seems more than likely that it will exhibit serial correlation since relatively few time series are random.

The traditional method of handling serially correlated disturbances is to model them by a stationary AR(1) process,

$$u_t = \phi u_{t-1} + \epsilon_t, \qquad |\phi| < 1 \tag{1.2}$$

Although such a model is not unreasonable for economic data, its main attraction lies in its ease of estimation. However, with modern computing facilities this is no longer a major consideration, and there is no reason why a general ARMA(p, q) disturbance should not be entertained. Such models are considered from Section 4 onwards, but from the point of view of exposition it is convenient to restrict attention to (1.2) initially.

Generalised Least Squares

The easiest way of carrying out GLS estimation is to find a matrix, L, with the property that $L'L = V^{-1}$. This may be used to transform the observations in the manner described in Section 2.4, and the

GLS estimator is then obtained by applying OLS. For a given V matrix — which in this case amounts to a given value of ϕ — this procedure yields the BLUE of β.

The V matrix may be readily constructed given the autocovariance properties of the AR(1) process. Since $E(u_t u_{t-\tau}) = \sigma^2 \phi^\tau / (1 - \phi^2)$, it follows that the ijth element of V is given by $\phi^{|i-j|}/(1 - \phi^2)$ for $i, j = 1, \ldots, T$. Inverting this matrix yields

$$
V^{-1} = \begin{bmatrix}
1 & -\phi & 0 & \cdots & 0 & 0 \\
-\phi & 1 + \phi^2 & -\phi & \cdots & 0 & 0 \\
\vdots & \vdots & \vdots & \vdots\vdots\vdots & \vdots & \vdots \\
0 & 0 & 0 & \cdots & 1 + \phi^2 & -\phi \\
0 & 0 & 0 & \cdots & -\phi & 1
\end{bmatrix}
\tag{1.3}
$$

and it is not difficult to verify that the $T \times T$ matrix

$$
L = \begin{bmatrix}
\sqrt{1 - \phi^2} & 0 & 0 & \cdots & 0 & 0 \\
-\phi & 1 & 0 & \cdots & 0 & 0 \\
\cdot & -\phi & 1 & \cdots & & \\
\vdots & \vdots & \vdots & \vdots\vdots\vdots & \vdots & \vdots \\
0 & 0 & 0 & \cdots & 1 & 0 \\
0 & 0 & 0 & \cdots & -\phi & 1
\end{bmatrix}
\tag{1.4}
$$

has the desired property $L'L = V^{-1}$. The transformed observations are therefore

$$
y_1^* = \sqrt{1 - \phi^2}\, y_1, \quad x_1^* = \sqrt{1 - \phi^2}\, x_1
\tag{1.5}
$$

$$
y_t^* = y_t - \phi y_{t-1}, \quad x_t^* = x_t - \phi x_{t-1}, \qquad t = 2, \ldots, T
\tag{1.6}
$$

For a given value of ϕ, regressing y_t^* on x_t^* for $t = 1, \ldots, T$, gives the GLS estimator of β.

Because the first observation is handled in a slightly different way from the remaining observations, it is often dropped from the computations. Thus OLS is based only on the $T - 1$ quasi-differences defined by (1.6). This is sometimes referred to as the *Cochrane–Orcutt* transformation. The rationale behind this transformation may be seen directly by multiplying (1.1) by ϕ at time $t - 1$. This yields

$$
\phi y_{t-1} = \phi x_{t-1}' \beta + \phi u_{t-1}
$$

Subtracting from (1.1) with u_t set equal to $\phi u_{t-1} + \epsilon_t$ then gives

$$
y_t - \phi y_{t-1} = (x_t - \phi x_{t-1})' \beta + \epsilon_t
\tag{1.7}
$$

The effect of dropping the first observation depends crucially on the structure of the explanatory variables. Under the regularity conditions generally imposed for deriving asymptotic properties, it can be shown that the effect of omitting (1.5) is negligible in large samples. On the other hand it is always possible to find sets of explanatory variables for which the efficiency of the Cochrane–Orcutt transform is very low. Maeshiro (1976) reports a series of calculations for a model with a constant term and a single explanatory variable equal to US real GNP. For all positive values of ϕ, the GLS estimator computed by the Cochrane–Orcutt transformation was actually less efficient than OLS. Furthermore its performance worsened the higher the value of ϕ; with $\phi = 0.98$ the variance of the Cochrane–Orcutt estimator of the coefficient of the explanatory variable was twice that of the OLS estimator.

Chipman (1979) provides analytic evidence on the above point. For the time trend model

$$y_t = \alpha + \beta t + u_t, \qquad t = 1, \ldots, T \tag{1.8}$$

he is able to show that the minimum efficiency of the Cochrane–Orcutt estimator is 0.54. On the other hand, the minimum efficiency of OLS is higher at 0.75. Unfortunately the actual efficiency is a complicated function of ψ and T, and the interested reader should refer to Chipman's paper for further details. However, a heuristic rationale for the poor performance of Cochrane–Orcutt in this particular case is obtained by noting that the transformed explanatory variable is $x_t^* = t - \phi(t - 1)$. When ϕ is close to one, $x_t^* \simeq 1$ and this is almost perfectly collinear with the constant term, $1 - \phi$. Hence the estimates of both α and β will be relatively imprecise.

Further discussion of the relative efficiency of the Cochrane–Orcutt transform with respect to both the full GLS transform and OLS will be found in Section 2. For the moment, however, the main point to note is a tendency for the Cochrane–Orcutt transformation to yield unsatisfactory results when one of the explanatory variables is trending.

Two-Step and Iterative Procedures

Since ϕ will not, in general, be known, it must be estimated. If a consistent estimator of ϕ can be found, a feasible GLS estimator of β may be constructed. Subject to regularity conditions of the type described in Chapter 2, the feasible GLS estimator may be shown to have the same large sample distribution as the GLS estimator proper.

There are basically two methods of estimating ϕ. The first is based on the OLS residuals, and is given by the regression of e_t on e_{t-1}, i.e.

$$\hat{\phi} = \sum_{t=2}^{T} e_t e_{t-1} \Big/ \sum_{t=2}^{T} e_{t-1}^2 \qquad\qquad (1.9)$$

The second approach, due to Durbin (1960), consists of regressing y_t on y_{t-1}, x_t and x_{t-1}. The coefficient of y_{t-1} then provides the required estimate. The rationale behind this may be seen by re-arranging (1.7). This yields

$$y_t = \phi y_{t-1} + x_t'\beta - x_{t-1}'(\phi\beta) + \epsilon_t, \qquad t = 1, \ldots, T \qquad (1.10)$$

Both of the above estimators of ϕ can be shown to be consistent under fairly general conditions. The inclusion of the first observation, (1.5), is immaterial as regards the asymptotic properties of the estimator of β, but as indicated in the previous sub-section this could be quite important in practice. If the first observation is omitted, the feasible GLS estimator based on (1.9) is known as two-step Cochrane—Orcutt. However, the nature of this estimation procedure lends itself naturally to iteration. Having computed a feasible GLS estimator of β, a new set of residuals may be used to construct another estimator of the form (1.9). This is then used to compute another feasible GLS estimator, and the process is repeated until convergence. The *Cochrane—Orcutt iterative procedure*, as it is called, implies no improvement as far as asymptotic properties are concerned, although it may have a significant effect in small samples. Similar considerations hold for an iterative procedure based on the full GLS transform.

Maximum Likelihood Estimation

If the disturbances in (1.1) are normally distributed, the log—likelihood function is of the form (3.5.12), but with y_t replaced by $y_t - x_t'\beta$. However, as in the case of the time series AR(1) model, considerable simplification results if y_1 is taken to be fixed. Conditional on β, the ML estimator of ϕ is then given by a straight-forward regression of $y_t - x_t'\hat{\beta}$ on $y_{t-1} - x_{t-1}'\hat{\beta}$. This suggests that if the ML estimator of β, conditional on ϕ, is also linear, a stepwise optimisation procedure is likely to be computationally effective. Now, maximising the approximate log—likelihood function is equivalent to minimising the residual sum of squares function

$$S(\beta; \phi) = \sum_{t=2}^{T} \{(y_t - x_t'\beta) - \phi(y_{t-1} - x_{t-1}'\beta)\}^2$$

$$= \sum_{t-2}^{T} \{y_t - \phi y_{t-1} - (x_t - \phi x_{t-1})'\beta\}^2$$

and using the definition (1.6) this becomes

$$S(\beta; \phi) = \sum_{t=2}^{T} (y_t^* - x_t^{*'}\beta)^2 \qquad (1.11)$$

The conditional ML estimator of β is therefore identical to the Cochrane–Orcutt estimator. The iterative Cochrane–Orcutt procedure can now be interpreted as a stepwise algorithm for computing the ML estimators of β and ϕ when y_1 is regarded as fixed.

By a similar argument to the one employed above, the full GLS estimator is given implicitly when the exact likelihood function is decomposed in terms of prediction errors as in (3.5.12). Indeed, the matrix, L, defined at the beginning of this section is identical to the matrix defined in (3.5.13). Thus, conditional on ϕ, the ML estimator of β is obtained by applying OLS to the complete set of transformed observations, (1.5) and (1.6), while the conditional ML estimator of ϕ is obtained by solving a certain cubic equation; see Beach and MacKinnon (1978).

Since the GLS estimator is the MVUE of β, the two-step estimator is asymptotically efficient under the normality assumption. Furthermore, because the OLS estimator of β is consistent, the estimator (1.9) is an asymptotically efficient estimator of ϕ.

Asymptotic Distribution of the ML Estimator

The large sample properties of the ML estimators of β, ϕ and σ^2 are unaffected by the treatment of the initial conditions. The simplest approach is to take y_1 as fixed. Taking the expectation of the matrix of second derivatives of $\log L$ yields a block diagonal matrix. The ML estimator of β is therefore distributed independently of the estimators of ϕ and σ^2. These in turn have exactly the same asymptotic distribution as in an AR(1) time series model; see (3.5.16). Thus, asymptotically, $(\tilde{\beta}', \tilde{\phi}, \tilde{\sigma}^2)'$ has a multivariate normal distribution with mean $(\beta', \phi, \sigma^2)'$ and covariance matrix

$$\text{Avar}(\beta, \phi, \sigma^2) = \begin{bmatrix} \sigma^2(X'V^{-1}X)^{-1} & 0 & 0 \\ 0 & (1-\phi^2)/T & 0 \\ 0 & 0 & 2\sigma^4/T \end{bmatrix} \qquad (1.12)$$

2. Comparison of Estimators

All the procedures described in the previous sections yield estimators with identical large sample properties. Nevertheless their behaviour in small samples can be very different. This section reviews some of the evidence on small sample properties. Before this is done, however, the effect of applying OLS when the disturbances are autocorrelated is considered in some detail. Two issues are examined. The first is the extent to which OLS is inefficient, while the second concerns the implications of making inferences when the autocorrelation in the disturbances is ignored.

Efficiency of OLS

If the value of the AR parameter, ϕ, is known, the OLS estimator of β cannot be more efficient than the GLS estimator. This follows directly from the Gauss–Markov theorem. Furthermore if ϕ is unknown, any consistent estimator of ϕ will lead to a feasible GLS estimator of β with the same asymptotic distribution as the GLS estimator proper. Thus a comparison between OLS and GLS for given values of ϕ provides a sensible measure of the relative efficiency of OLS in practice, provided that the·sample size is reasonably large. On the other hand if T is small, it is quite possible for OLS to yield a more efficient estimator of β than a method which 'allows' for serial correlation. If, for example, ϕ is close to zero the additional source of variation arising from the need to estimate ϕ may more than offset any loss in efficiency incurred by taking ϕ to be equal to zero. Unfortunately there is little analytic evidence available on this point, although the relatively small sample efficiency of OLS in certain circumstances does emerge quite clearly in Monte Carlo results. This sub-section therefore concentrates on examining the extent to which OLS is likely to be inefficient in large samples.

A single regressor model

$$y_t = \beta x_t + u_t, \qquad t = 1, \dots, T \tag{2.1}$$

is relatively straightforward to analyse, and provides a good deal of insight into the factors affecting the performance of OLS. If u_t is an AR(1) process of the form, (1.2), the GLS estimator is obtained by applying OLS to the transformed observations, (1.5) and (1.6). Therefore

$$\mathrm{Var}(\tilde{b}) = \sigma^2 \left[(1 - \phi^2)x_1^2 + \sum_{t=2}^{T} (x_t - \phi x_{t-1})^2 \right] \tag{2.2}$$

The variance of the Cochrane–Orcutt estimator is given by an expression of the form (2.2), but with the term involving x_1 omitted.

The variance of the OLS estimator of β may be obtained by evaluating the general expression (2.4.8). Thus

$$\text{Var}(b) = \sigma^2 \kappa \bigg/ \bigg[(1 - \phi^2)\bigg(\sum_{t=1}^{T} x_t^2\bigg)\bigg] \qquad (2.3)$$

where

$$\kappa = (1 - \phi^2)X'VX = \sum_{t=1}^{T} x_t^2 + 2\sum_{i=1}^{T-1}\sum_{t=1}^{T-i} \phi^i x_t x_{t+i} \qquad (2.4)$$

Now suppose that x_t itself is generated by a stationary AR(1) process of the form

$$x_t = \bar\phi x_{t-1} + \xi_t \qquad (2.5)$$

where $\xi_t \sim NID(0, \bar\sigma^2)$. Then

$$\text{plim } \kappa/\Sigma x^2 = 1 + 2\phi\bar\phi + 2(\phi\bar\phi)^2 + 2(\phi\bar\phi)^3 + \cdots$$
$$= 2/(1 - \phi\bar\phi) - 1 = (1 + \phi\bar\phi)/(1 - \phi\bar\phi)$$

and so

$$\text{Avar}(b) = T^{-1}\text{ plim } T\text{ Var}(b) = \frac{1}{T}\frac{\sigma^2}{\bar\sigma^2}\frac{(1-\bar\phi^2)(1+\phi\bar\phi)}{(1-\phi^2)(1-\phi\bar\phi)} \qquad (2.6)$$

For the GLS estimator

$$\text{Avar}(\tilde{b}) = \sigma^2 T^{-1}[\text{plim}(\Sigma x_t^2 - 2\phi\Sigma x_t x_{t-1} + \phi^2\Sigma x_{t-1}^2)/T]^{-1}$$
$$= \frac{1}{T}\frac{\sigma^2}{\bar\sigma^2}\frac{1 - \bar\phi^2}{1 + \bar\phi^2 - 2\phi\bar\phi} \qquad (2.7)$$

irrespective of whether the full or Cochrane–Orcutt transform is carried out.

The asymptotic efficiency of OLS is

$$\text{Eff}(b) = \frac{\text{Avar}(\tilde\beta)}{\text{Avar}(b)} = \frac{(1 - \phi^2)(1 - \phi\bar\phi)}{(1 + \phi\bar\phi)(1 + \phi^2 - 2\phi\bar\phi)} \qquad (2.8)$$

For $\phi = 0.9$ and $\bar\phi = 0.8$, $\text{Eff}(b) = 0.03$.

The above results suggest that the performance of OLS may be very poor in certain circumstances. As an alternative case, however, suppose that $x_t = 1$ for $t = 1, \ldots, T$, rather than being generated by (2.5). Then

$$\kappa = T + 2(T - 1)\phi + 2(T - 2)\phi^2 + \cdots + 2\phi^{T-1}$$
$$= \{T(1 - \phi^2) - 2\phi(1 - \phi^T)\}/(1 - \phi)^2$$

and on evaluating $\text{Avar}(\tilde\beta)$ and $\text{Avar}(b)$, it will be found that $\text{Eff}(b) = 1$. The efficiency of OLS in these circumstances is actually a reflection of a more general result, namely that for polynomial or trigonometric regressors OLS is asymptotically efficient for all values of ϕ.

Inferences Based on OLS

Although the true covariance matrix of the OLS estimator is given by (2.4.8), this formula will not, of course, be used in practice. Estimating $\text{Var}(b)$ in the usual way may give a misleading picture of the accuracy of the procedure.

As an example, consider model (2.1) with x_t generated by the AR(1) process (2.5). The variance of the OLS estimator will be estimated by

$$\text{var}(b) = s^2/\Sigma\, x_t^2 \tag{2.9}$$

and

$$T^{-1} \text{ plim } T \text{ var}(b) = T^{-1} \text{ plim } s^2/\text{plim } \Sigma x^2/T$$

$$= T^{-1} \frac{\sigma^2}{1 - \phi^2} \frac{(1 - \bar{\phi}^2)}{\bar{\sigma}^2} \tag{2.10}$$

A comparison of (2.10) with (2.6) shows that, in large samples, the ratio of the estimated variance to the true variance will be about $(1 - \phi\bar{\phi})/(1 + \phi\bar{\phi})$ on average. If both x_t and u_t exhibit positive serial correlation therefore, the analysis indicates that (2.9) will tend to *underestimate* the true variance.

Small Sample Properties

The results shown in table 6.1 are taken from a series of Monte Carlo experiments carried out by Harvey and McAvinchey (1978). They show the estimated root mean square error (RMSE) of a number of estimators in relation to the estimated RMSE of the full ML estimator. The model examined was

$$y_t = \beta_1 + \beta_2 x_t + u_t, \qquad t = 1, \ldots, T \tag{2.11}$$

with u_t generated by an AR(1) process. Two forms for the explanatory variable were considered. In the first, x_t displayed a strong upward *trend*, being generated by a process of the form

$$x_t = \exp(0.04t) + \xi_t, \qquad t = 1, \ldots, T \tag{2.12}$$

where $\xi_t \sim NID(0, 0.0009)$. In the second, x_t was *stationary* and random. The summary statistics reported in table 6.1 are based on two hundred independent replications for each model.

Table 6.1 Ratio of RMSEs of OLS, Cochrane–Orcutt and Two-Step Estimators to Full ML for $\phi = 0.6$

Model		Two-Step Estimators					Iterative Cochrane–Orcutt			OLS	
		Full Transform		ϕ	Cochrane–Orcutt						
		β_1	β_2		β_1	β_2	β_1	β_2	ϕ	β_1	β_2
Trending Data	$T=20$	99.8	100.0	101.0	113.0	109.2	183.9	133.5	100.9	103.3	103.9
	$T=100$	100.0	100.0	100.5	102.5	100.0	102.5	100.0	100.3	102.0	100.0
Stationary Data	$T=20$	100.0	100.9	105.3	109.6	100.5	112.4	99.8	100.0	100.0	115.7
	$T=100$	100.0	100.5	103.0	103.4	100.5	103.4	100.5	100.7	100.0	152.1

The design of the Monte Carlo experiments in Harvey and McAvinchey (1978) is exactly the same as that used by Beach and MacKinnon (1978). Unfortunately the Beach—MacKinnon study had been restricted to a comparison of full ML with iterative Cochrane—Orcutt. They had found that when the explanatory variable was trending, Cochrane—Orcutt could give very inefficient estimators of β. This point was confirmed by Harvey and McAvinchey as the results presented for $\phi = 0.6$ and $T = 20$ show. However, the main motive behind the Harvey—McAvinchey study was to extend the Beach—McKinnon results to include two-step estimators. Using the same initial consistent estimator of ϕ, (1.9), feasible GLS estimates of β were computed using the full transformation, (1.4), and Cochrane—Orcutt.

Several conclusions emerged from this exercise. Firstly, the two-step estimator of β based on the full transformation appears to be as efficient as full ML in small samples. Secondly, the results for the two-step Cochrane—Orcutt estimator show that it is quite possible that iterating a particular procedure until convergence will result in a deterioration in small sample properties. Thirdly, the results for trending data and $T = 20$ show the two-step Cochrane—Orcutt estimators of both β_1 and β_2 having a significantly higher RMSE than the corresponding estimators based on the full transformation. This lends support to the argument advanced in Section 1 around (1.7) and (1.8). However, the results for $T = 100$ lead to a fourth conclusion, namely that the effect of the Cochrane—Orcutt approximation is relatively unimportant for moderately large samples.

The main conclusion to emerge from table 6.1 — and this is supported by the results not reproduced here — is that a significant loss in efficiency may be incurred if the Cochrane—Orcutt transformation is used in small samples. However, this does not necessarily constitute an argument for the use of full ML estimation. As the results for the two-step estimators show, the key feature of a successful estimation procedure is the use of the full GLS transform.

The results in table 6.1 also shed some light on the small sample performance of OLS. This estimator was included in the simulations as a yardstick by which to judge the other estimators, and within this framework its performance for trending data is particularly striking. In fact the relatively high efficiency of OLS in these circumstances is perhaps not surprising in view of the result noted at the close of the sub-section on the efficiency of OLS. The performance of OLS for stationary data is, however, less impressive. This is what one would be led to expect from (2.8). However, this asymptotic formula does imply that OLS is much more inefficient than it typically is for a moderately sized sample. For $T = 100$, the estimated efficiency of OLS is $(100/152.1)^2 = 0.81$. By way of contrast, evaluating (2.8) for $\phi = 0.6$ and $\tilde{\phi} = 0.0$ gives $\text{Eff}(b) = 0.47$.

3. Testing for First-Order Autoregressive Disturbances

In Chapter 5 the construction of tests of specification based on the LR, Wald and LM principles was discussed. These procedures may all be applied to the problem of testing for an AR(1) disturbance term in (1.1), i.e. testing the null hypothesis $H_0 : \phi = 0$ against the alternative $H_1 : \phi \neq 0$. In order to simplify matters it will be assumed that y_1 is fixed. The remaining observations will be taken to be normally distributed.

If it is felt desirable to estimate the unrestricted model by an iterative procedure, such as Cochrane–Orcutt, a Wald test may be carried out quite easily. In terms of (5.3.8), $\psi = (\beta', \phi)'$, $R = (0 \ldots 1)'$ and $r = 0$. The Wald statistic (5.4.2) is therefore

$$W = \tilde{\phi}^2/\text{Avar}(\tilde{\phi}) = \tilde{\phi}^2/\{(1 - \tilde{\phi}^2)/T\} \tag{3.1}$$

and under H_0 this will have a χ_1^2 distribution. However, it will generally be preferable to base the test on $\tilde{\phi}/\sqrt{\text{Avar}(\tilde{\phi})}$ since this is $AN(0, 1)$ under H_0, and allows a one-sided test to be carried out. A slightly more convenient form of the test is given by noting that it is asymptotically equivalent to a t-test in the final regression of $y_t - x_t'\tilde{\beta}$ on $y_{t-1} - x_{t-1}'\tilde{\beta}$; cf. Example 4.6.1.

The need to estimate the unrestricted model may, however, be regarded as a disadvantage from the computational viewpoint. Because the restricted model is linear in this case, it may be estimated simply by OLS. This suggests a test based on the LM principle. For an AR(1) time series model it follows from (5.5.16) that the appropriate LM statistic is Tr_1^2, where r_1 is the first-order sample autocorrelation. Using exactly the same argument the LM principle in the regression case leads to a test statistic based on the first-order autocorrelation in the OLS residuals. A one-sided test may be carried out by treating r_1 as being normally distributed with mean 0 and variance $1/T$ in large samples.

In the time series AR(1) model a test of $\phi = 0$ is not usually based on r_1, but rather on the von Neumann Ratio. This statistic is a monotonic transformation of r_1 but its exact significance points have been tabulated. Unfortunately an exact test based on OLS residuals cannot be constructed in the same way, since the residuals will not, in general, be independently and identically distributed.

The Durbin–Watson d-statistic

Apart from a factor of $T/(T-1)$, the Durbin–Watson statistic

$$d = \frac{\sum\limits_{t=2}^{T} (e_t - e_{t-1})^2}{\sum\limits_{t=1}^{T} e_{t-1}^2} \tag{3.2}$$

is identical to the von Neumann Ratio, since the mean of the OLS residuals must be zero. However, the vector of OLS residuals is defined by $e = Mu$, where $M = I - X(X'X)^{-1}X'$. Although $E(e) = 0$ when $e \sim N(0, \sigma^2 I)$, the covariance matrix is given by $E(ee') = \sigma^2 M$ and so will not, in general, be scalar. Since the distribution of e depends on the set of regressors X, the distribution of d also depends on X. Therefore, exact significance points for d cannot be tabulated once and for all. However, as the sample size increases the effect of the regressors becomes less important. In fact, since $r_1 \sim AN(0, 1/T)$ and

$$d \simeq 2(1 - r_1) \tag{3.3}$$

it is not difficult to see that $d \sim AN(2, 4/T)$. The approximation in (3.3) arises because of the treatment of end points and these are unimportant in large samples.

The range of the d-statistic is $[0, 4]$, and again this ties in with its relationship with r_1 as given by (3.3). Furthermore, when the residuals are positively autocorrelated, r_1 is greater than zero and the value of d will tend to be low. On the other hand, d tends towards 4 for negative serial correlation. A one-sided test against positive serial correlation – the most common alternative – will therefore be based on a critical region for which the null is rejected when d falls *below* a certain value. Thus with $T = 100$, the critical region for a large sample one-sided test at the 5% level of significance comprises all values of d less than $2 - 1.64(0.2) = 1.67$.

Although the asymptotic distribution of d is independent of the regressors, the behaviour of the d-statistic for different regressors can vary quite markedly in small samples. A test based on the asymptotic distribution could therefore be misleading in certain circumstances. Hence the small sample distribution of d is of some interest. Although this distribution depends on X it may nevertheless be computed. In matrix terms, the d statistic is

$$d = \frac{e'Ae}{e'e} = \frac{u'MAMu}{u'Mu} \tag{3.4}$$

where A is a fixed $T \times T$ matrix defined in Durbin and Watson (1950). The expression on the far right-hand side of (3.4) is a quadratic form in normal variables, and the exact distribution of such a statistic may be found numerically using the method of Imhof or Pan Jie-Jan; see Durbin and Watson (1971) and Koerts and Abrahamse (1969). However, from the point of view of carrying out a test, all that is required is the appropriate significance point. Thus for a one-tailed test against positive serial correlations it is necessary to compute a value d^*, such that

$$Pr(d < d^*) = \alpha$$

where α is the size of the test. A procedure carried out on this basis may be referred to as the *exact Durbin–Watson test*.

The Bounds Test

Although the exact significance point for the d-statistic varies with X, it is subject to an upper and a lower bound. These bounds, denoted respectively by d_U and d_L were tabulated by Durbin and Watson in 1950. The critical values for a one-sided test at the 0.05 and 0.01 levels of significance are given in tables A.1 and A.2 (see pp. 360–361). Although these points apply to a test against positive serial correlation, a test against negative serial correlation may be carried out in exactly the same way using the statistic $4 - d$. The critical values d_U and d_L depend on k as well as T. It is important to note not only that k includes the constant term, but that a constant term must be included in the regression in order for the bounds to be valid.

The bounds test is carried out as follows. If $d < d_L$, the null hypothesis is rejected; if $d > d_U$ it is not rejected; while if $d_L \leqslant d \leqslant d_U$ the test is inconclusive. If the test is inconclusive, one might resort to calculating d^* numerically by one of the methods referred to above. Alternatively the 'Beta' or '$a + bd_U$' approximations may be used. Although these methods are unable to give d^* accurately they will usually give a value accurate enough for most practical purposes. This is a small price to pay in view of the fact that they are relatively inexpensive compared with the Imhof or Pan Jie-Jan algorithms. Further details of these approximate procedures will be found in Durbin and Watson (1951, 1971).

A totally different reaction to an inconclusive bounds test would simply be to treat it as a rejection of the null hypothesis. For economic data, which is typically 'slowly changing' this may not imply a particularly big increase in the size of the Type I error. The reason is that for this type of regressor, d^* is close to d_U. On the other hand this is less likely to be the case when first differences have been taken, and when k is large the bounds may be very wide. Nevertheless there is still a strong argument for going ahead and actually estimating a more general model. Any 'losses' incurred are likely to be much higher if the restricted model is accepted when it is inadequate.

Tests Based on LUS Residuals*

The problem of the inconclusive region in the bounds test may be by-passed by using LUS residuals. A test may then be based directly on the von Neumann Ratio. However, such tests will be less effective than the exact Durbin–Watson procedure. Given d^*, a test based on the d-statistic is uniformly the most powerful invariant against a one-sided alternative in the neighbourhood of $\phi = 0$. This theoretical evidence in favour of basing a test on OLS residuals is supported by the Monte Carlo results reported in Koerts and Abrahamse (1969) and Phillips and Harvey (1974). Thus a test against AR(1) disturbances using LUS residuals can only be justified on computational grounds. Since the computations involved in obtaining a set of BLUS residuals are of a similar order to those needed to compute d^* exactly the case for using them is very weak. Recursive residuals, on the other hand, are relatively easy to obtain and in Section 5.2 it was argued that a modified von Neumann Ratio statistic based on them would play an important role in any battery of misspecification tests. However, against an AR(1) alternative such a test is less attractive overall as compared with a Durbin–Watson procedure which utilises the *Beta* or $a + bd_U$ approximation.

Robustness of Other Test Statistics

It was pointed out in Section 5.2 that serial correlation in the residuals does not necessarily imply serial correlation in the disturbances. Therefore although the Durbin–Watson procedure is formally a test against an AR(1) disturbance term, it may well be indicative of other types of misspecification besides serial correlation. Indeed, it was in this spirit that it was proposed in Section 5.2 as a general test of misspecification, as was the von Neumann Ratio based on recursive residuals.

Bearing these considerations in mind it is important to consider supplementary statistics, which provide some indication of the nature of the misspecification when a significant d-statistic or von Neumann Ratio is obtained. Several tests were proposed in Section 5.2, and table 6.2 presents estimates of their respective powers for various

Table 6.2   Estimated Powers of Tests for $y_t = \alpha + \beta t + u_t$ with AR(1)
Disturbances

ϕ	0.5		0.8	
T	30	60	30	60
MVNR	0.78	0.96	0.94	1.00
$D-W$	0.68–0.79	0.95–0.97	0.93–0.94	1.00–1.00
CUSUM	0.16	0.24	0.34	0.50
ψ	0.15	0.18	0.34	0.36
CUSUMSQ	0.15	0.17	0.29	0.36
$H\dagger$	0.14	0.12	0.35	0.42
H	0.08	0.08	0.06	0.12

values of ϕ. The results, taken from Harvey and Phillips (1976), refer to model (2.11) with $x_t = t$. Apart from the von Neumann Ratio (with recursive residuals) and d-statistic, the tests are relatively robust to a moderate level of serial correlation such as $\phi = 0.5$. They are less robust when $\phi = 0.8$.

4. Higher Order Autoregressive Disturbances

An AR(p) disturbance term,

$$u_t = \phi_1 u_{t-1} + \cdots + \phi_p u_{t-p} + \epsilon_t \tag{4.1}$$

provides a natural extension of the model considered so far. The AR(p) model is rather more flexible than the AR(1) model, and specifying the disturbance in this way permits a wide variety of behavioural patterns to be represented. For example, pseudo-cyclical movements can often be modelled by an AR(2) process. In fact, it can be shown that any stationary stochastic process can be approximated to any degree of accuracy if p is made sufficiently large. Thus specifying an autoregressive process of sufficiently high order can be made the basis for a nonparametric approach to estimating the regression coefficients.

Higher order autoregressive representation is also important in the context of seasonal data. Just as seasonal effects need to be taken into account in modelling univariate time series, so they must be allowed for in specifying the form of the disturbance term in regression analysis. In economics, observations are generally available on a monthly or quarterly basis. This suggests the use of twelfth or fourth order processes. As in univariate time series analysis, these effects may be compounded with lower order autoregressive effects, so that in the quarterly case, for example, a typical specification for the disturbance might be

$$(1 - \phi_1 L)(1 - \phi_4 L^4)u_t = \epsilon_t \tag{4.2}$$

The seasonal effect is captured by the fourth-order coefficient, ϕ_4, while the serial correlation from one quarter to the next is accounted for by the first-order term. Combining the two processes leads to an eventual AR(5) specification, albeit with certain constraints on the parameters.

Maximum Likelihood Estimation

Methods for estimating regression models with AR(p) disturbances may be derived by generalising the results for the first order case. As with the AR(1) model, a unified approach is obtained by working with the likelihood function. The form of the GLS estimator then emerges quite naturally.

If $\epsilon_t \sim NID(0, \sigma^2)$, the full log–likelihood function may be decomposed as

$$\log L(y) = \sum_{t=p+1}^{T} \log l(y_t/y_{t-1}, \ldots, y_1) + \log L(y_1, \ldots, y_p);$$

cf. (3.5.5). This becomes

$$\log L(y; \phi, \beta, \sigma^2) = -\tfrac{1}{2}T \log 2\pi - \tfrac{1}{2}T \log \sigma^2 - \tfrac{1}{2} \log |V_p|$$

$$- \tfrac{1}{2}\sigma^2 (y_p - X_p\beta)' V_p^{-1}(y_p - X_p\beta)$$

$$- \tfrac{1}{2}\sigma^2 \sum_{t=p+1}^{T} \{(y_t - x_t'\beta) - \phi_1(y_{t-1} - x_{t-1}'\beta) -$$

$$\cdots - \phi_p(y_{t-p} - x_{t-p}'\beta)\}^2 \tag{4.3}$$

V_p denotes the covariance matrix of $y_p = (y_1, \ldots, y_p)'$, and $X_p = (x_1, \ldots, x_p)'$.

Maximising (4.3) with respect to β gives the GLS estimator for a given value of $\phi = (\phi_1, \ldots, \phi_p)'$. The form taken by this estimator is implicitly given by the last two terms in the expression. If L_p defines a $p \times p$ matrix with the property $L_p' L_p = V_p^{-1}$, the first p transformed observations are given by

$$y_p^* = L_p y_p \quad \text{and} \quad X_p^* = L_p X_p \tag{4.4}$$

The remaining $T - p$ transformed observations are effectively defined by the last term in (4.3) to be

$$y_t^* = y_t - \phi_1 y_{t-1} - \cdots - \phi_p y_{t-p},$$
$$x_t^* = x_t - \phi_1 x_{t-1} - \cdots - \phi_p x_{t-p}, \qquad t = p+1, \ldots, T \quad (4.5)$$

The transformed observations have a scalar covariance matrix and so the application of OLS yields the GLS estimator of β. The independence of the disturbances corresponding to the first p transformed observations follows by definition, while in the case of (4.5), it should be clear by analogy with (1.7) that

$$y_t^* = x_t^* \beta + \epsilon_t, \qquad t = p+1, \ldots, T \qquad (4.6)$$

The computation of L_p may be carried out by the Cholesky decomposition; cf. Section 1.3. The full GLS transform for $p = 1$ depends on (1.5) which is a special case of (4.4). The matrix L_p is simply a scalar taking the value $\sqrt{1 - \phi^2}$.

Given a method for computing the GLS estimator, the likelihood function may be concentrated with respect to ϕ. Substituting the conditional ML estimator

$$\tilde{\sigma}^2(\phi) = T^{-1}(y - X\tilde{b})' V^{-1}(y - X\tilde{b})$$

$$= T^{-1} \sum_{t=1}^{T} (y_t^* - x_t^{*'}\tilde{b})^2 \qquad (4.7)$$

into (4.3) yields

$$\log L_c(y; \phi) = -\tfrac{1}{2} T \log 2\pi - \tfrac{1}{2} T \log \tilde{\sigma}^2(\phi) + \tfrac{1}{2} \log |V_p^{-1}(\phi)| \qquad (4.8)$$

The function (4.8) may be maximised by a general numerical optimisation procedure, possibly incorporating constraints to ensure stationarity. However, a simplification of the ML procedure is obtained by taking y_p to be *fixed*. This allows a straightforward stepwise optimisation approach since the third and fourth terms in the log–likelihood function, (4.3) disappear. If $\hat{u}_t = y_t - x_t'\hat{\beta}$ denotes the tth residual from an estimator of β, the ML estimator of ϕ, conditional on $\hat{\beta}$, is given by a linear regression of $\hat{u}_t$ on $\hat{u}_{t-1}, \ldots, \hat{u}_{t-p}$. The form of the GLS estimator also simplifies somewhat since fixing y_p effectively removes the first p transformed observations, (4.4). The resulting transformation, based only on (4.5), is therefore a direct generalisation of Cochrane–Orcutt, and the stepwise optimisation procedure, involving successive estimation of β and ϕ, is the natural analogue of the Cochrane–Orcutt iterative procedure. However, given the undesirable properties exhibited by the Cochrane–Orcutt transformation for certain types of data, it may be advisable to base GLS estimation on the full transform, even though the estimation of ϕ is simplified. From the computational point of view this makes very little difference since the Cholesky decomposition of a $p \times p$ matrix should not be particularly time consuming. Furthermore, the extra computation is to some extent justified insofar as the Cholesky decomposition acts as a check on the stationarity of the AR process; when the process is non-stationary V_p will not be p.d. and the factorisation of V_p^{-1} breaks down. As a final point, V_p^{-1} need not be specifically computed since an estimate of it emerges directly from the inverse of the cross-product matrix in the regression of $\hat{u}_t$ on its lagged values.

Two-Step Estimators

Given an estimate of ϕ, a feasible GLS estimator of β may be computed using either the full transform or the approximation based on (4.5). Provided the estimator of ϕ is consistent, both estimators of β may be shown to have the same asymptotic distribution as the GLS estimator based on the true value of ϕ, subject, of course, to the usual conditions on the explanatory variables. However, insofar as the findings for the AR(1) model may be generalised, there is a strong argument for using the full GLS transform in small samples.

A consistent estimator of ϕ may be obtained from the OLS residuals by a regression of the type described in the previous subsection. This is rather convenient, since it makes it very easy to carry out tests of significance on the parameters in ϕ. In fact, it follows directly from the material presented in Section 5.5, that an 'F-test' of the hypothesis $\phi = 0$ represents a straightforward application of the modified LM procedure.

Asymptotic Distribution of Estimators

All the procedures considered in this section yield estimators of both β and ϕ which have identical large sample properties. Hence they may all be regarded as variants of what is sometimes termed autoregressive least square.

The asymptotic covariance matrix of ML estimators of β, ϕ and σ^2 takes exactly the same form as (1.12) but with $(1 - \phi^2)/T$ replaced by the general expression $T^{-1}V_p^{-1}$.

5. Moving Average and Mixed Disturbances

A general specification for the disturbance term in a regression model is the ARMA(p, q) model

$$u_t = \phi_1 u_{t-1} + \cdots + \phi_p u_{t-p} + \epsilon_t + \theta_1 \epsilon_{t-1} + \cdots + \theta_q \epsilon_{t-q} \quad (5.1)$$

An MA model is, of course, a special case of the ARMA model with $p = 0$. However, in terms of their statistical treatment it is convenient to consider ARMA and MA models together. Both give rise to estimation problems which do not occur with pure autoregressive disturbances.

An MA specification is plausible for a number of reasons. ARMA disturbances are perhaps even more important, firstly, because aggregation of a number of variables generally leads to an ARMA process, and, secondly, because any stationary stochastic process can be represented reasonably parsimoniously by an ARMA model.

As in the case of a pure time series model, the likelihood function for a regression model with ARMA disturbances simplifies considerably if certain assumptions are made about the initial values. This permits a straightforward decomposition of the likelihood in terms of prediction errors, and maximising the likelihood function is then equivalent to minimising a conditional sum of squares function. On the other hand, it may be felt that a full ML approach is desirable. A detailed discussion of full ML for ARMA disturbances is beyond the scope of this book, but in TSM a method based on the Kalman Filter is proposed.

Approaching estimation from an ML viewpoint effectively solves the problem of obtaining a GLS estimator. It is shown below that an approximation similar to Cochrane—Orcutt can be based on transformed observations built up from simple recursive formulae.

Approximate Maximum Likelihood Estimation

The principles underlying the CSS approach to estimating a regression model with ARMA disturbances may be developed by considering the case of an MA(1) disturbance term

$$u_t = \epsilon_t + \theta \epsilon_{t-1} \quad (5.2)$$

This embodies all the difficulties which are not encountered in the pure AR case, and the generalisation to other ARMA models is relatively straightforward.

If ϵ_0 is taken to be fixed and equal to zero, maximising the likelihood function is equivalent to minimising the conditional sum of squares. For a given value of θ, the residual is obtained by a recursion having exactly the same form as (3.5.14), i.e.

$$\epsilon_t = (y_t - x_t'\beta) - \theta \epsilon_{t-1}, \qquad t = 1, \ldots, T \quad (5.3)$$

with $\epsilon_0 = 0$. Minimising the sum of squares of these residuals with respect to β and θ may be carried out very easily by Gauss–Newton. The derivatives are computed from the recursions

$$\frac{\partial \epsilon_t}{\partial \beta} = -x_t - \theta \frac{\partial \epsilon_{t-1}}{\partial \beta}, \qquad t = 1, \ldots, T \tag{5.4}$$

with $\partial \epsilon_0 / \partial \beta = 0$, and

$$\frac{\partial \epsilon_t}{\partial \theta} = -\theta \frac{\partial \epsilon_{t-1}}{\partial \theta} - \epsilon_{t-1}, \qquad t = 1, \ldots, T \tag{5.5}$$

with $\partial \epsilon_0 / \partial \theta = 0$. As with autoregressive disturbances, the information matrix in this model is block diagonal, as

$$\text{plim} \frac{1}{T} \sum_{t=1}^{T} \frac{\partial \epsilon_t}{\partial \beta} \cdot \frac{\partial \epsilon_t}{\partial \theta} = 0 \tag{5.6}$$

This suggests a modified optimisation procedure whereby the Gauss–Newton scheme is broken down into two separate regressions. The first consists of regressing ϵ_t on $\partial \epsilon_t / \partial \beta$ while the second, which corresponds directly to the iterative procedure in the pure time series model, is a regression of ϵ_t on $\partial \epsilon_t / \partial \theta$. Although the function to be minimised is unchanged, this modified scheme should be more efficient computationally. It is worth noting that defining a Gauss–Newton scheme for an AR(1) disturbance term and then modifying it in this way, leads directly to Cochrane–Orcutt. However, there is a difference, in that for the MA(1) model a simple regression of ϵ_t on $\partial \epsilon_t / \partial \theta$ does not yield the ML estimator of θ conditional on β. On the other hand regressing ϵ_t on its derivatives with respect to β does yield an approximation to the GLS estimator. This therefore implicitly defines the equivalent of the Cochrane–Orcutt transform for the MA model.

The link between the Gauss–Newton recursion and GLS may be established as follows. Conditional on a value of θ, the sum of squares function, $S = S(\beta)$, is minimised when

$$\frac{\partial S}{\partial \beta} = 2 \sum_{t=1}^{T} \frac{\partial \epsilon_t}{\partial \beta} \cdot \epsilon_t = 0 \tag{5.7}$$

Repeatedly substituting in (5.3) and (5.4) yields

$$\epsilon_t = \sum_{j=0}^{t-1} (-\theta)^j (y_{t-j} - x'_{t-j}\beta) \tag{5.8}$$

and

$$\frac{\partial \epsilon_t}{\partial \beta} = -\sum_{j=0}^{t-1} (-\theta)^j x_{t-j} \tag{5.9}$$

respectively. Substituting from (5.8) and (5.9) into (5.7) gives

$$\sum_{t=1}^{T} x_t^*(y_t^* - x_t^{*\prime}\beta) = 0 \tag{5.10}$$

where

$$y_t^* = \Sigma(-\theta)^j y_{t-j} \quad \text{and} \quad x_t^* = \Sigma(-\theta)^j x_{t-j} = -\partial \epsilon_t/\partial \beta$$

Therefore, given the assumptions made for the initial value of the disturbance term, the GLS estimator of β is obtained by regressing y_t^* on x_t^*. This is identical to the estimate obtained by regressing ϵ_t on $\partial \epsilon_t/\partial \beta$ and subtracting the result from the current estimate of β. Note that the transformed dependent variable, y_t^*, can be obtained from a recursion with a similar form to (5.3), i.e.

$$y_t^* = y_t - \theta y_{t-1}^*, \qquad t = 1, \ldots, T \tag{5.11}$$

with $y_0^* = 0$.

Two-Step Estimators

A consistent estimator of β is given by OLS. A consistent estimator of θ is given by (4.5.4), where r_1 is now the first-order autocorrelation in the OLS residuals. From these starting values, asymptotically efficient estimators of both β and θ are produced by a single iteration of Gauss–Newton.

In view of (5.6), an alternative method for calculating an efficient estimator of β is simply to regress ϵ_t on $\partial \epsilon_t/\partial \beta$. This is equivalent to regressing y_t^* on x_t^* and so it can be regarded as a feasible GLS estimator. It is directly comparable with the two-step Cochrane–Orcutt estimator in the AR(1) case.

6. Tests Against Serial Correlation

The serial correlation tests described in Section 3 were constructed on the assumption that the disturbances followed an AR(1) process. However, in view of (3.3), a test procedure based on a statistic having the form of the von Neumann Ratio will be relatively effective against any disturbance term which exhibits strong first-order autocorrelation, irrespective of the form of the underlying stochastic process. Thus, for example, the probability that the Durbin–Watson test will reject the null hypothesis of white noise disturbances will be relatively high for an AR(2) process,

$$u_t = \phi_1 u_{t-1} + \phi_2 u_{t-2} + \epsilon_t \tag{6.1}$$

unless ϕ_1 is close to zero. In a similar way the test will also be effective against an MA(1) disturbance term. Indeed, in this case, the LM test is identical to the LM test against an AR(1) alternative.

As a general rule, therefore, a 'significant' value of a test statistic such as the Durbin–Watson should not necessarily be taken as an indication of an AR(1) disturbance term. Assuming that the structural part of the equation is correctly specified, the possibility of modelling the disturbance by a more general ARMA process should always be explored.

Although tests against AR(1) disturbances are effective against a wide variety of alternatives, they are nevertheless limited. In particular they may be totally inadequate when dealing with seasonal data, or when strong cyclical effects are present.

Testing for Seasonal Effects

Faced with quarterly observations, a reasonable model for the disturbance term may be

$$u_t = \phi_4 u_{t-4} + \epsilon_t, \qquad |\phi_4| < 1 \tag{6.2}$$

This is an AR(4) model but with ϕ_1, ϕ_2 and ϕ_3 constrained to be zero. Since successive observations in (6.2) are independent, any test based on the first-order autocorrelation of the residuals will obviously be completely inappropriate. As might be expected, the Lagrange multiplier principle suggests a test based on r_4, both for (6.2) and for the corresponding MA(4) seasonal process. Under the null hypothesis $r_4 \sim AN(0, T^{-1})$.

If an exact small sample test is required, a natural statistic to adopt is the generalisation of the Durbin–Watson d-statistic

$$d_4 = \frac{\sum\limits_{t=5}^{T} (e_t - e_{t-4})^2}{\sum\limits_{t=1}^{T} e_t^2} \tag{6.3}$$

Wallis (1972) presents upper and lower significance points for carrying out a bounds test based on d_4. The test procedure is essentially the same as in the first-order case, except that two tables are presented for d_4, depending on whether or not seasonal dummies are included in the regression. When the test is inconclusive, an approximation to the exact significance point may be obtained by a *Beta* or '$a + bd_u$' approximation. Alternatively the exact probability value of d_4 may be obtained at the outset by the method of Imhof or Pan Jie-Jan.

When using quarterly data it will usually be advisable to compute both d_1 and d_4. Statistically significant values of these statistics may then indicate that both first and fourth order autocorrelation effects should be allowed for in the estimation procedure. However, some care should be taken in interpreting what is meant by 'statistically significant' in this context. Suppose the exact significance points are computed for d_1 and d_4 at the 5% level, and the null hypothesis of white noise disturbances is rejected if either of the test statistics is 'significant'. The size of the Type I error of this test procedure will then be greater than 0.05.

Portmanteau Tests

The Q-statistic, (1.5.20) may be used to test the assumption of white noise disturbances. The fact that the autocorrelations are constructed from OLS residuals makes no difference to the large sample distribution of Q, which is asymptotically χ_P^2 under the null hypothesis. The modified statistic,

$$Q^* = T(T + \tau) \sum_{\tau = 1}^{P} (T - \tau)^{-1} r_\tau \qquad (6.4)$$

also has this distribution under the null, but the available evidence suggests that it is a better approximation.

The Q and Q^* statistics form the basis for 'portmanteau' tests since they are designed to pick up departures from randomness indicated by the first P residual autocorrelations. The alternative hypothesis might seem to be rather vague, but in fact the Q-test can be derived as the LM test against an $AR(P)$ or an $MA(P)$ process. However, there remains the problem of how to choose P. Too small a value of P may result in a failure to detect a process where high-order lags are important. On the other hand, too large a value of P may result in a test with a low power since the effect of the dominant autocorrelations is diluted by a series of autocorrelations close to zero.

Neither version of the portmanteau test is exact. Furthermore, the distribution of Q is difficult to compute since it cannot be expressed as a quadratic form in normal variables; cf. (3.4). An alternative approach is to construct a test statistic by generalising the d-statistic. Schmidt has suggested the statistic

$$d_{1,2} = \left[\sum_{t=2}^{T} (e_t - e_{t-1})^2 + \sum_{t=3}^{T} (e_t - e_{t-2})^2 \right] \bigg/ \sum_{t=1}^{T} e_t^2 \qquad (6.5)$$

Since $d_{1,2}$ may be expressed as a quadratic form in normal variables, its small sample distributions can be found by the method of Imhof or Pan Jie-Jan. A bounds test may also be developed, and the appropriate significance points are tabulated in Schmidt (1972). This approach could be generalised further, but it is not clear whether tests carried out on this basis would have desirable properties. The Q-statistic may at least be rationalised as the appropriate statistic for carrying out an LM test against an AR(P) or MA(P) alternative. This provides a sounder basis for constructing a test statistic than the requirement that its distribution be known exactly in small samples.

Tests when the Disturbance Term is Modelled as an ARMA Process

When a regression model has been estimated with a particular ARMA disturbance term, it will usually be desirable to carry out diagnostic checks on the specification. Insofar as the model has been estimated by a straightforward extension of a method used to fit an ARMA time series model, no new issues are raised. Thus the large sample distribution of the CSS residuals is unaffected by the addition of β to the parameter set, and the null hypothesis that the disturbances are serially independent can be tested by treating the Box–Pierce Q-statistic as having a χ^2 distribution with $P - p - q$ degrees of freedom.

7. Prediction

Given future values of the explanatory variables, the dependent variable in a regression model with ARMA disturbances, (5.1), may be predicted recursively. Multiplying both sides of (1.1) by the AR polynomial, $\phi(L)$, yields

$$\begin{aligned}
y_t &= \phi_1 y_{t-1} + \cdots + \phi_p y_{t-p} + (x_t - \phi_1 x_{t-1} - \cdots - \phi_p x_{t-p})' \beta \\
&\quad + \epsilon_t + \theta_1 \epsilon_{t-1} + \cdots + \theta_q \epsilon_{t-q} \\
&= x_t' \beta + \phi_1 (y_{t-1} - x_{t-1}' \beta) + \cdots + \phi_p (y_{t-p} - x_{t-p}' \beta) \\
&\quad + \epsilon_t + \cdots + \theta_q \epsilon_{t-q}
\end{aligned} \qquad (7.1)$$

This equation forms the basis for the recursive calculation of the optimal predictors of future values of y_t. Setting future disturbances equal to their expected value of zero yields:

$$\tilde{y}_{T+l/T} = \phi_1 \tilde{y}_{T+l-1/T} + \cdots + \phi_p \tilde{y}_{T+l-p/T}$$
$$+ (x_{T+l} - \phi_1 x_{T+l-1} - \cdots - \phi_p x_{T+l-p})' \beta$$
$$+ \tilde{\epsilon}_{T+l/T} + \cdots + \theta_q \tilde{\epsilon}_{T+l-q/T}, \qquad l = 1, 2, \ldots \quad (7.2)$$

where

$$\tilde{y}_{T+j/T} = \begin{cases} \tilde{y}_{T+j/T} & \text{for} \quad j > 0 \\ y_{T+j} & \text{for} \quad j \leqslant 0 \end{cases}$$

$$\tilde{\epsilon}_{T+j/T} = \begin{cases} 0 & j > 0 \\ \epsilon_{T+j} & j \leqslant 0; \end{cases}$$

cf. (1.5.22).

Mean Square Error of Predictions

If β, ϕ and θ are known, the MSEs of predictions may be obtained in exactly the same way as for a time series ARMA model; see TSM (Section 6.4). Thus

$$\text{MSE}(\tilde{y}_{T+l/T}) = (1 + \psi_1^2 + \cdots + \psi_{l-1}^2)\sigma^2 \qquad (7.3)$$

where ψ_j is the coefficient of ϵ_{t-j} in the MA representation of the ARMA process. For an AR(1) model it follows from (1.5.5) that

$$\text{MSE}(\tilde{y}_{T+l/T}) = \{1 + \phi^2 + \cdots + \phi^{2(l-1)}\}\sigma^2$$
$$= \sigma^2(1 - \phi^{2l})/(1 - \phi^2) \qquad (7.4)$$

When β is unknown, the predictions in the regression model are subject to an additional source of variation as compared with the pure time series case. This arises from the estimation of β. The effect may be illustrated for AR(1) disturbances. Here, predictions are computed from the recursion

$$y_{T+l/T} = \phi \tilde{y}_{T+l-1/T} + (x_{T+l} - \phi x_{T+l-1})' \tilde{b}, \qquad l = 1, 2, \ldots$$
$$(7.5)$$

where $\tilde{b}$ is the GLS estimator of β and $\tilde{y}_{T/T} = y_T$. Repeated substitution for predicted values of y_t gives

$$\tilde{y}_{T+l/T} = (x_{T+l} - \phi^l x_T)' \tilde{b} + \phi^l y_T \qquad (7.6)$$
$$= x_{T+l}' \tilde{b} + \phi^l \tilde{u}_T \qquad (7.7)$$

where $\tilde{u}_T = y_T - x_T' \tilde{b}$. The last expression shows how the predictor of $\tilde{y}_{T+l}$ is made up of two components: one the BLUE of $E(y_{T+l})$, and the other the BLUP of u_{T+l}.

The true value of y_{T+l} is given by

$$y_{T+l} = x'_{T+l}\beta + u_{T+l}$$

$$= x'_{T+l}\beta + \sum_{j=0}^{l-1} \phi^j \epsilon_{T+l-j} + \phi^l(y_T - x'_T\beta)$$

$$= (x_{T+l} - \phi^l x_T)'\beta + \sum_{j=0}^{l-1} \phi^j \epsilon_{T+l-j} + \phi^l y_T \qquad (7.8)$$

Subtracting (7.8) from (7.5) gives

$$y_{T+l} - \tilde{y}_{T+l/T} = \sum_{j=0}^{l-1} \phi^j \epsilon_{T+l-j} + (x_{T+l} - \phi^l x_T)'(\beta - \tilde{b}) \qquad (7.9)$$

The two components in (7.9) are independent, and so the MSE of the prediction error consists of a contribution from the estimation error, $\beta - \tilde{b}$, plus a contribution from future values of the disturbance term, i.e.

$$\text{MSE}(\tilde{y}_{T+l/T}) = E(y_{T+l} - \tilde{y}_{T+l/T})^2$$

$$= \sigma^2 \frac{1 - \phi^{2l}}{1 - \phi^2}$$

$$+ \sigma^2(x_{T+l} - \phi^l x_T)'(X'V^{-1}X)^{-1}(x_{T+l} - \phi^l x_T)$$

$$\qquad (7.10)$$

Estimation of ARMA Parameters

The fact that the parameters in ϕ and θ will generally need to be estimated raises additional problems in the computation of the MSE, although the actual predictions, $\tilde{y}^*_{T+l/T}$, are obtained exactly as in the case when ϕ and θ are known. Baillie (1979) has considered this problem and has evaluated the asymptotic mean square error (AMSE) of predictions made for regression models with AR disturbances. In the case of AR(1) disturbances

$$\text{AMSE}(\tilde{y}^*_{T+l/T}) \simeq \sigma^2 \frac{1 - \phi^{2l}}{1 - \phi^2}$$

$$+ \sigma^2(x_{T+l} - \phi^l x_T)'(X'V^{-1}X)(x_{T+l} - \phi^l x_T)$$

$$+ \frac{\sigma^2}{T} l^2 \phi^{2(l-1)} \qquad (7.11)$$

terms of $O(T^{-1})$ having been excluded from the formula. The first two terms in (7.11) give the MSE of $\tilde{y}_{T+l/T}$ when ϕ is known while the last term arises because of the estimation of ϕ. Thus the AMSE depends on three things: the unknown disturbances $\epsilon_{T+1}, \ldots, \epsilon_{T+l}$, the error in estimating β, and the error in estimating ϕ. However, it must be stressed that the justification for simply adding up these components is not as simple as might seem at first sight.

The most important question raised by expression (7.11) concerns the likely relative importance of the three components in practice. If T is large the last term will become relatively small. However, the extent to which the first term dominates the second will depend very much on the future values of x_t. If these fall outside the range of the sample x_t's the error arising from the estimation of β could be very large. Furthermore, as l increases the first term in (7.11) tends towards a constant value of $\sigma^2/(1 - \phi^2)$. If the explanatory variables are trending the contribution from the second term could well exceed that of the first term even if T is relatively large.

Some useful evidence on the likely order of magnitude of the three terms in (7.11) is given in the paper by Baillie (1979). He considers an equation estimated by Hendry on the relationship between UK imports and income. The model contained a constant term and quarterly dummy variables in addition to income, and ϕ was estimated as 0.53. The sample size was 46. For a one period ahead forecast ($l = 1$) the proportion of the total AMSE accounted for by each of the three terms in (7.11) was 0.860, 0.121 and 0.019 respectively. For $l = 4$ the corresponding figures were 0.748, 0.248 and 0.004. Thus the contribution from the estimation of ϕ is small, and for $l = 4$ it could quite happily be ignored. The estimation of β, however, becomes increasingly important as the lead time increases.

The OLS predictor is simply $\tilde{y}_{T+l/T} = x'_{T+l}b$. This ignores any serial correlation in the disturbances, both in estimation and prediction, and the MSE is given by

$$\mathrm{MSE}(\tilde{y}_{T+l/T}) = \sigma^2/(1 - \phi^2)$$
$$+ \sigma^2 \cdot x'_{T+l}(X'X)^{-1}X'VX(X'X)^{-1}x_{T+l}$$

However, the relative inefficiency of the OLS predictor decreases as the lead time increases, since the disturbance term becomes more difficult to predict even though the form of the underlying process is known. This point is illustrated in Baillie's results where the MSE for the OLS predictor exceeds the AMSE of the BLUP by 140% for $l = 1$ and by 55% for $l = 4$.

8. Systems of Equation

The SURE system introduced in Section 2.9 may be extended to allow the disturbances to follow a vector ARMA process. Using the notation of (2.9.15), such a model may be written as

$$y_t = X_t\beta + u_t, \qquad t = 1, \dots, T \tag{8.1a}$$

$$u_t = \Phi_1 u_{t-1} + \cdots + \Phi_p u_{t-p} + \epsilon_t + \Theta_1 \epsilon_1 + \cdots \Theta_q \epsilon_{t-q} \tag{8.1b}$$

where $\epsilon_t \sim NID(0, \Omega)$. Since the explanatory variables in all the equations are fixed, estimation and testing may be carried out by a straightforward extension of the methods appropriate for dealing with a multivariate ARMA time series model.

Estimation

Given the specification of the vector ARMA disturbance term, the computation of ML estimates is, in principle, straightforward. By making appropriate assumptions about the initial observations and disturbances, maximisation of the likelihood reduces to minimisation of the multivariate conditional sum of squares function

$$S = \Sigma \epsilon_t' \Omega^{-1} \epsilon_t + T \log |\Omega| \tag{8.2}$$

The residuals are evaluated from the recursion

$$\epsilon_t = (y_t - X_t\beta) - \cdots - \Phi_p(y_t - X_{t-p}\beta)$$
$$- \Theta_1 \epsilon_{t-1} - \cdots - \Theta_q \epsilon_{t-q}, \qquad t = p+1, \dots, T \tag{8.3}$$

and the derivatives of these residuals with respect to $\Phi_1, \dots, \Phi_p$, $\Theta_1, \dots, \Theta_q$ and β may be used to minimise (8.2) by the multivariate Gauss–Newton procedure described in Section 4.4. However, because the explanatory variables are fixed, the ML estimator of β is distributed independently of the ML estimators of the ARMA parameters. The block diagonality of the information matrix can therefore be exploited by breaking each iteration down into two parts, one based on a SURE regression of ϵ_t on $\partial\epsilon_t/\partial\beta'$ and the other based on a SURE regression of ϵ_t on its derivatives with respect to the elements of $\Phi_1, \dots, \Theta_q$. For given values of $\Phi_1, \dots, \Phi_p, \Theta_1, \dots, \Theta_q$, and Ω, the first SURE regression, ϵ_t on $\partial\epsilon_t/\partial\beta'$, yields the GLS estimator of β when subtracted from the initial estimates. An estimator of the asymptotic covariance matrix of $\tilde{\beta}$ is directly available since

$$\text{Avar}(\tilde{\beta}) = \left[\sum_{t=p+1}^{T} \frac{\partial\epsilon_t'}{\partial\beta} \Omega^{-1} \frac{\partial\epsilon_t}{\partial\beta'} \right]^{-1} \tag{8.4}$$

For a vector MA(1) disturbance term, the recursion in (8.3) is

$$\epsilon_t = (y_t - X_t\beta) - \Theta\epsilon_{t-1}, \qquad t = 1, \ldots, T \tag{8.5}$$

with $\epsilon_0 = 0$; cf. (3.5.14). The matrix of derivatives with respect to β is given by a recursion which is direct generalisation of (5.4) i.e.

$$\frac{\partial\epsilon_t}{\partial\beta'} = -X_t - \Theta\frac{\partial\epsilon_{t-1}}{\partial\beta'}, \qquad t = 1, \ldots, T \tag{8.6}$$

with $\partial\epsilon_0/\partial\beta' = 0$. The derivatives with respect to θ are obtained by generalising (4.4.19).

Matters are simplified considerably if the disturbance term is assumed to follow a pure autoregressive process. In the first order case, when $u_t = \Phi u_{t-1} + \epsilon_t$, the recursion (8.3) becomes

$$\epsilon_t = y_t - \Phi y_{t-1} - (X_t - \Phi X_{t-1})\beta \tag{8.7}$$

and so

$$\frac{\partial\epsilon_t}{\partial\beta'} = -(X_t - \Phi X_{t-1}), \qquad t = 2, \ldots, T \tag{8.8}$$

Conditional on Φ, the GLS estimator of β is given directly by a SURE regression of $y_t - \Phi y_{t-1}$ on $X_t - \Phi X_{t-1}$. At the same time, the recursions for estimating Φ are simplified considerably. If $\hat{u}_t$ denotes the tth residual vector at a particular iteration, the ith row in Φ is estimated by an OLS regression of the ith element in $\hat{u}_t$ on the complete vector $\hat{u}_{t-1}$. Following the argument in Examples of Section 3.5 these estimates will be the ML estimates of Φ conditional on β. Thus for an AR(1) disturbance term the multivariate Gauss–Newton scheme is a stepwise optimisation procedure. It could have been derived as a natural generalisation of Cochrane–Orcutt since subtracting Φy_{t-1} from both sides of (8.1) gives

$$y_t - \Phi y_{t-1} = (X_t - \Phi X_{t-1})\beta + \epsilon_t, \qquad t = 2, \ldots, T \tag{8.9}$$

Conditional on Φ and Ω, the GLS estimator of β in (8.9) is given by a SURE regression of $y_t - \Phi y_{t-1}$ on $X_t - \Phi X_{t-1}$.

Asymptotically efficient two-step estimators of β may be constructed in much the same way as for single equation models. The first step is to obtain consistent estimators of the ARMA parameters and Ω from the OLS residuals. These estimators are then used to construct a two-step estimator of β by carrying out a SURE regression of ϵ_t on $\partial\epsilon_t/\partial\beta'$. As in the single equation model, it will usually be more convenient to carry out the second step in a slightly different way. Thus in the MA(1) case, a transformed set of observations on the dependent variables may be obtained from the recursion of the form (8.5), i.e.

$$y_t^\dagger = y_t - \Theta y_{t-1}^\dagger, \qquad t = 1, \ldots, T \tag{8.10}$$

with $y_t^\dagger = 0$. A SURE regression of $y_t^\dagger$ on $-\partial \epsilon_t / \partial \beta'$ then yields the two-step estimator of β directly; there is no need to subtract from the initial OLS estimates.

For an AR(1) disturbance term, the two-step procedure is particularly appealing. The estimator, Φ, obtained from the OLS residuals is not only consistent, but asymptotically efficient. Furthermore, the second step is simply a SURE regression of $y_t - \Phi y_{t-1}$ on $X_t - \Phi X_{t-1}$.

Example   Pagan and Byron (1977) estimate the investment model of Section 2.9, under the assumption that the disturbances are generated by a vector MA(1) process. Both the unrestricted MA(1) model and the seemingly unrelated MA(1) process in which $\Theta_{12} = \Theta_{21} = 0$ are considered. In the latter case they report

$$\hat{y}_1 = 0.041 x_{11} + 0.132 x_{21}$$
$$\hat{y}_2 = 0.048 x_{12} + 0.065 x_{22} \tag{8.11}$$

with

$$\tilde{\Theta} = \begin{bmatrix} 0.638 & 0 \\ 0 & 0.647 \end{bmatrix} \quad \text{and} \quad |\tilde{\Omega}| = 13189.$$

The LR statistic for testing $H_0 : \Theta_{11} = \Theta_{22} = 0$ is

$$LR = 20 \log \frac{25768}{13189} = 13.40$$

This is significant at the 1% level, since under H_0, LR is asymptotically distributed at χ^2 with two degrees of freedom.

Common Regressors

In the multivariate regression model, described in Section 2.10, the explanatory variables in each equation are identical. As a result OLS applied to each equation separately will give fully efficient estimators of the regression parameters when the disturbances are normally distributed and serially independent. However, once serial correlation is introduced into the model, asymptotically efficient estimators of the parameters can no longer be constructed by considering each equation in isolation.

The above point may be illustrated by a relatively simple example.

Suppose that the disturbance term, u_t, follows a vector AR(1) process in which Φ is diagonal. The transformed system, corresponding to (8.9), may be written as

$$y_{it} - \phi_i y_{i,t-1} = (x_t - \phi_i x_{t-1})' \beta_i + \epsilon_i,$$

$$i = 1, \ldots, N, \, t = 2, \ldots, T \quad (8.12)$$

where ϕ_i is the ith diagonal element of Φ. However, unless the value of ϕ_i is the same in all the equations, the regressors in (8.12) will be different. Thus even with Φ known, the estimation of the model must, in general, be carried out within the SURE framework developed in this section. It will often be convenient to cast the model in the form (2.10.5) for this purpose.

Notes

Section 2   Griliches and Rao (1969).
Section 5   Pierce (1971a), Harvey and McAvinchey (1980).
Section 6   Pierce (1971b), Ljung and Box (1978). Durbin (1969) describes a test based on the cumulative periodogram.
Section 7   Box and Jenkins (1976, Ch. 5, pp. 267–269).
Section 8   Monte Carlo evidence on various systems estimators will be found in Kmenta and Gilbert (1970).

Exercises

1. Consider the AR(1) model, (1.2.10), with $|\phi| < 1$ and $\epsilon_t \sim NID(0, \sigma^2)$ and show that the LM principle leads to a test based on the first order sample autocorrelation, r_1. Show that similar reasoning leads to a test based on the corresponding sample autocorrelation for the OLS residuals in a linear regression model.

2. Compare the Gauss–Newton, Newton–Raphson and method of scoring procedures for model (1.1), (1.2). How are these related to the Cochrane–Orcutt iterative procedure?

3. (a) For the single regressor model (2.1) with $x_t = 1$ for $t = 1, \ldots, T$, show that for a given value of ϕ in the range $0 < \phi < 1$,

$$\frac{\text{Var}(\bar{b}^\dagger)}{\text{Var}(b)} = \frac{T(1 - \phi^2) - 2\phi(1 - \phi^T)}{T^2(1 - \phi)^2} \cdot \frac{(T-1)(1-\phi)}{(1+\phi)} \geqslant T/(T-1)$$

where $\bar{b}^\dagger$ is the Cochrane–Orcutt approximation to GLS.

(b) Repeat the above calculations with $x_t = 1$ for t odd and $x_t = 0$ for t even. Comment on the results.

4. For model (2.1) with $x_t = 1$, determine whether the usual estimator of the variance of the OLS estimator, b, will tend to underestimate or overestimate the true variance in large samples.

5. Consider an AR(2) model

$$y_t = \phi_1 y_{t-1} + \phi_2 y_{t-2} + \epsilon_t, \qquad \epsilon_t \sim NID(0, \sigma^2)$$

Explain: (a) how you would test the hypothesis $\phi_1 = \phi_2 = 0$ using LR and Wald . statistics; (b) how you would test the hypothesis $\phi_2 = 0$ using LR, Wald and LM statistics. How would you apply these tests if the AR(2) process were a disturbance term in a regression model?

6. The following is a sequence of residuals from a regression on two exogenous variables and a constant term: +1.4, +1.6, +0.4, −0.9, −0.2, −0.6, +0.1, −1.9, −0.7, +0.2, +1.8, +1.2, +0.4, −0.5, −1.2, −1.1. Test the hypothesis that the disturbances are serially independent against the alternative that they are positively autocorrelated using the Durbin—Watson test. State any assumptions you make and comment on the significance of your findings.

Carry out a large sample test of the same hypothesis based on r_1, the first order autocorrelation in the residuals.

7. Consider the model (1.1), in which the disturbance term follows an ARMA(1, 1) process

$$u_t = \phi u_{t-1} + \epsilon_t + \theta \epsilon_{t-1}$$

(a) Show, that under certain simplifying assumptions, maximising the likelihood function is equivalent to minimising a sum of squares function. Hence show how maximum likelihood estimates may be obtained by the Gauss—Newton procedure. Write down expressions for any derivatives employed in this procedure. Indicate briefly how you would test the hypothesis that $\theta = 0$ within this framework.

(b) Suppose that $x'_{T+1}\tilde{\beta} = 10$ and $x'_{T+2}\tilde{\beta} = 11$, $y_T - x'_T\tilde{\beta}$ and $e_T = 1.0$, where e_T is the maximum likelihood residual corresponding to ϵ_T. Obtain predictions for y_{T+1} and y_{T+2} given that $\phi = \theta = 0.5$.

8. Show that the portmanteau test based on OLS residuals may be derived as an LM test against either an AR(P) or an MA(P) disturbance term.

9. Formulate LR, Wald and LM statistics for testing that the disturbance vector in a SURE model is serially autocorrelated against the alternative that it is generated by a vector AR(1) process. If the AR matrix, Φ, is taken to the diagonal how would this affect the estimation and test procedures?

10. (i) Explain how you would extend the Gauss—Newton algorithm in (4.4.20) if the disturbance term were generated by an AR(1) process, (1.2). (ii) How are the estimators distributed in large samples? (iii) Discuss any modifications you might consider making to the Gauss—Newton algorithm.

7
Dynamic Models I

1. Introduction

quarterly

The regression models considered up to this point have been *static*, in the sense that the dependent variable is taken to be a function of a set of explanatory variables observed at the same point in time. A *dynamic* model, on the other hand, involves non-contemporaneous relationships between variables. A simple example of such a model is therefore

$$y_t = \delta x_{t-3} + u_t \tag{1.1}$$

In this case the effect of a change in x is only felt on y after three time periods have elapsed. However, (1.1) does not really capture the spirit of a dynamic model since once x begins to influence y its full effect is immediately realised. If, on the other hand, several lagged values of x are included in the model, any change in the level of x will work its way through to y in stages. Thus suppose that

$$y_t = 0.4x_{t-1} + 1.0x_{t-2} + 0.6x_{t-3} + u_t \tag{1.2}$$

If x_t is running at a constant level, i.e. $x_t = \bar{x}$, then

$$E(y_t) = 0.4\bar{x} + 1.0\bar{x} + 0.6\bar{x} = 2\bar{x}$$

An increase in the level of x, from $\bar{x}$ to $\bar{x} + 1$; clearly increases the expected value of y by 2 units. However, this is accomplished gradually. There is no immediate change in the expected value of y, but after one and two time periods respectively, 20% and 70% of the adjustment will have taken place. The full effect is realised when three time periods have elapsed.

The model in (1.2) is known as a *distributed lag*. Such a model may be written more generally as

$$y_t = \delta_0 x_t + \delta_1 x_{t-1} + \cdots + \delta_m x_{t-m} + u_t \qquad (1.3)$$

where $\delta_0, \ldots, \delta_m$ are the *lag coefficients*. These determine the way in which y will respond to a change in x. Models of this kind are important in econometrics since there are many situations where it is plausible to assume that the effect of a change in an explanatory variable takes time to work through to the dependent variable. A capital appropriation in a given quarter, for example, may result in capital expenditure in several subsequent quarters, as real investment in buildings and machinery cannot be realised instantaneously. Similarly, there will be time delays of some sort before changes in prices and output are induced by a change in the money supply.

In principle, a finite distributed lag model of the form (1.3) causes no new estimation problems. If $u_t \sim NID(0, \sigma^2)$, OLS will yield efficient estimators of all the parameters, while if u_t is subject to serial correlation or heteroscedasticity it may be handled by the GLS techniques already introduced. However, if the object of the analysis is to obtain reasonably precise estimates of the lag coefficients, such an approach may be unsatisfactory. The reason is that economic time series are typically slowly changing and so the x_t's on the right hand side of (1.3) are likely to be highly correlated with each other. This multicollinearity makes it difficult to determine the lag structure very precisely.

There are a number of ways of tackling this problem. However, they all have the same objective, which is to impose some *a priori* structure on the form of the lag, thereby reducing the number of parameters to be estimated. One approach, originally suggested by Almon (1965), is to approximate the pattern of lag coefficients by a polynomial of a fairly low degree. An alternative approach, which fits more naturally into a time series framework, is to represent the lag structure by the ratio of two polynomials in the lag operator. This is known as a *rational distributed lag*. Approximating the lag structure in this way is motivated by the same principle which leads a stationary stochastic process to be modelled by a mixed ARMA process.

One of the simplest examples of a rational lag structure is the *Koyck* or *geometric* distributed lag. In this model the coefficients are constrained to decline exponentially as the length of the lag increases. Thus, in terms of (1.3), $\delta_j = \beta \alpha^j$, $j = 0, 1, 2, \ldots, m$, where β and α are parameters and $0 < \alpha < 1$. By imposing this constraint, a more parsimonious model is obtained since $m + 1$ lag coefficients are represented by only two parameters. Although the lag length, m, could take any value, it is actually most convenient to set it equal to infinity. This will not, in general, involve any serious distortion in a finite lag structure since the implied lag coefficients will tend to be very small for high values of j. However, from the technical point of view an infinite lag structure is far easier to handle. The model may be written as

$$y_t = \beta \sum_{j=0}^{\infty} \alpha^j x_{t-j} + u_t \tag{1.4}$$

and multiplying both sides of (1.4) at time $t - 1$ by α, subtracting from (1.4) and rearranging gives

$$y_t = \alpha y_{t-1} + \beta x_t + v_t \tag{1.5}$$

see Thiel. Smith Schmidt

where

$$v_t = u_t - \alpha u_{t-1} \tag{1.6}$$

MA(1) with lagged variable $\quad Cov(y_{t-1}, v_t) \neq 0$ *inconsistent*

The simple form of equation (1.5) is very appealing, although the simplicity is, to some extent, illusory. If the disturbance term in (1.4) is white noise, v_t will be serially correlated and OLS will *not* be an appropriate estimation technique. However, provided the problem is approached within a maximum likelihood framework, rather than on an *ad hoc* basis, there is no difficulty in constructing efficient estimators of α and β.

See Mann-Wald - 47-49

By defining the polynomial

$$D(L) = \delta_0 + \delta_1 L + \delta_2 L^2 + \cdots + \delta_m L^m \tag{1.7}$$

model (1.3) may be expressed in the form

$$y_t = D(L)x_t + u_t \tag{1.8}$$

For the Koyck model,

$$
\begin{aligned}
D(L) &= \beta(1 + \alpha L + \alpha^2 L^2 + \alpha^3 L^3 + \cdots) \\
&= \beta\{1 + \alpha L + (\alpha L)^2 + (\alpha L)^3 + \cdots\} \\
&= \beta/(1 - \alpha L) \quad \text{like} \quad AR(1) \quad y_t = \frac{\varepsilon}{1 - \phi L} \tag{1.9}
\end{aligned}
$$

Thus (1.4) is indeed a member of the rational lag family since $D(L)$ is equal to the ratio of two polynomials in the lag operator. (Although the numerator is simply β.) More generally we may define

$$B(L) = \beta_0 + \beta_1 L + \cdots + \beta^s L^s \quad \text{[Koyck with } \beta \text{]} \tag{1.10}$$

$$A(L) = 1 - \alpha_1 L - \alpha_2 L^2 - \cdots - \alpha_r L^r \tag{1.11}$$

A general rational distributed lag structure is then given by setting $D(L) = B(L)/A(L)$. If u_t is assumed to be generated by an ARMA(p, q) process, the resulting model,

rational distributed lag

$$y_t = \frac{B(L)}{A(L)} x_{t-v} + \frac{\theta(L)}{\phi(L)} \epsilon_t \qquad \text{rational distributed} \qquad (1.12)$$

is known as a *transfer function*. The first term on the right hand side of (1.12) encompasses the *systematic dynamics* of the model, while the second term reflects the *disturbance dynamics*.

Augmenting the right hand side of a regression equation with lagged values of the explanatory variables is not the only way of making the model dynamic. An alternative approach is to introduce lagged values of the dependent variable into the regression. A simple example of such a model is

$$y_t = \alpha y_{t-1} + \beta x_t + u_t \qquad (1.13)$$

where α and β are parameters and u_t is a disturbance term. There are a number of reasons why a model may be cast directly in this form. However, the important point to note for the moment is the close similarity between (1.13) and (1.5). This implies that if $|\alpha| < 1$, (1.13) has an implicit rational lag structure. By substituting repeatedly for lagged values of y_t, or by writing the model as

$$(1 - \alpha L)y_t = \beta x_t + u_t \qquad (1.14)$$

and multiplying both sides by $1/(1 - \alpha L)$, it will be seen that (1.13) is equivalent to

$$y_t = \sum_{j=0}^{\infty} \alpha^j x_{t-j} + \sum_{j=0}^{\infty} \alpha^j u_{t-j} \qquad (1.15)$$

The systematic part of (1.15) is therefore a Koyck distributed lag, but the model differs from (1.14) in that the disturbance term depends on the same parameter, α, as does the lag structure itself.

The duality between stochastic difference equations and transfer function models persists in more general cases. Provided a lagged dependent variable model is stable, its mean path i.e. $E(y_t)$, will be a rational distributed lag function of the explanatory variable or variables. Defining $A(L)$ and $B(L)$ as in (1.11) and (1.10) respectively leads to the following general formulation for a single explanatory variable: ARMAX — ARMA with x in y_t

$$A(L)y_t = B(L)x_t + u_t \qquad (1.16)$$

The model is *stable* if the roots of $A(L)$ lie outside the unit circle. Provided that this condition is met, the lag structure is given by $B(L)/A(L)$, and so the systematic dynamics may be deduced from the corresponding results derived for (1.12). As with the transfer function model, u_t may be modelled by an ARMA(p, q) process.

$$y_t = \frac{B(L)}{A(L)}$$

Estimation of Dynamic Regression Models

If the disturbance term in a stochastic difference equation is a white noise process, the Mann–Wald result of Section 2.3 is applicable. The parameters in $A(L)$ and $B(L)$ of (1.16) will be estimated consistently by OLS and if u_t is normally distributed these estimators will be asymptotically efficient. However, if u_t is serially correlated, OLS may not even be consistent. Thus consider (1.13) and suppose that $E(u_t u_{t-1}) \neq 0$. Let $z_t = (y_{t-1} x_t)'$. The OLS estimator of α and β may then be written

$$\begin{bmatrix} a \\ b \end{bmatrix} = \begin{bmatrix} \alpha \\ \beta \end{bmatrix} + \left[\frac{\Sigma z_t z_t'}{T} \right]^{-1} \frac{\Sigma z_t u_t}{T} \tag{1.17}$$

If plim $T^{-1} \Sigma x_t^2$ is a positive number, $T^{-1} \Sigma z_t z_t'$ converges to a p.d. matrix. The estimators a and b will therefore be consistent if plim $T^{-1} \Sigma z_t u_t = 0$. Since x is exogenous plim $T^{-1} \Sigma x_t u_t = 0$ by definition, but because u_t and u_{t-1} are correlated

$$\text{plim } T^{-1} \Sigma y_{t-1} u_t \neq 0 \tag{1.18}$$

Expression (1.18) captures the basic difficulty with this model, namely that one of the regressors is correlated with the disturbance term, and from (1.17) it will be seen that this is sufficient to render the OLS estimators of both α and β inconsistent.

The Koyck distributed lag model, (1.4), may be transformed to (1.5), which is very similar to (1.13). However if the disturbance term in (1.4) is white noise, the disturbance term in the transformed model is generated by an MA(1) process. Thus regressing y_t on y_{t-1} and x_t will *not* yield consistent estimators of the parameters in the model.

In a classical regression model the OLS estimator remains unbiased (and consistent) even though the disturbances are serially correlated. Once a lagged dependent variable is introduced into a model, either directly as in (1.13) or by a transformation as in (1.5), serial correlation has more serious consequences. Furthermore, attempting to overcome these problems by constructing a feasible GLS estimator will not, in general, be successful. Fortunately these problems are not difficult to overcome provided that estimation is approached within a maximum likelihood framework rather than on an *ad hoc* basis. This point is worth stressing since in econometrics there has, in the past, been a tendency to approach estimation from a least squares point of view. This is not usually a successful strategy in dynamic models insofar as it rarely leads to fully efficient estimators. In addition the argument that such procedures have the compensating advantage of simplicity is not generally valid since two-step maximum likelihood estimators are usually very easy to compute while still being fully efficient.

Partial Adjustment and Adaptive Expectations

Although the systematic part of the stochastic difference equation, (1.13), is a Koyck distributed lag, the structure of the disturbance term in (1.15) places the model in a different class from (1.4). The decision as to which formulation is appropriate in a given situation will depend on a number of factors. These include pragmatic considerations such as ease of estimation and goodness of fit. More fundamentally, however, both formulations may be rationalised in terms of a behavioural theory. The transfer function specification, (1.4), can be derived from a theory of adaptive expectations, while the stochastic difference equation (1.13) arises as a consequence of a process of partial adjustment.

The adaptive expectations hypothesis is best described in a specific context. Suppose we are dealing with the estimation of an agricultural supply function in which quantity, y, is related to price, x. A farmer must make his decisions regarding how much of a given crop to plant before he knows the price the crop will actually fetch on the market. Thus he will plan production on the basis of *expected* price in the next time period. It may therefore be reasonable to write

$$y_t = \beta x_{t+1}^* + u_t \tag{1.19}$$

where u_t is a disturbance term which represents factors such as the weather, as well as individual differences between farmers in the way they react to a particular expected price.

Since x_{t+1}^* is not directly observable, a reasonable hypothesis concerning the manner in which expectations are generated must be formulated. One possibility is to put

$$x_{t+1}^* - x_t^* = \gamma(x_t - x_t^*), \qquad 0 < \gamma < 1 \tag{1.20}$$

This is the *adaptive expectations* hypothesis, and it states that decision makers revise their expectations on the basis of the discrepancy between their prediction of the current value of x_t and the actual outcome. The parameter γ measures the extent to which they react to this discrepancy.

Rearranging (1.20) and rewriting it using the lag operator gives

$$(1 - \alpha L)x_{t+1}^* = (1 - \alpha)x_t \tag{1.21}$$

where $\alpha = 1 - \gamma$. Substituting for x_{t+1}^* in (1.19) then yields

$$y_t = \beta(1-\alpha)\,\frac{x_t}{1-\alpha L} = \beta(1-\alpha)\sum_{j=0}^{\infty}\alpha^j x_{t-j} + u_t$$

which is of the same form as (1.4).

The *partial adjustment* mechanism operates in a slightly different way. Suppose that y_t^*, the *desired* level of y at time t, is related to x by the equation

$$y_t^* = \beta x_t \tag{1.22}$$

In an economic context x_t might be income and y_t^* desired expenditure. Alternatively x_t may be the price of a good, while y_t^* is the quantity the firm wishes to produce at that price. In both cases, however, there are plausible reasons why the adjustment to the desired level cannot take place immediately. In the first example, the consumer is likely to take time to adjust to a new set of spending habits associated with his new income, while in the second example the firm may not be able to switch production very easily from one good to another, or undertake the investment required to increase production. Because of these rigidities only a partial adjustment is possible and this may be modelled by a *reaction function* of the form

$$y_t - y_{t-1} = \gamma(y_t^* - y_{t-1}) + u_t \tag{1.23}$$

where $0 < \gamma \leqslant 1$. Thus (1.23) asserts that in the current period the consumer or firm only moves part of the way towards the optimal position, the speed of adjustment being determined by the parameter γ.

Combining (1.22) and (1.23) gives

$$y_t = (1-\gamma)y_{t-1} + \beta\gamma x_t + u_t \tag{1.24}$$

which is a stochastic difference equation of the form (1.13).

Model Selection

The choice between a transfer function model and a stochastic difference equation cannot always be resolved on the basis of *a priori* theory in the way suggested above. In a situation where very little information regarding a suitable specification is available, both a transfer function, (1.12), and a stochastic difference equation, (1.16), represent valid ways of approaching dynamic modelling. There is, however, some divergence in the literature regarding the appropriate way to proceed. The strategy adopted by Box and Jenkins (1976) is based on the transfer function formulation, and a suitable dynamic specification is reached by using techniques similar to those employed in the identification of an ARMA time series model. Econometricians, on the other hand, have tended to favour stochastic difference equations, at least in macroeconomics.

At one time the predilection of econometricians for stochastic difference equations could be attributed to pragmatic motives. If the disturbance term in (1.16) is a normally distributed white noise process, OLS is the appropriate estimation technique. However, while ease of estimation might have been a sensible criterion for model selection ten or fifteen years ago, this is no longer the case now. If a particular specification is felt to be appropriate, there seems to be general agreement that the correct course of action is to attempt to estimate it by maximum likelihood.

The fundamental reasons for the differences in the literature concern specification. These differences are a reflection of different attitudes to model building, which stem mainly from the type of problems under consideration. In the examples considered by Box and Jenkins (1976, Chapter 11), the explanatory variable is an input to a physical system. The observations are generated by a controlled experiment, and so the question of whether other explanatory variables should be included in the model does not arise. In econometrics the problem is to estimate a behavioural relationship from nonexperimental data. Thus specification takes on two dimensions. As well as determining the dynamics of the model, there is also the more fundamental question of which variables should be included in the model in the first place. This favours an approach in which the first stage in model selection is to estimate a very general model. If such an approach is adopted, there are important technical reasons for preferring to work within the framework of a stochastic difference equation. These points are discussed in detail in Section 5 of the next chapter.

There are two other reasons why stochastic difference equations have been favoured in econometrics. The first is that generalisation to systems of equations, and in particular to systems of simultaneous equations, is somewhat easier. Secondly, economic theory sometimes suggests a priori constraints on the model specification because of long-run solutions. These constraints are generally easier to handle within a stochastic difference equation framework.

The next section describes ways in which the systematic dynamics of a model may be characterised. This material is common to both transfer functions and stochastic difference equations. The same could be said of the material on seasonality in Section 6. However, the bulk of the chapter is concerned with the estimation and testing of transfer function models. The last section is devoted to polynomial, or Almon, distributed lags. This represents an alternative

approach to approximating the lag structure in a model of the form
(1.8). Hence it fits more naturally into this chapter than the next,
which is primarily concerned with stochastic difference equations.
However, from the point of view of specification, the polynomial
distributed lag possesses some of the advantages of the stochastic
difference equation, insofar as it is relatively easy to move from a
general specification to a more specific one.

2. Systematic Dynamics *(same for transfer + stochastic diffce)*

The systematic dynamics of a model concern the behaviour of the
mean path of y_t. This is defined as the expected value of y_t, condi- *Forget*
tional on the current and past values of the exogenous variables. For *stoch random*
both the transfer function (1.12), and the stochastic difference *parts*
equation, (1.16), the mean path is given by

$$\bar{y}_t = E(y_t) = A^{-1}(L)B(L)x_t \tag{2.1}$$

Hence the systematic dynamics for the two classes of model can be
considered together.

The nature of the expected response of y to changes in x depends
on the pattern of lag coefficients implied by the polynomial ratio
$B(L)/A(L)$. Thus the first step in an examination of the systematic
dynamics of the model is the derivation of the lag coefficients. Once
this has been done, the properties of the lag may be characterised by
summary statistics, such as the mean lag, and by various multipliers.

Throughout the discussion it will be assumed that the model is
stable. This will be the case if the roots of the polynomial $A(L)$ lie
outside the unit circle. An *equilibrium* solution is then possible since
if x_t is constant at $\bar{x}$, the mean path of y_t is also constant at $\bar{y}$,
where

$$\bar{y} = \frac{\beta_0 + \beta_1 + \cdots + \beta_s}{1 - \alpha_1 - \cdots - \alpha_r}\bar{x} = \frac{B(1)}{A(1)}\bar{x} \tag{2.2}$$

Note that setting $L = 1$ in the lag polynomials $A(L)$ and $B(L)$ has the
effect of reducing them to the sums of their respective coefficients.

Derivation of the Lag Coefficients

In the geometric lag model, the relationship between the ratio
$\beta/(1 - \alpha L)$ and the lag coefficients emerges as a consequence of the
result on summing a geometric series. This approach may be extended
to cases where $r > 1$. Suppose that

$$\bar{y}_t = \frac{2.0}{1 - 0.8L + 0.15L^2} x_t$$

The roots of $A(L)$ are obtained by factorising it as

$$A(L) = (1 - 0.5L)(1 - 0.3L)$$

Both roots are real and distinct, and so expanding the ratio $B(L)/A(L)$ in partial fractions yields

$$\frac{B(L)}{A(L)} = 2\left\{\frac{0.5}{0.2}\frac{1}{1 - 0.5L} - \frac{0.3}{0.2}\frac{1}{1 - 0.3L}\right\}$$

Each of the terms in this expression may be treated as the sum of an infinite number of terms in a geometric progression. Therefore

$$D(L) = 2\left(2.5 \sum_{j=0}^{\infty} 0.5^j L^j - 1.5 \sum_{j=0}^{\infty} 0.3^j L^j\right)$$

$$= 2 \sum_{j=0}^{\infty} \{2.5(0.5^j) - 1.5(0.3)^j\}L^j$$

and so $\delta_0 = 2$, $\delta_1 = 1.6$, $\delta_2 = 0.98$ and so on. The same technique may be applied even if the numerator polynomial, $B(L)$, is of a higher order.

An alternative method of obtaining the lag coefficients is based on equating coefficients of powers of L in the expression

$$B(L) = A(L)D(L)$$

This gives

$$\delta_j = \sum_{i=1}^{\min(j,r)} \alpha_i \delta_{j-i} + \beta_j, \qquad 1 \leqslant j \leqslant s \qquad (2.3a)$$

$$\delta_j = \sum_{i=1}^{\min(j,r)} \alpha_i \delta_{j-i}, \qquad j > s \qquad (2.3b)$$

Note that in all cases $\delta_0 = \beta_0$.

In the numerical example above, the lag coefficients emerge as follows:

$$\delta_0 = 2.0$$

$$\delta_1 = 0.8(2.0) = 1.6$$

$$\delta_2 = 0.8(1.6) - 0.15(2.0) = 0.98$$

and so on. Thus the weights are determined by a second-order difference equation from $j = 2$ onwards. In this example the roots of $A(L)$ are real. However with complex roots, the possibility of the lag coefficients following a damped cyclical pattern arises. This implies that certain of the δ_j's will be negative, a situation which may, in some cases, be ruled out by prior considerations.

Total, Interim and Impact Multipliers

The concept of a multiplier is defined within the context of an equilibrium solution to the model. Suppose that x_t is constant at a level $\bar{x}$, so that the equilibrium solution, $\bar{y}$, is given by (2.2). If this level increases by unity, i.e. from $\bar{x}$ to $\bar{x} + 1$, the new equilibrium solution will be

$$\bar{y}^* = \frac{B(1)}{A(1)}(\bar{x} + 1) = \bar{y} + \frac{B(1)}{A(1)}$$

This reflects a rise in the level of the mean path from $\bar{y}$ to $\bar{y} + B(1)/A(1)$. The change is known as the _total multiplier_, and since

$$B(1)/A(1) - D(1) = \sum_{j=0}^{\infty} \delta_j \tag{2.4}$$

it may be interpreted as the sum of the lag coefficients.

The new equilibrium solution, $\bar{y}^*$, represents the long-run effect of a unit increase in $\bar{x}$. At the other extreme is the immediate effect on $\bar{y}_t$. The increase in $\bar{y}_t$ induced by a unit increase in $\bar{x}$ is known as the _impact multiplier_, and this is simply equal to δ_0.

The effect of a unit change in $\bar{x}$ after J time periods is measured by the Jth _interim multiplier_,

$$\delta_J^* = \sum_{j=0}^{J} \delta_j, \qquad J = 0, 1, 2, \ldots \tag{2.5}$$

It is sometimes convenient to standardise the interim multipliers by dividing through by the total multiplier. The resulting quantities,

$$\delta_J^{\dagger} = \delta_J^*/\delta_{\infty}^*, \qquad J = 0, 1, 2, \ldots \tag{2.6}$$

then give the proportion of the total change completed after J time periods.

Example 1   The pattern of interim multipliers for the Koyck lag, (1.4), with $\alpha = 0.5$ is shown implicitly in figure 7.1. The diagram traces out the mean path of y_t resulting from a unit change in x_t at

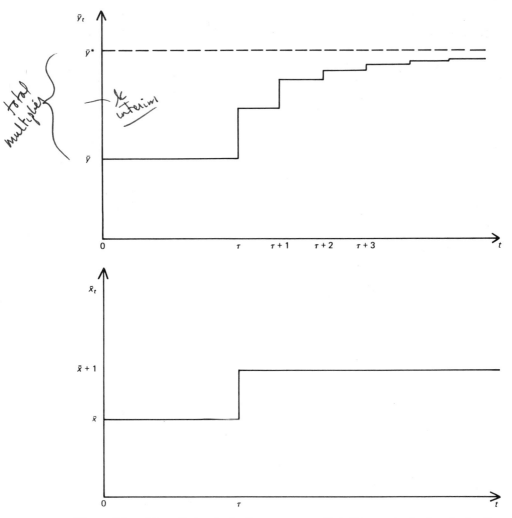

Figure 7.1   *Mean Path of y in Response to a Unit Change in the Level of x for a Koyck Distributed Lag*

the time $t = \tau$. Since the lag is infinite, the new equilibrium level of y is only approached asymptotically.

Example 2   Consider a model in which the systematic part is

$$\bar{y}_t = 0.6\,\bar{y}_{t-1} - 0.4\,\bar{y}_{t-2} + 1.2x_t$$

The roots of the polynomial $A(L) = 1 - 0.6L + 0.4L^2$ are complex with the result that the pattern of lag coefficients shows a damped cyclical movement. This is illustrated in figure 7.2(a). Figure 7.2(b)

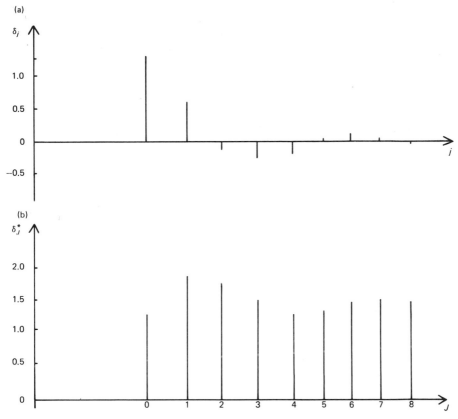

Figure 7.2   *Lag Coefficients and Interim Multipliers for Example 2*

shows the interim multipliers. The initial response in $\bar{y}_t$ is to over-shoot the new equilibrium level. Thereafter a gradual adjustment takes place by a series of cyclical movements.

Mean and Median Lags

When all the lag coefficients are non-negative, the *mean or average-lag* is defined as

$$\text{Mean lag} = \sum_{j=0}^{\infty} j\delta_j \bigg/ \sum_{j=0}^{\infty} \delta_j \tag{2.7}$$

Higher order moments may also be defined although these are not often employed. However, another summary statistic which does often prove useful is the *median lag*. This gives the number of time periods it takes for 50% of the total adjustment to be completed.

Example 3 In the finite distributed lag model

$$\bar{y}_t = 0.2x_{t-2} + 0.4x_{t-3} + 0.1x_{t-4} \tag{2.8}$$

the impact multiplier is 0, the total multiplier 0.7, and

$$\text{mean lag} = \frac{0.2 \times 2 + 0.4 \times 3 + 0.1 \times 4}{0.7} = 2.86$$

The standardised interim multipliers at $J = 2$ and $J = 3$ are 0.29 and 0.86 respectively. The median lag is therefore given by[1]

$$\text{median lag} = 2.0 + \frac{0.50 - 0.29}{0.86 - 0.29} = 2.37$$

Example 4 Consider the Koyck distributed lag model, (1.4). The total multiplier is $\beta/(1 - \alpha)$ while the standardised interim multipliers are

$$\delta_J^\dagger = 1 - \alpha^{J+1}, \qquad J = 0, 1, \ldots \tag{2.9}$$

The median lag is obtained by finding a value of J such that $\delta_J^\dagger = 0.5$. Solving the equation

$$1 - \alpha^{J+1} = 0.5$$

gives the following expression:

$$\text{median lag} = (\log \alpha)^{-1} \log 0.5 - 1 \tag{2.10}$$

This will be negative if $\alpha < 0.5$, in which case it may be sensible to round it to zero.

The mean lag may be obtained as a special case of a general formula which is applicable to all rational lag distributions. Differentiating the lag polynomial, $D(L)$, with respect to L gives

$$D'(L) = \delta_1 + 2\delta_2 L + 3\delta_3 L^2 + \cdots + j\delta_j L^{j-1} + \cdots$$

The mean lag, (2.7), is then given by $D'(1)/D(1)$. However,

$$D'(L) = \frac{A(L)B'(L) - A'(L)B(L)}{A^2(L)}$$

and so

$$\text{mean lag} = \frac{B'(1)}{B(1)} - \frac{A'(1)}{A(1)} \tag{2.11}$$

Expression (2.11) is readily evaluated. In the Koyck case

[1]There is a case for adding 0.5 to the median lag. This would ensure that the mean and median lags were identical for a symmetric pattern of lag coefficients.

$A'(1) = -\alpha$ while $B'(1) = 0$ and so mean lag $= \alpha/(1 - \alpha)$. The values typically taken by the mean and median lags in the Koyck model provide a useful contrast. The long tail on the distribution means that the mean lag will sometimes exceed the median lag by a fairly large amount. Thus for $\alpha = 0.9$, the mean lag is 9.00 while the median lag is only 5.58.

In many applications the variables in the distributed lag equation are in logarithms, i.e.

$$\log y_t = \sum_{j=0}^{\infty} \delta_j \log x_{t-j} + u_t \tag{2.12}$$

A 1% increase in the level of x will mean that after J periods the percentage increase in the mean path of y will be

$$100 \sum_{j=0}^{J} \delta_j \log 1.01 \simeq \sum_{j=0}^{J} \delta_j$$

since $\log 1.01 \simeq 0.00995 \simeq 0.01$. Thus the interim and total multipliers in a double log model may be interpreted as giving percentage increases in the level of y when x changes by one per cent.

3. Estimation of Transfer Function Models with Independent Disturbances

This section examines the way in which ML estimation of a transfer function model may be carried out when the disturbances are normally distributed white noise. Attention is focused on the geometric distributed lag

$$y_t = \beta \sum_{j=0}^{\infty} \alpha^j x_{t-j} + \epsilon_t, \qquad t = 1, \ldots, T \tag{3.1}$$

with $\epsilon_t \sim NID(0, \sigma^2)$. Although this is a very simple model, it embodies all the important features present in more general formulations.

The first sub-section considers an instrumental variable estimator of the parameters in the geometric distributed lag. Although it is not efficient, the IV estimator is consistent and easy to compute. It therefore provides a suitable initial estimate for a two-step or iterative ML procedure. The ML estimator is then developed and its properties examined. The section closes by showing how the results may be extended to more general models, including those with several explanatory variables.

IV Estimation

[handwritten: $y_t = \alpha\, y_{t-1} + \beta x_t + v_t$ $v_t = u_t - \alpha u_{t-1}$ Koyck]

The transformed equation for the geometric distributed lag model is (1.5). However, as indicated in Section 1, the OLS estimators of both α and β will be inconsistent, because the disturbance term, v_t, is correlated with y_{t-1}. This may be shown formally by noting that

$$\text{plim } T^{-1}\Sigma\, y_{t-1}v_t = \text{plim } T^{-1}\Sigma\, y_{t-1}(\epsilon_t - \alpha\epsilon_{t-1})$$

$$= 0 - \alpha\, \text{plim } T^{-1}\Sigma\,(\alpha y_{t-2} + \beta x_{t-1} + \epsilon_{t-1})\epsilon_{t-1}$$

$$= -\alpha\sigma^2 \tag{3.2}$$

[handwritten: ✳] The parameters may be estimated consistently by constructing an IV estimator in which x_{t-1} is used as an instrument for y_{t-1}. In terms of the notation of Section 2.11, $z_t = (y_{t-1}\, x_t)'$ and $w_t = (x_{t-1}\, x_t)'$ and so the estimator may be written as

$$\begin{bmatrix} \tilde{a} \\ \tilde{b} \end{bmatrix} = (\Sigma w_t z_t')^{-1}\Sigma w_t y_t$$

$$= \begin{bmatrix} \Sigma x_{t-1}y_{t-1} & \Sigma x_{t-1}x_t \\ \Sigma x_t y_{t-1} & \Sigma x_t^2 \end{bmatrix}^{-1} \begin{bmatrix} \Sigma x_{t-1}y_t \\ \Sigma x_t y_t \end{bmatrix} \tag{3.3}$$

[handwritten: MSE - lousy procedure] Since x is exogenous, x_{t-1} is, by definition, uncorrelated with v_t since it is independent of both ϵ_t and ϵ_{t-1}. Furthermore, it is not necessary to place particularly restrictive conditions on x in order to ensure that plim $T^{-1}\Sigma w_t z_t'$ is positive definite, and so the conditions necessary for consistency are easily satisfied.

The structure of the model ensures that x_{t-1} will show a reasonably high degree of correlation with y_{t-1}. The actual correlation depends on a number of factors including the value of α and the nature of the process generating x_t. However, the IV estimator will not, in general, be asymptotically efficient. Furthermore, even though it is consistent, it is quite conceivable that its small sample performance will be poorer than that of OLS. In fact the Monte Carlo results reported in the Appendix to Dhrymes (1971) indicate that in small samples OLS is 'a remarkably good estimator' when α is large, say around 0.9, and the x_t sequence exhibits strong positive autocorrelation. This point should be borne in mind in constructing initial estimates for an ML procedure.

Maximum Likelihood Estimation

Maximising the likelihood function of (3.1) is equivalent to minimising the sum of squares function

$$S(\alpha, \beta) = \sum_{t=1}^{T} \left(y_t - \beta \sum_{j=0}^{\infty} \alpha^j x_{t-j} \right)^2 \tag{3.4}$$

However, this poses an immediate problem, since the right hand side of (3.4) contains an infinite number of explanatory variables. Furthermore, $x_0, x_{-1}, x_{-2}, \ldots$, are, in any case, assumed to be unobservable.

One approach to the problem is to decompose the infinite lag on x into two parts, one observable and the other unknown. Equation (3.1) then becomes

$$y_t = \beta \sum_{j=0}^{t-1} \alpha^j x_{t-j} + \alpha^t \left\{ \beta \sum_{j=0}^{\infty} \alpha^j x_{-j} \right\} + \epsilon_t$$

$$= \beta x_t^* + \alpha^t \xi + \epsilon_t, \qquad t = 1, \ldots, T \tag{3.5}$$

The term $\xi = E(y_0)$ represents the unobservable component in the model. There are two ways of dealing with it, the first of which is simply to set $\xi = 0$. The estimation problem then reduces to an exercise in nonlinear least squares, minimising

$$S(\alpha, \beta) = \sum_{t=1}^{T} (y_t - \beta x_t^*)^2 \tag{3.6}$$

ok with stationary X's + large sample

with respect to α and β. There are a number of ways of going about this, but since (3.6) is linear in β for a given value of α, carrying out a direct search over the interval $0 \leqslant \alpha \leqslant 1$ is one possibility. Note that x_t^* may be built up recursively from the relationship

$$x_t^* = \alpha x_{t-1}^* + x_t, \qquad t = 1, \ldots, T, \tag{3.7}$$

with $x_0^* = 0$.

The second way of dealing with ξ is to treat it as an unknown parameter which is to be estimated along with α and β. As before, the likelihood function may be concentrated with respect to α and a grid search carried out. In other words, for a given value of α, y_t is regressed on x_t^* and α^t to yield a residual sum of squares which forms the criterion function to be minimised.

Both methods of handling the unobservable component in (3.5) have disadvantages. Although the second solution makes full use of all the available information, adding extra parameters to a model is not usually regarded as desirable. On the other hand setting $\xi = 0$ may cause considerable distortion, particularly if x_t is non-stationary.

A third solution, which is, to some extent, a compromise between the first two, is to write the model as

$$y_t = \beta \sum_{j=0}^{t-2} \alpha^j x_{t-j} + \alpha^{t-1} \left(\beta \sum_{j=0}^{\infty} \alpha^j x_{1-j} \right) + \epsilon_t$$

$$= \beta x_t^\dagger + \alpha^{t-1} \xi_1 + \epsilon_t, \qquad t = 2, \ldots, T \qquad (3.8)$$

Apart from x_1, all the lagged values of x_t in ξ_1 are unobservable. However, because $\xi_1 = E(y_1)$, it seems reasonable to replace it by y_1. The parameters in the model may then be estimated by nonlinear least squares in much the same way as before. If ϵ_1 were fixed and equal to zero, this procedure would be identical to full ML.

The construction of the likelihood function may be approached from a completely different direction if the transformed model, (1.5), is taken as the starting point. A set of prediction errors may be computed from the recursion,

$$\epsilon_t = y_t - \alpha y_{t-1} - \beta x_t + \alpha \epsilon_{t-1}, \qquad t = 2, \ldots, T \qquad (3.9)$$

Taking y_1 to be fixed provides a theoretical justification for setting $\epsilon_1 = 0$ and, from the prediction error decomposition of Section 3.5, maximising the likelihood function is equivalent to minimising the sum of squares function,

$$S(\alpha, \beta) = \sum_{t=2}^{T} \epsilon_t^2 \qquad (3.10)$$

The recursion in (3.9) is similar to the recursion used to construct the residual function in an ARMA time series model. It was argued in Section 3.5 that setting pre-sample values of y_t equal to zero is not a good way of approximating the likelihood function. This consideration is even more important in the present context because of the presence of the explanatory variable. Thus only $T - 1$ residuals are computed in (3.9), thereby avoiding the need to set any pre-sample values equal to zero. However, if we were prepared to set y_0 (and ϵ_0) equal to zero a full set of T residuals could be obtained. On substituting repeatedly for lagged values of ϵ_t, it can be seen that the T residuals obtained in this way are identical to the T residuals obtained from (3.5) with $\xi = 0$. Similarly, replacing ξ_1 by y_1 in (3.8) leads to a set of $T - 1$ residuals which are identical to the residuals obtained from (3.9). This correspondence is not surprising, since the assumption that ϵ_1 is fixed and equal to zero is implicit in both (3.9) and (3.8).

Unless full ML estimation is contemplated, therefore, the most satisfactory approach is by (3.8) or (3.9). Although these will yield identical sum of squares functions, the recursive equation (3.9) is more convenient since it implicitly includes the recursion for $x_t^\dagger$; cf. (3.7). Furthermore the derivatives of ϵ_t may also be computed recursively. Differentiating (3.9) with respect to α and β yields

$$\frac{\partial \epsilon_t}{\partial \alpha} = -y_{t-1} + \epsilon_{t-1} + \alpha \frac{\partial \epsilon_{t-1}}{\partial \alpha}, \qquad t = 2, \ldots, T \qquad (3.11)$$

and

$$\frac{\partial \epsilon_t}{\partial \beta} = -x_t + \alpha \frac{\partial \epsilon_{t-1}}{\partial \beta}, \qquad t = 2, \ldots, T \qquad (3.12)$$

Both derivatives are equal to zero for $t = 1$ since $\epsilon_1 = 0$. The ease with which analytic derivatives may be obtained suggests that they be exploited in a Gauss–Newton optimisation procedure.

Asymptotic Properties of the ML Estimator

The three ML estimators discussed above differ in the assumptions made about starting values. However, these differences are unimportant in large samples, and all three procedures yield estimators with the same asymptotic distribution. On examining (3.8) it will be apparent that we are essentially dealing with a classical regression model, since $x_t^\dagger$ is fixed and the disturbances are white noise. The asymptotic efficiency of the ML estimator therefore follows immediately from standard results.

One minor problem concerns the regularity conditions which must be imposed on x_t in order that the ML estimators of α and β have a well defined asymptotic distribution. To see what this entails, consider the definition of the asymptotic information matrix in (3.5.18). On examining (3.11) and (3.12), it will be seen that this involves the evaluation of terms containing y_t and ϵ_t as well as x_t. However, (3.12) may be written as

$$(1 - \alpha L) \frac{\partial \epsilon_t}{\partial \beta} = -x_t, \qquad t = 2, \ldots, T \qquad (3.13)$$

and relaxing the assumption that y_1 is fixed, yields

$$z_{2t} = -\frac{\partial \epsilon_t}{\partial \beta} = \frac{x_t}{1 - \alpha L} = \sum_{j=0}^{\infty} \alpha^j x_{t-j} \qquad (3.14)$$

Similarly,

$$z_{1t} = -\frac{\partial \epsilon_t}{\partial \alpha} = \frac{-(y_{t-1} - \epsilon_{t-1})}{1 - \alpha L} = \frac{-\beta \sum\limits_{j=0}^{\infty} \alpha^j x_{t-j-1}}{1 - \alpha L}$$

$$= -\beta \sum_{j=0}^{\infty} j\alpha^j x_{t-j-1} \tag{3.15}$$

These formulae make it possible to write down an expression for the asymptotic information matrix which involves only x_t, α and β. Any conditions which must be imposed on x_t may therefore be obtained explicitly.

An Efficient Two-Step Estimator

An asymptotically efficient two-step estimator of α and β may be based on a single iteration of the Gauss–Newton scheme. The first step is to estimate α and β consistently, say by the method of instrumental variables. These estimates are then used to compute the residual function, ϵ_t, and its derivatives, using (3.9), (3.11) and (3.12) respectively. Regressing ϵ_t on its derivatives and subtracting the resulting coefficients from the initial estimates then yields the two-step estimator.

Although consistent initial estimates of α and β are necessary for the two-step estimator to be asymptotically efficient, there is a case for using OLS rather than instrumental variables at the first stage. This arises because OLS appears to have rather better small sample properties than the IV estimator, even though it is inconsistent. However, it must be stressed that if OLS were used as the basis for a two-step procedure, the resulting estimator would not, in general, be asymptotically efficient. A thorough Monte Carlo study is therefore needed before such an estimator can be recommended.

Estimation of a General Rational Lag Structure

The principles of estimation established for the geometric lag extend quite naturally to the more general model

$$y_t = \frac{B(L)}{A(L)} x_t + \epsilon_t, \qquad t = 1, \ldots, T \tag{3.16}$$

with $\epsilon_t \sim NID(0, \sigma^2)$. Multiplying through by $A(L)$ gives

$$A(L)y_t = B(L)x_t + A(L)\epsilon_t \tag{3.17}$$

and so the prediction errors may be evaluated from the recursion,

$$\epsilon_t = y_t - \alpha_1 y_{t-1} - \cdots - \alpha_r y_{t-r} - \beta_0 x_t - \cdots - \beta_r x_{t-r}$$

$$+ \alpha_1 \epsilon_{t-1} + \cdots + \alpha_r \epsilon_{t-r}, \qquad t = t^* + 1, \ldots, T \qquad (3.18)$$

where $t^* = \max(s, r)$ and $\epsilon_t = 0$ for $t = t^* + 1 - s, \ldots, t^*$. The sum
of squares function is then minimised with respect to $\alpha_1, \ldots, \alpha_r$,
$\beta_0, \ldots, \beta_s$, with or without the use of analytic derivatives. Although
t^* residuals are 'lost' by starting the recursion at $t = t^* + 1$, no
distortion is introduced by setting pre-sample values of y_t or x_t equal
to zero.   *Also assume $\epsilon_{t-k} = zero$*

A two-step estimation procedure could be based on the Gauss—
Newton scheme as before. Consistent estimators of the parameters
may be obtained by an IV estimator which uses $x_t, x_{t-1}, \ldots, x_{t-r-s}$
as the set of instruments.

Several Explanatory Variables

The discussion so far has been restricted to the case of a single
regressor. However, there is no difficulty in extending the model to
include additional variables which are not dynamic. Adding a second
explanatory variable to (3.1) yields

$$y_t = \beta_1 \sum_{j=0}^{\infty} \alpha^j x_{1,t-j} + \beta_2 x_{2t} + \epsilon_t \qquad (3.19)$$

The transformed equation is

$$y_t = \alpha y_{t-1} + \beta_1 x_{1t} + \beta_2 x_{2t} - \alpha \beta_2 x_{2,t-1} + \epsilon_t - \alpha \epsilon_{t-1},$$

$$t = 2, \ldots, T \qquad (3.20)$$

and a recursive expression for the residuals, together with corresponding formulae for the derivatives may be built up in the same way as
before.

The estimation problem becomes more complex if there is a
dynamic relationship between x_2 and y. Suppose that the lag
structure for x_2 is again of the Koyck form, so that (3.19) becomes

$$y_t = \beta_1 \sum_{j=0}^{\infty} \alpha_1^j x_{1,t-j} + \beta_2 \sum_{j=0}^{\infty} \alpha_2^j x_{2,t-j} + \epsilon_t$$

$$= \frac{\beta_1 x_{1t}}{1 - \alpha_1 L} + \frac{\beta_2 x_{2t}}{1 - \alpha_2 L} + \epsilon_t \qquad (3.21)$$

The residuals may be computed from the recursion.

$$\epsilon_t = y_t - (\alpha_1 + \alpha_2)y_{t-1} + \alpha_1\alpha_2 y_{t-2} - \beta_1 x_{1t} + \alpha_2\beta_1 x_{1,t-1}$$
$$- \beta_2 x_{2t} + \alpha_1\beta_2 x_{2,t-1} + (\alpha_1 + \alpha_2)\epsilon_{t-1} - \alpha_1\alpha_2\epsilon_{t-2},$$
$$t = 3, \ldots, T \quad (3.22)$$

This is started at $t = 3$ in order to avoid setting any pre-sample values of y_t equal to zero. For a more general model,

$$y_t = \frac{B_1(L)}{A_1(L)} x_{1t} + \frac{B_2(L)}{A_2(L)} x_{2t} + \epsilon_t \qquad (3.23)$$

can use asymptotic D.W. for samples large [handwritten annotation]

such an approach implies the loss of t^* residuals where $t^* = \max (s_1 + r_2, s_2 + r_1, r_1 + r_2)$. However, this is preferable to the alternative strategy of setting pre-sample values of the observations equal to zero. Nevertheless, this second approach is the one adopted by Box and Jenkins (1976, p. 389). They write (3.21) in the form

$$y_t = \beta_1 x_{1t}^* + \beta_2 x_{2t}^* + \epsilon_t \qquad (3.24)$$

and suggest that x_{1t}^* and x_{2t}^* be computed from the recursion

$$x_{it}^* = \alpha_i x_{i,t-1}^* + x_{it}, \qquad t = 1, \ldots, T \qquad (3.25)$$

with $x_{i,0}^* = 0$ for $i = 1, 2$; cf. (3.7). For a single explanatory variable, this is the estimation procedure in which ξ of expression (3.5) is set equal to zero.

4. Serial Correlation

In the general transfer function model, (1.12), the disturbance term is assumed to be generated by an ARMA(p, q) process. The $p + q$ parameters of this process must therefore be estimated along with the parameters of the rational distributed lag. It is shown below how the method of maximum likelihood may be applied to a geometric distributed lag with AR(1) disturbances. Estimation of (1.12) raises no new issues of principle. Testing is considered in the final sub-section.

Estimation

Consider (1.4) with disturbances generated by an AR(1) process (6.1.2), i.e.

$$y_t = \frac{\beta x_t}{1 - \alpha L} + \frac{\epsilon_t}{1 - \phi L} \qquad (4.1)$$

Multiplying by $(1 - \alpha L)(1 - \phi L)$ gives the transformed equation, and this leads to the following recursion for the residuals:

$$\epsilon_t = y_t - (\alpha + \phi)y_{t-1} + \alpha\phi y_{t-2} - \beta x_t + \beta\phi x_{t-1} + \alpha\epsilon_{t-1},$$

$$t = 3, \ldots, T \quad (4.2)$$

with $\epsilon_2 = 0$. Recursive expressions for the analytic derivatives are also available. As before, these may be exploited in a Gauss–Newton procedure, although for more complicated models there is an argument for evaluating the derivatives numerically if programming time is at a premium.

Minimising the residual sum of squares function is equivalent to maximising the likelihood function when the disturbances are normally distributed and y_1 and y_2 are fixed. However, this last assumption is relatively unimportant, and minimising the sum of squares is a reasonable approximation to ML. An explicit expression for the information matrix may be derived in much the same way as when the disturbances are independent. However, the important point to note is that the ML estimator of ϕ is independent of the ML estimators of α and β in large samples. In other words the information matrix has the same block diagonal form exhibited by the information matrix of a static regression model with AR(1) disturbances; cf. (6.1.12). One implication of this is that the lag structure will be estimated consistently even if the disturbance term is misspecified.

Test Procedures

The block diagonality of the information matrix in a transfer function model means that all the test procedures discussed in the previous chapter can be applied in large samples. Thus a large sample test against an AR(1), or MA(1), disturbance term may be based on the Durbin–Watson d-statistic. If an ARMA disturbance term has been fitted to the model, the Box–Pierce Q-statistic may be used. This will have a χ^2 distribution with $P - p - q$ degrees of freedom under the null hypothesis, provided that P is sufficiently large.

5. Model Selection

Finding efficient estimates of the parameters in a transfer function model (1.12) is primarily a matter of computation. However, actually arriving at a suitable specification is a far more difficult task, requiring a good deal of skill and judgement. Box and Jenkins (1976, Chapter 11) approach the problem in the same spirit as in univariate time series model building, pursuing the overall strategy of identification, estimation and diagnostic checking. In the absence of any *a priori* knowledge of the system, this is a sensible strategy, although there are a number of different ways of handling the most difficult aspect of the problem, which is identification.

The identification procedure consists of three main stages. The first is to obtain some idea of the shape of the lag structure. This entails computing rough estimates of a finite number of the lag coefficients, $\delta_0, \delta_1, \delta_2, \ldots, \delta_m$. Given these estimates, the second stage is to find a suitable approximation to the lag polynomial, $D(L)$, in terms of the ratio, $B(L)/A(L)$. This means assigning numbers to r and s, the orders of the polynomials in the rational lag structure, as well as determining the extent of any delay, v. Finally, a suitable ARMA(p, q) process must be found for the disturbance term.

Estimates of the parameters in $A(L)$ and $B(L)$ may be derived from the lag coefficients, and these estimates may be used as starting values in an ML algorithm. An ARMA(p, q) model may then be identified from the residual autocorrelations as discussed in Section 1.5. Having chosen suitable values of r, s, v, p and q, the full model is estimated by ML. The residuals resulting from this procedure provide the basis for diagnostic checking of the model. As indicated in Section 4, the main test statistic is the Box–Pierce Q-statistic, although this could be supplemented by tests based on the Lagrange multiplier principle and by graphical procedures.

In univariate modelling it is important to consider the possibility of common factors in the ARMA polynomials $\phi(L)$ and $\theta(L)$. The same question arises in a transfer function model, but with respect to the rational lag polynomials, $A(L)$ and $B(L)$, as well as the ARMA polynomials in the disturbance term. However, it was demonstrated in Section 3.6 that one aspect of common factors, or near common factors, is the large standard errors associated with the estimates. The implications of this point are summarised by Box and Jenkins (1976, p. 387) as follows: 'In practice we shall be dealing with estimated coefficients which may be subject to rather large errors, so that only approximate factorization can be expected, and considerable imagination may need to be exerted to spot a possible factorization.' Another pertinent point concerns the possible instability of ML algorithms. Difficulties in convergence may well be experienced in these circumstances, thereby giving an early indication of an over-parameterised model.

Determining the Shape of the Lag Structure

The first stage in the identification procedure is to obtain rough estimates of the lag coefficients $\delta_0, \delta_1, \ldots, \delta_m$. The *direct* approach is to fix the maximum lag length, m, at a suitable high value and to regress y_t on $x_t, x_{t-1}, \ldots, x_{t-m}$. The main difficulty with this procedure is in determining a suitable value for m. The sequence of x_t's will typically be slowly changing, and so setting m too high could result in very imprecise estimates. On the other hand, including too few lagged values of x_t could give a totally false impression of the shape of the lag distribution. Some experimentation with different values of m will therefore be necessary.

It is possible to circumvent the need to run several regressions with different lag lengths by *pre-whitening* the explanatory variable. Suppose x_t can be modelled by an ARIMA process,

$$\varphi_x(L)x_t = \theta_x(L)\xi_t \tag{5.1}$$

where ξ_t is white noise and $\varphi_x(L) = \Delta^d \phi_x(L)$. Applying the transformation $\varphi_x(L)/\theta(L)$ to both sides of (1.8) gives

$$y_t^* = D(L)x_t^* + u_t^* \tag{5.2}$$

where $y_t^* = \theta^{-1}(L)\varphi_x(L)y_t$, $u_t^* = \theta^{-1}(L)\varphi_x(L)u_t$ and $x_t^* = \theta^{-1}(L)\varphi_x(L)x_t = \xi_t$. The advantage of working with (5.2) as opposed to (1.8) is that when y_t^* is regressed on $x_t^*, \ldots, x_{t-m}^*$, the cross-product matrix of the explanatory variables is approximately diagonal. Therefore

$$d_j^* \simeq \sum_{t=m+1}^{T} (x_{t-j}^* - \bar{x}^*)(y_t - \bar{y}) \Big/ \sum_{t=m+1}^{T} (x_{t-j}^* - \bar{x}^*)^2,$$

$$j = 0, \ldots, m \tag{5.3}$$

No matrix inversions are required and the coefficients remain the same if m is increased. In the time series literature, d_j is normally expressed as the ratio of the cross-covariance between x_t and y_t at lag j to the variance of x_t.

Although pre-whitening has some computational attractions it suffers from the disadvantage that x_t must be modelled as an ARIMA process. Furthermore, there are no statistical reasons for preferring (5.2) to (1.8). If u_t is white noise, OLS applied to (1.8) will, in the absence of any restrictions on $D(L)$, yield the BLUE of δ. On the other hand, (5.3) will be inefficient. Only if x_t and u_t are generated by *identical* ARIMA processes will u_t^* be white noise. Applying OLS to the transformed model, (5.2), is then equivalent to computing a GLS estimator for (1.8).

Example   If x_t follows an AR(1) process with parameter ϕ_x, the pre-whitening transformation is $x_t^* = x_t - \phi_x x_{t-1}$. Applying the same transformation to y_t yields a disturbance term of the form

$$u_t^* = u_t - \phi_x u_{t-1}$$

If $u_t \sim NID(0, \sigma^2)$, u_t^* will follow on MA(1) process. Only if u_t is an AR(1) process with $\phi = \phi_x$ will (5.3) be efficient.

When the model contains more than one explanatory variable, pre-whitening is no longer a viable alternative to the direct approach. The only other possibility is to employ a spectral estimator. This has a good deal to recommend it, but a full discussion of the technique is beyond the scope of this book.

Identification of a Rational Lag Structure

Once the shape of the lag coefficients has been determined, suitable rational lag structures may be considered. As an example, suppose that the estimated coefficients take the values shown in figure 7.3.

The first point to note is that both $\hat{\delta}_0$ and $\hat{\delta}_1$, are very close to zero. This suggests that there is a delay of two time periods, i.e. $v = 2$. The next problem is to find polynomials $A(L)$ and $B(L)$ such that the pattern exhibited by the estimated lag coefficients is captured by a relatively small number of parameters. On examining figure 7.3 it will be observed that $\hat{\delta}_2$, $\hat{\delta}_3$ and $\hat{\delta}_4$ appear to follow no set pattern, but that from $\hat{\delta}_4$ onwards there is the suggestion of an exponential decline. It will be recalled that exponential decline is a characteristic of the geometric distributed lag, (1.4), and so setting $r = 1$ would seem to be an appropriate course of action. The irregular behaviour of the first three coefficients in the lag distribution may be accommodated by setting $s = 3$.

These results suggest that the rational lag structure

$$\frac{B(L)L^2}{A(L)} = \frac{\beta_0 L^2 + \beta_1 L^3 + \beta_2 L^4}{1 - \alpha L} \tag{5.4}$$

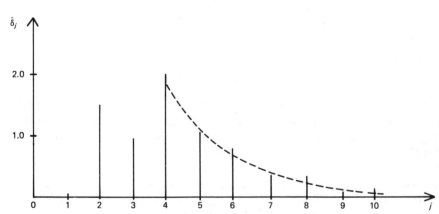

Figure 7.3   *Estimated Lag Coefficients for Rational Lag Structure in (7.5.4)*

should provide a suitable approximation to $D(L)$. Preliminary estimates of the parameters in $A(L)$ and $B(L)$ may be obtained by equating coefficients of powers of L in the expression

$$(1 - \alpha L)D(L) = \beta_0 L^2 + \beta_1 L^3 + \beta_2 L^4$$

This yields

$$\delta_2 = \alpha\delta_1 + \beta_0$$

$$\delta_3 = \alpha\delta_2 + \beta_1$$

$$\delta_4 = \alpha\delta_3 + \beta_2$$

$$\delta_j = \alpha\delta_{j-1}, \qquad j \geqslant 5 \tag{5.5}$$

The parameter α may be estimated by

$$\hat{\alpha} = \hat{\delta}_5/\hat{\delta}_4 = 1.2/2 = 0.6$$

Substituting estimated values for α, δ_2, δ_3 and δ_4, and setting $\delta_1 = 0$ then gives

$$\hat{\beta}_0 = \hat{\delta}_2 = 1.5$$

$$\hat{\beta}_1 = \hat{\delta}_3 - \hat{\alpha}\hat{\delta}_2 = 1.0 - 0.6(1.5) = 0.1$$

$$\hat{\beta}_2 = \hat{\delta}_4 - \hat{\alpha}\hat{\delta}_3 = 2.0 - 0.6(0.1) = 1.94$$

use $\delta_0, \delta_1, \delta_2$
for starting
values.

Identification of the Disturbance Term

The final stage in the identification process is to determine a suitable model for the disturbance term. This is done by treating the residuals as a univariate time series and applying the techniques described in Section 1.5.

A set of residuals may be constructed either from the estimated finite lag model or from the rational lag. The second approach is probably preferable, and in the example given in the previous subsection the residuals would be obtained from the recursion,

③

$$\hat{u}_t = \hat{\alpha}\hat{u}_{t-1} + y_t - \hat{\alpha}y_{t-1} + \hat{\beta}_0 x_{t-2} + \hat{\beta}_1 x_{t-3} + \hat{\beta}_2 x_{t-4},$$

$$t = 5, \ldots, T \tag{5.6}$$

ould go from ① to ③

Differencing

Variables are often differenced before a transfer function model is constructed. This is done so that the two series can be regarded as being jointly stationary, and a suitable model identified on the basis of the cross-correlation function; cf. (5.3). However, if the true model is in terms of levels, differencing will introduce an additional MA term into the disturbance. For example, taking first differences in the model

$$y_t = \beta x_t + \epsilon_t, \qquad t = 1, \ldots, T \qquad (5.7)$$

gives

$$\Delta y_t = \beta \Delta x_t + v_t, \qquad t = 2, \ldots, T \qquad (5.8)$$

where v_t is the non-invertible MA(1) process,

$$v_t = \epsilon_t - \epsilon_{t-1}, \qquad t = 2, \ldots, T \qquad (5.9)$$

This strict non-invertibility is a characteristic feature of 'over-differencing'.

  If transfer functions have been specified and estimated in both levels and first differences, it is possible to discriminate between them on the basis of their residual sums of squares; see Harvey (1980). When the number of parameters in the two specifications is different, some allowance must be made for parsimony by a criterion such as the AIC.

6. Seasonality

When monthly or quarterly data are being used, seasonal effects must usually be incorporated into the model. Two methods are described below, one based on seasonal dummies, the other based on seasonal differences. The question of seasonal adjustment is then examined.

  As in univariate time series modelling, it is preferable to work with unadjusted data wherever possible. If this is not done, there is the possibility of introducing considerable distortion into the estimated relationship between the series.

Modelling Seasonal Behaviour

A deterministic seasonal component may be incorporated into a transfer function model by introducing a set of dummy variables. The model takes the form

$$y_t = A^{-1}(L)B(L)x_t + \sum_{i=1}^{s} \gamma_i z_{it} + u_t \qquad (6.1)$$

where the z_{it}'s and γ_i's are the seasonal dummies and their respective coefficients, and s is the number of 'seasons'. A model of this

kind poses no new problems as regards estimation and specification.

A stochastic seasonal pattern may be introduced into the disturbance term by modelling it as an ARMA process with non-zero coefficients at the seasonal lags. More generally, u_t may be modelled by
a seasonal ARMA process which combines seasonal and non-
seasonal components in a multiplicative fashion. Expression (6.4.2)
is a simple example.

The use of seasonal dummies can be avoided by differencing. This
may or may not be combined with first differencing. The general
formulation is

$$\Delta^d \Delta_s^D y_t = A^{-1}(L)B(L)\Delta^d \Delta_s^D x_t + u_t \tag{6.2}$$

where $\Delta_s = 1 - L^s$ is the seasonal difference operator and D denotes
the number of times it is applied. The disturbance term will
normally be modelled by a seasonal ARMA process.

The model

$$\Delta_4 y_t = \beta \Delta_4 x_t + e_t \tag{6.3}$$

provides a simple illustration of how seasonality may be handled in a
relationship based on quarterly observations. The crucial point about
(6.3), and indeed any model of the form (6.2), is that the seasonal
variation in y_t arises directly from the seasonal variation in x_t. This
would be the case even if D were equal to zero. On the other hand,
adding a seasonal component to the model, as in (6.1), implies that
the seasonality in y_t is at least partly independent of any seasonal
movement in x_t.

Seasonal Adjustment

Most seasonal adjustment procedures yield a series of observations
which can be approximated by a weighted average of the original
observations. If the superscript (a) denotes an adjusted series,

$$y_t^a = \sum_{j=-m}^{n} w_j y_{t-j} \tag{6.4}$$

where the w_j's are weights; see TSM (Chapter 3). A transformation of
this kind is known as a moving average filter, and the filter itself may
be expressed as a polynomial in the lag operator by writing

$$S(L) = \sum_{j=-m}^{n} w_j L^j \tag{6.5}$$

Suppose that two time series, y_t and x_t, are subject to seasonal adjustment. The filters employed need not be the same and so

$$y_t^a = S_y(L)y_t \quad \text{and} \quad x_t^a = S_x(L)x_t \tag{6.6}$$

If the relationship between the adjusted variables is a dynamic model of the form (3.16), substituting from (6.6) implies the following model for the adjusted variables:

$$y_t^a = S_x^{-1}(L)S_y(L)D(L)x_t^a + S_y(L)u_t \tag{6.7}$$

There are two points to note about (6.7). The first is that by subjecting the variables to different adjustment procedures, the systematic dynamics are altered. The lag function becomes $S_y(L)D(L)/S_x(L)$, and in general this will be two-sided. It may therefore appear that y_t^a depends on future as well as past values of x_t^a. When the model is static, the effect of different seasonal filters is to introduce a spurious dynamic element into the relationship.

If the variables are adjusted in the same way, i.e. $S_y(L) = S_x(L)$, the lag structure is preserved. However, a second feature of (6.7), namely the change in the disturbance term, remains. If u_t is white noise, $S_y(L)u_t$ will be autocorrelated. Least squares will no longer be efficient, and it becomes necessary to make some allowance for serial correlation in the estimation procedure.

7. Prediction and Forecasting

Optimal predictions of future values of y_t in a transfer function model may be computed recursively. The required expression is obtained by multiplying (1.12) by $A(L)\phi(L)$ to yield

$$\phi(L)A(L)y_t = \phi(L)B(L)x_{t-v} + A(L)\theta(L)\epsilon_t \tag{7.1}$$

and rearranging so that only y_t appears on the left-hand side.

Example 1   For the geometric distributed lag, (3.1),

$$\tilde{y}_{t+l/T} = \alpha\tilde{y}_{T+l-1/T} + \beta x_{T+l} + \tilde{\epsilon}_{T+l/T} - \alpha\tilde{\epsilon}_{T+l-1/T},$$

$$l = 1, 2, \ldots, \tag{7.2}$$

where $\tilde{y}_{T+l/T}$ and $\tilde{\epsilon}_{T+l/T}$ are defined in (6.7.2).

There is no problem in extending these results to handle models cast in differenced form. If the model is

$$\Delta^d y_t = \frac{B(L)}{A(L)}\Delta^d x_{t-v} + \frac{\theta(L)}{\phi(L)}\epsilon_t \tag{7.3}$$

$\Delta^d y_t \sim ARMA(p,q)$

$y_t \sim ARIMA(p,d,q)$

$ARIMA(1,1,0)$

$\Delta y_t \overset{?+t}{=} \phi \Delta y_{t-1} + \epsilon_t$

expression (7.1) is modified to

$$\phi(L)A(L)\Delta^d y_t = \phi(L)B(L)\Delta^d x_{t-v} + A(L)\theta(L)\epsilon_t \qquad (7.4)$$

Example 2 If y_t and x_t in (3.1) are replaced by Δy_t and Δx_t, predictions are computed from the recursion

$$\tilde{y}_{T+l/T} = (1+\alpha)\tilde{y}_{T+l-1/T} - \alpha\tilde{y}_{T+l-2/T} + \beta x_{T+l} - \beta x_{T+l-1}$$
$$+ \tilde{\epsilon}_{T+l/T} - \alpha\tilde{\epsilon}_{T+l-1/T} \qquad (7.5)$$

MSE of Predictions

Suppose that the following assumptions hold: (i) the parameters in $A(L)$, $B(L)$, $\phi(L)$ and $\theta(L)$ are known; and (ii) future values of x_t are known. The MSE of $\tilde{y}_{T+l/T}$ may then be obtained by a straight-forward extension of the result given in (6.7.3). If differences have been taken, ψ_j is the jth power of L in the polynomial

$$\Psi(L) = \theta(L)/(\Delta^d \phi(L)) \qquad (7.6)$$

Example 3 In example 2 above, $\Psi(L) = 1/(1-L)$. Therefore,

$$\Psi(L) = L\Psi(L) + 1$$

and so

$$\psi_j = \psi_{j-1} + 1, \qquad j = 1, 2, \ldots$$

with $\psi_0 = 0$. Hence

$$\text{MSE}(\tilde{y}_{T+l/T}) = \sigma^2 l, \qquad l = 1, 2, \ldots, \qquad (7.7)$$

Assumption (i) will not hold in practice. The parameters will be replaced by their ML estimates, while the disturbances are set equal to the residuals. However, the additional contribution to the MSE will be of $0(1/T)$.

MSE of Forecasts

The assumption that future values of x_t are known will not usually be reasonable. Nevertheless, the evaluation of the MSE conditional on future x_t's is still very useful. The user of the model will often have a particular scenario in mind with regard to future values of the exogenous variables. The conditional MSE is then a relevant piece of information, since it enables the user to gauge the accuracy of the model's predictions given his particular view of the world.

On the other hand, in making unconditional predictions or *forecasts*, some account must be taken of the error in predicting

[handwritten notes:] ...ing d - If a series is not stationary the correlogram will not die away

y_t [sketch]

Δy_t [sketch]

future values of x_t. If $\hat{x}_{T+l/T}$ denotes a prediction of x_t, l periods ahead, the variance of the prediction error is defined by

$$\sigma_{x,l}^2 = E(x_{T+l} - \hat{x}_{T+l/T})^2 \tag{7.8}$$

An optimal forecast of a future value of y_t is computed in exactly the same way as if future values of x_t were known with certainty and it will be denoted by the same symbol, $\tilde{y}_{T+l/T}$. In order to derive the MSE of $\tilde{y}_{T+l/T}$, it is first written in the form

$$\tilde{y}_{T+l/T} = \sum_{j=1}^{l} \delta_{l-j}\hat{x}_{T+j/T} + \sum_{j=0}^{\infty} \delta_{l+j}x_{T-j} + \sum_{j=0}^{\infty} \psi_{l+j}\epsilon_{T-j},$$
$$l = 1, 2, \ldots ; \tag{7.9}$$

As in the previous sub-section the parameters, together with present and past disturbances, are assumed to be known. Subtracting (7.9) from

$$y_{T+l} = \sum_{j=1}^{l} \delta_{l-j}x_{T+j} + \sum_{j=0}^{\infty} \delta_{l+j}x_{T-j} + \sum_{j=1}^{l} \psi_{l-j}\epsilon_{T+j}$$

$$+ \sum_{j=0}^{\infty} \psi_{l+j}\epsilon_{T-j} \tag{7.10}$$

squaring, and taking expectations yields

$$\text{MSE}(\tilde{y}_{T+l/T}) = \sum_{j=1}^{l} \sigma_{x,j}^2 \delta_{l-j}^2 + \sigma^2 \sum_{j=0}^{l-1} \psi_j^2, \qquad l = 1, 2, \ldots \tag{7.11}$$

The first term on the right hand side of (7.11) measures the contribution arising from the uncertainty associated with future values of x_t. The second term is identical to (6.7.3), the MSE of a conditional prediction.

A natural way in which to forecast future values of x_t within the present framework is to model the series by an ARIMA process, i.e.

$$\phi_x(L)\Delta^d x_t = \theta_x(L)\zeta_t \tag{7.12}$$

where $\zeta_t \sim WN(0, \sigma_\zeta^2)$. Substituting in (7.3) yields

$$y_t = \frac{B(L)\theta_x(L)L^v\zeta_t}{A(L)\phi_x(L)\Delta^d} + \frac{\theta(L)}{\phi(L)\Delta^d}\epsilon_t$$

$$= K(L)x_t + \Psi(L)\epsilon_t \tag{7.13}$$

The argument used to obtain (7.11) leads directly to the expression

Seasonal ARIMA of order $(p, d, q) \times (P, D, Q)$

$$\text{MSE}\,(\tilde{y}_{T+l/T}) = \sigma_\xi^2 \sum_{j=0}^{l-1} \kappa_j^2 + \sigma^2 \sum_{j=0}^{l-1} \psi_j^2 \tag{7.14}$$

where κ_j is the coefficient of L^j in the polynomial $K(L)$.

Example 4   Box and Jenkins (1976, pp. 409–412) build the following model for forecasting sales, y:

$$\Delta y_t = 0.035 + \frac{4.82\Delta x_{t-3}}{1 - 0.72L} + (1 - 0.54L)\epsilon_t \tag{7.15}$$

The exogenous variable, x, is a *leading indicator* since it is known with certainty three time periods in advance of y.

The forecast function is

$$\tilde{y}_{T+l/T} = 1.72\tilde{y}_{T+l-1/T} - 0.72\tilde{y}_{T+l-2/T} + 0.0098$$
$$+ 4.82\hat{x}_{T+l-3/T} - 4.82\hat{x}_{T+l-4/T} + \tilde{\epsilon}_{T+l/T}$$
$$- 1.26\tilde{\epsilon}_{T+l-1/T} + 0.39\tilde{\epsilon}_{T+l-2/T}, \qquad l = 1, 2, \ldots \tag{7.16}$$

However for $l \leqslant 3$, the values of $\hat{x}_t$ appearing in (7.16) are known with certainty and so the MSE is given by (6.7.3). The weights are obtained by equating coefficients in the expression

$$\Psi(L)(1 - L) = 1 - 0.54L$$

On expanding this becomes

$$\psi_0 + (\psi_1 - \psi_0)L + (\psi_2 - \psi_1)L^2 + \cdots = 1 - 0.54L$$

Hence $\psi_0 = 1$, while $\psi_1 - \psi_0 = -0.54$, implying that $\psi_1 = 0.46$. Thereafter $\psi_j = \psi_{j-1} = 0.46$, and so, for example,

$$\text{MSE}(\tilde{y}_{T+3/T}) = \sigma^2(1 + 0.46^2 + 0.46^2) = 1.42\sigma^2$$

For making predictions beyond $l = 3$, forecasts of future values of x_t are needed. Box and Jenkins fit the model

$$\Delta x_t = (1 - 0.32L)\zeta_t$$

and so from (7.13),

$$K(L) = \frac{4.82(1 - 0.32L)L^3}{(1 - 0.72L)(1 - L)}$$

Therefore

$$(1 - 0.72L)(1 - L)K(L) = 4.82(1 - 0.32L)L^3$$

Equating coefficients in powers of L yields $\kappa_3 = 4.82$, $\kappa_4 = 6.75$ and so on.

Leading Indicators

There are basically two reasons for constructing a dynamic regression model for the variable y_t. The first is to obtain an understanding of the behaviour of the system, and to relate this to the theory of the subject. The second is to obtain better forecasts.

If the goal of model building is forecasting, the exercise can only be judged successful if, by relating y to x, improved forecasts are obtained. The yardstick against which improvements are judged is the accuracy of forecasts derived from a univariate model for y. Such a model must be a serious contender, since if the bivariate model is of the form (7.3) and $\Delta^d x_t$ is modelled by an ARMA process, y_t must have an ARIMA representation. This follows because $\Delta^d y_t$ is the sum of two independent ARMA processes; cf. TSM (Chapter 2). The question of whether there is any gain in forecasting performance from constructing a bivariate model is therefore a pertinent one.

Nelson (1975) was the first to establish one of the basic results in this area. For a correctly specified model he was able to show that when x is forecast, the one-step ahead prediction error variance of y cannot exceed that which would result from forecasting y on the basis of its past history alone. This result may be generalised to forecasts l steps ahead. However, it does not answer the question of when the forecasts from the bivariate model are actually better. One obvious case is when some values of x are known in advance of observing y. This corresponds to the pure delay model, where v is positive.

When there is a pure delay in the model, x is said to be a leading indicator of y. However, Pierce (1975) provides a more general definition of this term. Let $\Delta(l)$ denote the reduction in the MSE of the forecast l periods ahead, when y is modelled in terms of present and past values of x. Then x is a *leading indicator* of y whenever $\Delta(l)$ is strictly positive. Pierce then proceeds to establish conditions under which x is a leading indicator in this sense. In other words, he establishes conditions under which there is a gain in forecasting accuracy from utilising the information contained in x. The pure delay model emerges only as a special case of a leading indicator.

The extent to which x is useful in predicting y is measured by the *predictive*$-R^2$. This is a scale free quantity defined by

$$R_\dagger^2 = \Delta(1)/\sigma_\dagger^2 \tag{7.17}$$

where $\sigma_\dagger^2$ is MSE of one-step ahead predictions in the univariate model for y. A positive value of $R_\dagger^2$ indicates that x is a leading indicator for y. However, if the transfer function model is misspecified, it is quite conceivable that $R_\dagger^2$ will be less than zero. In these circumstances a 'naive' univariate model is actually giving better forecasts than a dynamic regression model. This is something which has often been observed in practice.

8. Polynomial Distributed Lags

An alternative approach to modelling a distributed lag is to approximate its structure by a polynomial. The lag is assumed to be finite, with a maximum lag length of m, and so the basic model is

$$y_t = \sum_{\tau=0}^{m} \delta_\tau x_{t-\tau} + u_t, \qquad t = m+1, \ldots, T \qquad (8.1)$$

Rather than attempting to estimate all $(m+1)$ coefficients, it is assumed that they lie on a polynomial of degree $n < m$. If this supposition is valid, the systematic part of (8.1) will depend on only $n+1$ parameters, and by imposing the appropriate structure, more efficient estimators of the lag coefficients will be obtained.

Let $P(\tau)$ be a polynomial of degree n which is a continuous function of τ. This depends on $n+1$ coefficients, $\gamma_0, \ldots, \gamma_n$, and is defined by

$$P(\tau) = \sum_{i=0}^{n} \gamma_i \tau^i \qquad (8.2)$$

Under the polynomial lag hypothesis, the lag coefficients in (8.1) are defined by

$$\delta_\tau = P(\tau) \qquad (8.3)$$

for the integers $\tau = 0, 1, \ldots, m$. Note that $\delta_0 = P(0) = \gamma_0$.

A simple illustration should convey the flavour of the approach. Suppose that $P(\tau)$ is a second degree polynomial,

$$P(\tau) = 2 + 5\tau - \tau^2 \qquad (8.4)$$

The maximum lag length is five, and the six lag coefficients in (8.1) are obtained by evaluating (8.4) at $\tau = 0, \ldots, 5$. The pattern of the coefficients is shown in figure 7.4. However, because the coefficients all lie on a second degree polynomial, the lag structure depends on only three parameters $\gamma_0 = 2$, $\gamma_1 = 5$ and $\gamma_2 = -1$.

Estimation

Expressions (8.2) and (8.3) imply that

$$\delta_\tau = \sum_{i=0}^{n} \gamma_i \tau^i, \qquad \tau = 0, \ldots, m \tag{8.5}$$

Substituting into (8.1) yields

$$y_t = \sum_{\tau=0}^{m} \left(\sum_{i=0}^{n} \gamma_i \tau^i \right) x_{t-\tau} + u_t$$

$$= \sum_{i=0}^{n} \gamma_i w_{it} + u_t, \qquad t = m+1, \ldots, T \tag{8.6}$$

where

$$w_{it} = \sum_{\tau=0}^{m} \tau^i x_{t-\tau}, \qquad i = 0, \ldots, n \tag{8.7}$$

If (8.1) obeys the classical assumptions, OLS applied to (8.6) will yield the BLUE of $\gamma = (\gamma_0, \ldots, \gamma_n)'$. An estimator of $\delta = (\delta_0, \ldots, \delta_m)'$ may then be constructed directly from (8.5). This estimator, denoted by d, is the BLUE of δ, a result which follows directly from the Gauss–Markov theorem.

The covariance matrix of d may be obtained most easily by writing the transformation (8.5) in matrix terms as

$$\delta = S\gamma \tag{8.8}$$

where S is an $(m+1) \times (n+1)$ matrix defined by $S = (s_0, \ldots, s_m)'$, where

$$s_0 = (1, 0, \ldots, 0)'$$

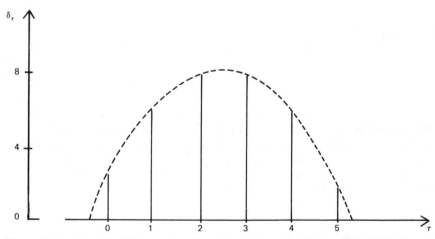

Figure 7.4   *Lag Coefficients Based on Second Degree Polynomial, (7.8.4)*

$$\mathrm{Var}\,(\hat{\gamma}) = \sigma^2(W'W)^{-1}$$
$$\delta = S\gamma$$

and

$$s_\tau = (\tau^0, \tau^1, \ldots, \tau^n)', \qquad \tau = 1, \ldots, m \tag{8.9}$$

If W is the $(T - m + 1) \times (n + 1)$ matrix of transformed observations, (8.7), then

$$E[(d - \delta)(d - \delta)'] = \sigma^2 S(W'W)^{-1}S' \tag{8.10}$$

Thus if $u_t \sim NID(0, \sigma^2)$, d will have a multivariate normal distribution with a mean of δ and a covariance matrix, (8.10).

An alternative approach is obtained by recognising that constraining the m lag coefficients to lie on an nth degree polynomial is equivalent to imposing a set of $(m - n)$ linear restrictions on the model. These may be written

$$R\delta = 0 \tag{8.11}$$

where R is an $(m - n) \times (m + 1)$ matrix. Differencing a polynomial reduces its degree by one. The $(n + 1)$th difference of (8.5) is therefore zero and so

$$\Delta^{n+1}\delta_\tau = 0, \qquad \tau = n + 1, \ldots, m \tag{8.12}$$

Since $\Delta^{n+1} = (1 - L)^{n+1}$, each row of R consists of the coefficients of L in the difference polynomial $(1 - L)^{n+1}$. For a second degree polynomial such as (8.4), each row of R is equal to $(1 - 3, 3, -1)$, and (8.12) becomes

$$\delta_\tau = 3\delta_{\tau-1} - 3\delta_{\tau-2} + \delta_{\tau-3}, \qquad \tau = n + 1, \ldots, m$$

Thus, taking the figures in (8.4), the coefficient at lag 5 is

$$\delta_5 = 3(6) - 3(8) + 1(8) = 2$$

This agrees with the direct calculation.

The lag coefficients may now be estimated by restricted least squares, i.e.

$$d = (X'X)^{-1}X'y - (X'X)^{-1}R'\{R(X'X)^{-1}R'\}^{-1}R(X'X)^{-1}X'y \tag{8.13}$$

where X is the $(T - m + 1) \times (m + 1)$ matrix of observations on $x_t, \ldots, x_{t-m}$, and $y = (y_{m+1}, \ldots, y_T)'$; cf. (2.2.21).

Test of the Polynomial Lag Hypothesis

The validity of the polynomial lag hypothesis, (8.3), may be tested by regarding the unrestricted lag coefficients as lying on a polynomial of degree m. The model may then be written in the form

$$y = XS^*\gamma^* + u = W^*\gamma^* + u \qquad \text{(8.14)}$$

[handwritten: testing model $n < m$ is valid]

where S^* is an $(m + 1) \times (m + 1)$ matrix defined by (8.8). This may be partitioned as $S^* = [S \vdots S^\dagger]$ where S is the $(m + 1) \times (n + 1)$ matrix introduced earlier. In a similar way γ^* and W^* are partitioned as $\gamma^* = (\gamma', \gamma^{\dagger\prime})'$ and $W^* = [W : W^\dagger]$ respectively. Expression (8.14) may then be written as

[handwritten: original ↓ ; all exc. $m-n$ ↙]

$$y = XS\gamma + X^\dagger S^\dagger \gamma^\dagger + u = W\gamma + W^\dagger \gamma^\dagger + u \qquad \text{(8.15)}$$

If $u_t \sim NID(0, \sigma^2)$, a subset F-test may be used to test the hypothesis $H_0 : \gamma^\dagger = 0$. This is obviously a test of the validity of a polynomial lag of degree n. However, in practice it is unnecessary to estimate the unrestricted model in the form (8.14). The residual sum of squares from regressing y on W^* is

$$e^{*\prime}e^* = u'\{I - W^*(W^{*\prime}W^*)^{-1}W^{*\prime}\}u$$

$$= u'\{I - XS^*(S^{*\prime}X'XS^*)^{-1}S^{*\prime}X'\}u$$

$$= u'\{I - X(X'X)^{-1}X'\}u \qquad \text{(8.16)}$$

Thus $e^{*\prime}e^*$ is identical to the residual sum of squares from regressing y on X, a result which is hardly surprising since taking the lag coefficients to lie on a polynomial of degree $(m + 1)$ imposes no restrictions on the original model. The subset F-test, (2.8.18), is therefore carried out by computing the residual sum of squares from the OLS regression of y on X and comparing this with the residual sum of squares from the polynomial distributed lag model. The resulting statistic has an F-distribution with $(m - n, T - 2m - 1)$ degrees of freedom.

[handwritten: $\dfrac{(SSE_0 - SSE)(m-n)}{SSE/(T - 2m - 1)}$]

The test may be carried out without actually estimating the restricted model. In the previous sub-section it was shown that the polynomial distributed lag formulation imposes $(m - n)$ linear restrictions of the form $R\delta = 0$. If d is the unrestricted OLS estimator of δ, the validity of these restrictions may be tested using the statistic

$$\frac{d'R'\{R(X'X)^{-1}R'\}^{-1}Rd}{s^2(m - n)} \qquad \text{(8.17)}$$

where s^2 is the unbiased estimator of σ^2 in (8.1). Under $H_0 : R\delta = 0$, (8.17) has an F-distribution with $(m - n, T - 2m - 1)$ degrees of freedom; cf. (5.4.5). The null hypothesis being tested is the same as that in the test discussed above, as $R\delta = 0$ and $\gamma^\dagger = 0$ are equivalent. A little algebraic manipulation reveals that the two statistics are, in fact, identical.

[handwritten left margin: $SSE = y = XS + u$; $SSE =$ restricted, to unrestricted ; h_2: ; (if rejected, ...)]

Testing the Polynomial Lag Specification

Given the maximum lag m, testing the validity of a polynomial of a given degree is straightforward. When n is unknown, this suggests carrying out a series of tests in the manner described in Section 5.8. We first test for a polynomial of degree $n - 1$. If this hypothesis is not rejected, we test for a polynomial of degree $n - 2$ and so on.

Although the above procedure is optimal in the sense defined in Section 5.8, it does hinge on prior specification of the length of the lag, m. It may therefore be prudent to carry out the procedure with a number of different values of m, possibly using the AIC or R^2 as a final check on the goodness of fit of the preferred specification in each series of tests.

Akaike Information Criterion

Sargan
RSS (8.0)

Polynomial Distributed Lags and the Rational Lag Structure *

Although the polynomial and rational lag formulations have traditionally been viewed as alternatives, Pagan (1978) has recently pointed out that they are related. Consider a rational lag model of the form (1.12). The lag associated with such a model will normally be infinite. However, suppose that for comparability with (8.1) it is decided that the underlying lag distribution should be finite with a maximum lag of m. This leads to a *finite rational distributed lag* formulation. The recurrence relations (2.3) are still assumed to hold, but only for $\delta_0, \ldots, \delta_s$. Thereafter, $\delta_r = 0$. The relationship between this model and the polynomial distributed lag then emerges as follows. If $A(L) = (1 - L)^{n+1}$ and $B(L)$ is of order n, the estimates of $\delta_0, \ldots, \delta_m$ will be identical to those obtained by a polynomial distributed lag estimator in which the approximating polynomial is of degree n. The rationale behind this result follows almost immediately from (8.12). On examining the recurrence relations (2.3a) and (2.3b) it will be seen that (2.3a) imposes no constraints on the lag structure since for the $n + 1$ equations there are $n + 1$ unknown parameters $\beta_0, \ldots, \beta_n$. However, the relations (2.3b) impose $m - n$ constraints, and these constraints are identical to those in (8.12). The restrictions imposed on the model therefore correspond exactly to those imposed by the polynomial distributed lag.

> *Example*   In (8.4), the corresponding finite rational distributed lag model is
>
> $$y_t = \frac{(2 - 4L^2)}{(1 - L)^3} x_t + u_t \qquad (8.18)$$

Perhaps the most important implication of this result concerns the interpretation of estimates of rational distributed lag models in which the roots of $A(L)$ lie close to, or on, the unit circle.

Notes

Section 1   Expectations models are developed further in the literature on 'rational expectations'; see, for example, Shiller (1978) and Wallis (1980).

Sections 3 and 4   Maddala and Rao (1971), Pesaran (1973), and Schmidt (1975).

Section 5   The treatment here differs from that in Box and Jenkins (1976, Chapter 11) in that more stress is placed on the 'direct approach' to model identification. Box and Jenkins place much more emphasis on the role of the cross-correlation function; see Haugh and Box (1977) as well. In addition, differencing is seen primarily as a way of achieving joint stationarity. Further discussion on 'overdifferencing' will be found in Plosser and Schwert (1977).

Section 6   Wallis (1974).

Section 8   Polynomial lags were introduced by Almon (1965), and are sometimes known as 'Almon' lags. Specification is considered in Godfrey and Poskitt (1975) and Sargan (1980b).

Exercises

1. Find the mean and median lags in (1.1).
2. Explain how you would construct an efficient two-step estimator in a model of the form (3.1) with $s = 0$ and $r = 2$.
3. Consider (1.4) with an MA(1) disturbance term. Explain how you would compute consistent estimates of α, β and the MA parameter θ. Derive recursive expressions for ϵ_t and its derivatives, and explain how you would compute efficient estimators of all three parameters by least squares regression. How would you test the hypothesis that $\theta = 0$ on the basis of this regression?
4. Construct an LM test of the hypothesis that $\phi = 0$ in (4.1). Show that it is equivalent to a test based on the DW statistic in large samples.
5. A regression of y_t on $x_t, x_{t-1}, \ldots, x_{t-6}$ produced the following coefficients:

$$d_0 = 2.00, d_1 = 0.80, d_2 = -0.72, d_3 = 0.64, d_4 = -0.58, d_5 = 0.52,$$

$$d_6 = -0.46.$$

Explain why a rational distributed lag with $r = s = 1$ might be an appropriate way of modelling the relationship between x and y. Compute preliminary estimates of the parameters in such a model and indicate how you would proceed to identify a suitable model for the disturbance term.

6. How would you estimate the model

$$\Delta_4 y_t = (1 - \alpha L)^{-1} \Delta_4 x_t + u_t$$

if u_t was generated by the seasonal ARMA process, (6.4.2)?

7. If the explanatory variable in (3.16) is modelled by the ARIMA(1, 1, 0) process

$$\Delta x_t = 0.6\Delta x_{t-1} + \zeta_t$$

and $r = s = 1$ with $\tilde{\alpha} = 0.3$, $\tilde{\beta} = 2.0$, find an expression for the forecast function and compute $\text{MSE}(\bar{y}_{T+l/T})$ for $l = 1$, 2 and 3.

8. Given that

$$\Sigma x_t^2 = 50 \qquad \Sigma x_t x_{t-1} = 48 \qquad \Sigma x_t y_t = 10.2$$

$$\Sigma x_t y_{t-1} = 10 \qquad \Sigma x_{t-1} y_t = 9.9 \qquad \Sigma x_{t-1} y_{t-1} = 9.8$$

where all summations are from $t = 2$ to T, calculate consistent estimates of α and β in (1.4), showing why the method you have chosen does, in fact, yield consistent estimates. State any assumptions you make. Find the time taken for the overall change in y resulting from a change in x to be 90% complete.

End point restrictions -

Restricted through end pts.

$$\delta_{m+1} = \delta_{-1} = 0$$

unrestricted

a) $\quad \delta_{T} = \gamma_0 + \gamma_1 \tau + \gamma_2 \tau^2$

$0 = \gamma_0 - \gamma_1 + \gamma_2 \quad (8.20a)$

$0 = \gamma_0 + \gamma_1 (m+1) + \gamma_2 (m+1)^2 \, (8.20\,b) \qquad$ substitute

$\gamma_0 = -\gamma_2 (m+1)$

$\gamma_1 = -\gamma_2 m \qquad\qquad$ then $\quad y_t = \gamma_0 \omega_{0t} + \gamma_1 \omega_{1t} + \gamma_2 \omega_{2t} + u_t$

and substituting

$y_t = \gamma_2 \omega_t^* + u_t$

$\omega_t^* = \omega_{2t} - m\omega_{1t} - (m+1)\omega_{0t}$

$\quad = \sum_{j=0}^{m} \left(j^2 - mj - m - 1 \right) x_{t-j} \qquad t = m+1, \cdots T$

8
Dynamic Models II: Stochastic Difference Equations

1. Introduction

In modelling dynamic relationships, econometricians have tended to favour stochastic difference equations over transfer functions. There are cases where this preference reflects a strong prior belief in a partial adjustment theory of behaviour. However, as a general rule, economic theory cannot be relied upon to provide any firm guidance on this matter. Given that a model is only an approximation to reality, questions of model specification are, to a large extent, determined on pragmatic grounds.

The advantages of a stochastic difference equation formulation only really become apparent when the model contains more than one explanatory variable. Suppose that y depends on k explanatory variables, and that the influence of each of these variables is felt through a geometric distributed lag. Such a relationship could be modelled by the 'multiple input' transfer function

$$y_t = \sum_{i=1}^{k} \frac{\beta_i x_{ti}}{1 - \alpha_i L} + \frac{\epsilon_t}{1 - \phi L} \tag{1.1}$$

Although the disturbance term in (1.1) has been specified as an AR(1) process it could, in principle, be any ARMA process. However, assuming it to be an AR(1) process is a convenient simplification in the argument that follows.

As it stands, (1.1) is highly nonlinear, and contains $2k + 2$ unknown parameters. Furthermore, in constructing the sum of squares function, kr residuals are 'lost' if setting pre-sample values of x and y to zero is to be avoided; cf. Section 7.3. However, matters are simplified considerably by imposing the constraints $\alpha_i = \alpha$, $i = 1, \ldots, k$ and $\phi = \alpha$. On re-arranging this gives the stochastic difference equation,

$$y_t = \alpha y_{t-1} + \sum_{i=1}^{k} \beta_i x_{ti} + \epsilon_t \tag{1.2}$$

There are two attractions to (1.2), the first obvious, the second less so. The obvious attraction concerns estimation. If ϵ_t is normally distributed, OLS will be fully efficient. The second point in favour of a formulation like (1.2), however, is more fundamental. It is much easier, within a stochastic difference equation framework, to adopt a model selection procedure which moves from the general to the specific. Such an approach is particularly appropriate when there is uncertainty regarding the explanatory variables to be included in the model. The technical aspects of model specification are discussed in some detail in Section 5.

The price paid for imposing the restrictions leading to (1.2) is a less flexible lag structure. However, this can, to some extent, be compensated for by adding lagged values of the x_i's. If (1.2) is extended to include each explanatory variable lagged one period, the model will contain the same number of unknown parameters as (1.1). This may, or may not, give a better approximation to the underlying dynamic structure. However, the advantages of (1.2) with regard to estimation and specification still remain.

in engineer...

when $\dfrac{B_i(L)}{A_i(L)} \longrightarrow A_i(L) = A(L)$

Autoregressive Distributed Lag and ARMAX Models

Equation (1.2) may be generalised to

$$A(L)y_t = \sum_{i=1}^{k} B_i(L)x_{ti} + \epsilon_t \tag{1.3}$$

this preferred to transfer function — reduce down   *model selection*   *can*

where $A(L)$ is a polynomial in the lag operator defined by (7.1.11), and $B_i(L)$ is a polynomial of order s_i defined as in (7.1.10) for $i = 1, \ldots, k$. A necessary condition for stability is that the roots of $A(L)$ should be outside the unit circle. Such a model is sometimes known as an *autoregressive distributed lag*, and the precise specification is given by the abbreviation $AD(r, s_1, \ldots, s_k)$. Using this terminology (7.1.13) is $AD(1, 0)$. *order of lag*

The specification of the disturbance term in (1.3) may be extended to allow it to follow any ARMA process. There are, however, some advantages in restricting attention to pure AR processes, and these advantages will become apparent in Section 5. On the other hand, an MA specification for the disturbance term does have a certain appeal. Such a model may be written in the form

$$A(L)y_t = \sum_{i=1}^{k} B_i(L)x_{ti} + \theta(L)\epsilon_t \qquad\qquad (1.4)$$

and referred to as ARMAX $(r, s_1, \ldots, s_k, q)$. Any transfer function may be expressed as an ARMAX model with restrictions on the lag polynomials. For example, (7.3.16) has $A(L) = \theta(L)$.

The properties of the disturbance term are irrelevant as regards the systematic dynamics of the model. As was indicated in Section 7.1, the material on transfer functions is directly applicable to stochastic difference equations. Thus the mean path of y_t in (1.3), or (1.4), is given by

$$E(y_t) = \sum_{i=1}^{k} \frac{B_i(L)x_{ti}}{A(L)}$$

and the systematic dynamics may be deduced by regarding the model as a special case of a transfer function.

Example 1   Consider the model

$$y_t = \alpha y_{t-1} + \beta_1 x_{t1} + \beta_2 x_{t2} + u_t \qquad\qquad (1.5)$$

The expected effect of a change in the vector $x_t = (x_{t1}\, x_{t2})'$ may be assessed additively. Thus the impact multiplier, the immediate effect of a unit change in both x_1 and x_2, is simply $\beta_1 + \beta_2$. After one time period this increases to $(\beta_1 + \beta_2)(1 + \alpha)$ and so on.

Seasonality and Prediction

The systematic dynamics of a stochastic difference equation can be deduced from the properties of the corresponding transfer function model. Thus the material of the previous section is directly relevant. In a similar way, the treatment of seasonality and the handling of predictions essentially follows from what has gone before.

It was argued in Section 7.6 that dynamic relationships should, wherever possible, be estimated using series which have not been seasonally adjusted. This conclusion continues to hold for models formulated as stochastic difference equations. By and large the same techniques for introducing deterministic and stochastic seasonal components into the model are appropriate, and seasonal differencing and seasonal dummies are widely used. In addition, seasonal effects are often introduced into the lag structure itself in a stochastic difference equation, e.g.

$$y_t = \alpha y_{t-1} + \beta_0 x_t + \beta_4 x_{t-4} + \epsilon_t \tag{1.6}$$

Seasonality could also be introduced into a transfer function model in this way, although the approach seems to be a more natural one within the context of a stochastic difference equation. When there is no option but to use seasonally adjusted data, it should be borne in mind that seasonal effects may still be present.

Predictions from a stochastic difference equation can be made recursively, since the model is already in a form corresponding to (7.7.1). The MSE of a prediction may be deduced by casting the model in transfer function form. Consider the general formulation, (1.4). If the parameters in the model are known, the prediction MSE is the same as the MSE of a prediction from the time series model, $y_t = A^{-1}(L)\theta(L)\epsilon_t$.

Example 2   The predictions of future values of y_t in the model

$$y_t = \alpha y_{t-1} + \beta x_t + \epsilon_t \tag{1.7}$$

are made directly from the recursion

$$\tilde{y}_{T+l/T} = \alpha \tilde{y}_{T+l-1/T} + \beta x_{T+l}, \qquad l = 1, 2, \ldots \tag{1.8}$$

with $y_{T/T} = y_T$. The transfer function formulation of (1.7),

$$y_t = \frac{\beta x_t}{1 - \alpha L} + \frac{\epsilon_t}{1 - \alpha L}$$

implies that for given values of α and β the MSE of $\tilde{y}_{T+l/T}$ is identical to the MSE of the l-step ahead predictor in an AR(1) time series model. When α and β are estimated by OLS, formula (2.5.3) may be applied directly. This provides an estimate of the MSE of $\tilde{y}_{T+l/T}$, which takes account of the fact that α and β are unknown.

2. Estimation

The main issues which arise in estimating stochastic difference expectations can be discussed with respect to the simple model introduced in (7.1.13)

$$y_t = \alpha y_{t-1} + \beta x_t + u_t, \qquad t = 2, \ldots, T \text{ with } |\alpha| < 1 \tag{2.1}$$

The extensions to the more general model, (7.1.16), are straight-forward and will not be discussed explicitly.

OLS and Instrumental Variables

The simplest way to estimate α and β in (2.1) is to regress y_t on y_{t-1} and x_t. If the disturbance term is white noise and y_1 is regarded as being fixed, the Mann–Wald theorem is directly applicable. This result was discussed in Section 2.3, and its main implication is that a model of the form (2.1) may be treated in exactly the same way as a classical regression model, provided the sample size is large. Thus if a and b denote the OLS estimators of α and β respectively,

$$\begin{bmatrix} a \\ b \end{bmatrix} \sim AN\left[\begin{pmatrix} \alpha \\ \beta \end{pmatrix}, \ \sigma^2 \begin{pmatrix} \Sigma y_{t-1}^2 & \Sigma y_{t-1}x_t \\ \Sigma y_{t-1}x_t & \Sigma x_t^2 \end{pmatrix}^{-1} \right] \tag{2.2}$$

If, in addition to being white noise, u_t is normally distributed, the OLS estimator is the ML estimator. It is therefore asymptotically efficient.

Although OLS has very desirable properties when u_t is white noise, it becomes less attractive in the presence of serial correlation. If $E(u_t u_{t-1}) \neq 0$, u_t and y_{t-1} will be correlated even in large samples, cf. (7.1.18). As was noted in Section 7.1, the result of this is that OLS is no longer consistent.

The IV solution to problems of this kind was discussed in Section 7.3, and the results established there are directly applicable in this case as well. The IV estimator of (2.1) is identical to the estimator introduced for the Koyck distributed lag. Thus taking x_{t-1} as an instrument for y_{t-1} leads to estimators of α and β which are consistent.

ML Estimation with AR(1) Disturbances

Suppose that the disturbances in (2.1) follow a stationary AR(1) process as defined by (6.1.2). If y_1 and y_2 are fixed, the joint density of $y_3, \ldots, y_T$ is determined by $\epsilon_3, \ldots, \epsilon_T$. For given values of α, β and ϕ, a set of $T - 2$ residuals is defined by

$$\epsilon_t = y_t - \alpha y_{t-1} - \beta x_t - \phi(y_{t-1} - \alpha y_{t-2} - \beta x_{t-1}),$$
$$t = 3, \ldots, T \tag{2.3}$$

and ML estimation is carried out by minimising the sum of squares function

$$S(\alpha, \beta, \phi) = \sum_{t=3}^{T} \epsilon_t^2 \tag{2.4}$$

A natural algorithm for minimising $S(\alpha, \beta, \phi)$ is Gauss–Newton, particularly as analytic first derivatives are readily available. Differentiating (2.3) with respect to α, β and ϕ gives

$$\frac{\partial \epsilon_t}{\partial \alpha} = -(y_{t-1} - \phi y_{t-2}) \qquad\qquad (2.5)$$

$$\frac{\partial \epsilon_t}{\partial \beta} = -(x_t - \phi x_{t-1}) \qquad\qquad (2.6)$$

$$\frac{\partial \epsilon_t}{\partial \phi} = -(y_{t-1} - \alpha y_{t-2} - \beta x_{t-1}), \qquad t = 3, \ldots, T \qquad (2.7)$$

respectively. Note that the last derivative is equal to $-u_{t-1}$ when α and β are set at their true values.

A Cochrane–Orcutt procedure could also be adopted by minimising (2.4). This would be applied in much the same way as in Chapter 6, with α and β being estimated for a given value of ϕ, and ϕ being estimated conditional on α and β. For a fixed regressor model, it was shown that Cochrane–Orcutt is equivalent to Gauss–Newton in large samples. However, this equivalence no longer holds for a model with a lagged dependent variable, the reason being that

$$\text{plim } T^{-1} \sum \frac{\partial \epsilon_t}{\partial \alpha} \frac{\partial \epsilon_t}{\partial \phi} \neq 0 \qquad\qquad (2.8)$$

The Cochrane–Orcutt scheme will eventually converge as it is a stepwise optimisation procedure. However, it is arguably less attractive than Gauss–Newton insofar as information about the structure of the model is suppressed.

There are a number of ways of obtaining starting values for an iterative ML procedure in this model. One possibility is to estimate α and β by the method of instrumental variable, and then compute an estimate of ϕ from the residuals in the usual way; cf. (6.1.9). This has the advantage that the estimates of all three parameters will be consistent. On the other hand, OLS may be more attractive in practice, even though it is inconsistent. Note that the Gauss–Newton iterations cannot be started by setting all three parameters equal to zero. On examining (2.5) and (2.7), it can be seen that this would invite immediate disaster in view of the perfect multi-collinearity between $\partial \epsilon_t / \partial \alpha$ and $\partial \epsilon_t / \partial \phi$.

Asymptotic Properties

Under suitable conditions the ML estimator of $\psi = (\alpha, \beta, \phi)'$ is asymptotically normally distributed with a mean of ψ and a covariance matrix equal to the inverse of the asymptotic information matrix, $IA(\psi)$, divided by T. A formal demonstration of this result will be found in Hatanaka (1974). All that will be done here is to derive an expression for $\mathrm{Avar}(\check{\psi})$. This may be carried out most easily by using the first derivatives, (2.5) to (2.7), in formula (3.5.18). As is usual in this type of problem, the ML estimator of σ^2 is distributed independently of the other estimators in large samples with a variance of $2\sigma^4/T$. It can therefore be ignored in any discussion of the large sample properties of the ML estimator of ψ.

The terms which only involve derivatives with respect to α or β are relatively easy to evaluate. Thus, for example,

$$\mathrm{plim}\ \frac{1}{T}\Sigma\left(\frac{\partial \epsilon_t}{\partial \beta}\right)^2 = q_x^2$$

where

$$q_x^2 = \mathrm{plim}\ T^{-1}\Sigma(x_t - \phi x_{t-1})^2 \tag{2.10}$$

Similar expressions may be obtained for the other elements in the 2×2 sub-matrix in the top left hand corner of $IA(\alpha, \beta, \phi)$. The terms q_y^2 and q_{xy} are defined in the same way as q_x^2 with y interpreted as y_{t-1}. Note that these definitions imply that certain assumptions must be imposed on the sequence of x_t's.

The remaining terms in $IA(\psi)$ may be evaluated as follows:

$$\mathrm{plim}\ \frac{1}{T}\Sigma\left(\frac{\partial \epsilon_t}{\partial \phi}\right)^2 = \mathrm{plim}\ \frac{1}{T}\Sigma u_{t-1}^2 = \frac{\sigma^2}{1 - \phi^2} \tag{2.11}$$

$$\mathrm{plim}\ \frac{1}{T}\Sigma\ \frac{\partial \epsilon_t}{\partial \phi}\cdot\frac{\partial \epsilon_t}{\partial \beta} = 0 \tag{2.12}$$

and

$$\mathrm{plim}\ \frac{1}{T}\Sigma\ \frac{\partial \epsilon_t}{\partial \alpha}\cdot\frac{\partial \epsilon_t}{\partial \phi} = \frac{\sigma^2}{1 - \alpha\phi} \tag{2.13}$$

The last of these expressions is evaluated by noting that $y_t = \{\beta x_t + u_t\}/(1 - \alpha L)$ and so

$$\mathrm{plim}\ \frac{1}{T}\Sigma u_t y_{t-j} = \frac{\sigma^2 \phi^j}{1 - \phi^2}\ (1 + \alpha\phi + \alpha^2\phi^2 \ldots) = \frac{\sigma^2}{1 - \phi^2}\cdot\frac{\phi^j}{1 - \alpha\phi} \tag{2.14}$$

Using this result for $j = 0$ and $j = 1$ then gives

$$\text{plim}\,\frac{1}{T}\,\Sigma\,\frac{\partial\epsilon_t}{\partial\alpha}\cdot\frac{\partial\epsilon_t}{\partial\phi} = \text{plim}\,T^{-1}\sum(y_{t-1}-\phi y_{t-2})u_{t-1}$$

$$= \frac{\sigma^2}{1-\phi^2}\cdot\frac{1-\phi^2}{1-\alpha\phi} = \frac{\sigma^2}{1-\alpha\phi}$$

The full asymptotic covariance matrix is

$$\text{Avar}(\tilde{\psi}) = T^{-1}IA^{-1}(\alpha,\beta,\phi)$$

$$= T^{-1}\begin{bmatrix} \sigma^{-2}q_y^2 & \sigma^{-2}q_{xy} & (1-\alpha\phi)^{-1} \\ . & \sigma^{-2}q_x^2 & 0 \\ . & . & (1-\phi^2)^{-1} \end{bmatrix}^{-1} \tag{2.15}$$

The crucial feature of (2.15) as compared with (6.1.12) is the appearance of the term $1/(1-\alpha\phi)$. Thus, when a lagged dependent variable appears in a model, its estimated coefficient is not independent of the coefficient of the AR process driving the disturbance term. One implication of this point has already been noted in discussing ML iterative procedures. However, the consequences of the result are far more important in the context of two-step estimation and testing for serial correlation.

Two-Step Estimators

Given initial consistent estimators of α, β and ϕ, one iteration of Gauss–Newton will yield a set of asymptotically efficient estimators; see Section 4.5. A feasible Aitken estimator, based on a regression of $y_t - \hat{\phi}y_{t-1}$ on $y_{t-1} - \hat{\phi}y_{t-2}$ and $x_t - \hat{\phi}x_{t-1}$, where $\hat{\phi}$ is consistent, will *not*, however, be fully efficient. This is basically because of the term $(1-\alpha\phi)^{-1}$ in the information matrix which establishes a link between y_{t-1} and the disturbance.

Hatanaka (1974) was the first to point out how an efficient two-step estimator might be obtained for this model. His estimator may be constructed as follows:

(i) Estimate α and β consistently using the method of instrumental variables. Denoting these estimators by $\tilde{\alpha}$ and $\tilde{\beta}$, form the residuals $\hat{u}_t = y_t - \hat{\alpha}y_{t-1} - \hat{\beta}x_t$, and estimate ϕ by regressing $\hat{u}_t$ on $\hat{u}_{t-1}$. Denote this estimator by $\hat{\phi}$.

(ii) Regress $y_t - \hat{\phi}y_{t-1}$ on $y_{t-1} - \hat{\phi}y_{t-2}$, $x_t - \hat{\phi}x_{t-1}$ and $\hat{u}_{t-1}$. The two-step estimators of α and β are given directly by the regression coefficients of $y_{t-1} - \hat{\phi}y_{t-2}$ and $x_t - \hat{\phi}x_{t-1}$.

respectively. If $\phi^\dagger$ denotes the regression coefficient of $\hat{u}_{t-1}$, the two-step estimator of ϕ is equal to $\hat{\phi} + \phi^\dagger$.

Hatanaka refers to this estimator as the *residual adjusted Aitken estimator*. It is the presence of $\hat{u}_{t-1}$ in the second stage regression which differentiates it from the feasible Aitken estimator, and ensures asymptotic efficiency.

Unfortunately, the form of the residual adjusted Aitken estimator does not make its relationship to Gauss—Newton immediately apparent. Given $\hat{\alpha}$, $\hat{\beta}$ and $\hat{\phi}$ as in (i) above, one iteration of Gauss—Newton involves regressing ϵ_t on the three derivatives, (2.5) to (2.7). However, if ϵ_t is split up into three parts,

$$\epsilon_t = (y_t - \phi y_{t-1}) - \beta(x_t - \phi x_{t-1}) - \alpha(y_{t-1} - \phi y_{t-2}) \qquad (2.16)$$

it will be observed from (2.5) and (2.6) that it may be written

$$\epsilon_t = (y_t - \phi y_{t-1}) + \beta \cdot \partial\epsilon_t/\partial\beta + \alpha \cdot \partial\epsilon_t/\partial\alpha \qquad (2.17)$$

Now $\partial\epsilon_t/\partial\beta$ and $\partial\epsilon_t/\partial\alpha$ are regressors in the Gauss—Newton iteration, and when a regressor is identical to the dependent variable, its estimated coefficient will be unity while all the other coefficients are zero; see for example, Theil (1971, p. 549). Given this, it follows almost immediately that the two-step Gauss—Newton estimator is identical to Hatanaka's residual adjusted Aitken estimator.

Treatment of the 'First Observation'*

In Section 6.1, the full GLS transformation was contrasted with the Cochrane—Orcutt transform. It was argued there that omitting the first set of transformed observations, as is done in Cochrane—Orcutt, can have an adverse effect on small sample properties. The same argument applies in dynamic models, and it therefore seems reasonable to define the likelihood function with only y_1 regarded as being fixed. This means that y_2 is taken to be normally distributed with a mean of $y_2 - \alpha y_1 - \beta x_2$ and a variance of $\sigma^2/(1 - \phi^2)$. The set of residuals given in (2.3) are therefore augmented by

$$\epsilon_2 = \sqrt{1 - \phi^2}\,(y_2 - \alpha y_1 - \beta x_2) \qquad (2.18)$$

and the sum of squares function, (2.4), is redefined to include ϵ_2.

Unfortunately, the way in which y_2 is handled introduces a determinantal term into the likelihood. The function to be minimised is therefore

$$S^*(\alpha, \beta, \phi) = S(\alpha, \beta, \phi)/|1 - \phi^2|^{1/T} \qquad (2.19)$$

However, it was argued in Section 6.1, that the inclusion of the determinant in the likelihood function is usually unimportant in practice. It therefore seems reasonable to construct an estimator based on minimising the sum of squares, $S(\alpha, \beta, \phi)$. This simplifies matters considerably since the Gauss—Newton algorithm can again be applied and the structure of the two-step estimator remains the same. The derivatives of (2.18) are

$$\frac{\partial \epsilon_2}{\partial \alpha} = -\sqrt{1 - \phi^2} \cdot y_1 \tag{2.20}$$

$$\frac{\partial \epsilon_2}{\partial \beta} = -\sqrt{1 - \phi^2} \cdot x_2 \tag{2.21}$$

and

$$\frac{\partial \epsilon_2}{\partial \phi} = -\phi(y_2 - \alpha y_1 - \beta x_2)/\sqrt{1 - \phi^2} \tag{2.22}$$

respectively.

Given the evidence from the fixed regressor model it would seem advisable to include ϵ_2 in both ML and two-step estimation procedures, particularly when one or more of the exogenous variables is subject to a trend. The only relevant Monte Carlo work for the lagged dependent variable model is provided, somewhat incidentally, in an article by Maddala and Rao (1973). They examine an ML estimator for (2.1) with AR(1) disturbances and, with US GNP data used for x_t, find that ML is in many cases inferior to OLS on an MSE comparison; see Maddala and Rao (1973, Table V, p. 767). The estimates they used did not allow for the first set of transformed observations implicit in (2.18), and since GNP data is trending, the poor results for the ML estimator exactly parallel those for the Cochrane—Orcutt procedure in the fixed regressor case.

Estimation of ARMAX Models

In more general formulations of stochastic difference equation models, the disturbances are generated by an ARMA(p, q) process. However, the main principles involved in estimating such models can be illustrated by taking (2.1) and allowing the disturbance term to follow an MA(1) process. The model in question,

$$y_t = \alpha y_{t-1} + \beta x_t + \epsilon_t + \theta \epsilon_{t-1} \tag{2.23}$$

falls within the class of ARMAX models defined in (1.4).

If y_1 is assumed to be fixed and ϵ_1 is assumed to be zero, maximising the likelihood function is equivalent to minimising the sum of squares function

$$S(\alpha, \beta, \theta) = \sum_{t=2}^{T} \epsilon_t^2 \tag{2.24}$$

where

$$\epsilon_t = y_t - \alpha y_{t-1} - \beta x_t - \theta \epsilon_{t-1}, \qquad t = 2, \ldots, T \tag{2.25}$$

with $\epsilon_1 = 0$. As is usual with MA disturbances, the derivatives of the residuals may be obtained recursively. In this case

$$\frac{\partial \epsilon_t}{\partial \alpha} = -y_{t-1} - \theta \frac{\partial \epsilon_{t-1}}{\partial \alpha} \tag{2.26}$$

$$\frac{\partial \epsilon_t}{\partial \beta} = -x_t - \theta \frac{\partial \epsilon_{t-1}}{\partial \beta} \tag{2.27}$$

$$\frac{\partial \epsilon_t}{\partial \theta} = -\epsilon_{t-1} - \theta \frac{\partial \epsilon_{t-1}}{\partial \theta} \tag{2.28}$$

All three derivatives are evaluated by setting them equal to zero for $t = 1$.

The analytic derivatives can be exploited in an iterative ML scheme such as Gauss—Newton. However, from the theoretical point of view the most interesting aspect of (2.26) to (2.28) lies in the construction of a two-step estimator. Initial consistent estimators of α and β can be obtained by the method of instrumental variables, as in Section 7.3, and the residuals used to construct a consistent estimator of θ using (4.5.4). Asymptotically efficient estimators of all three parameters can then be obtained by regressing ϵ_t on (2.26) to (2.28) and subtracting the resulting coefficients from the corresponding initial estimates. This procedure exactly parallels the residual adjusted Aitken estimator in that it is based on a single iteration of Gauss—Newton. However, its small sample properties have yet to be investigated.

The Effect of Misspecification

It was pointed out earlier that applying OLS to (2.1) in the presence of serially correlated disturbances generally yields inconsistent estimators of α and β. However, while underparameterising the disturbance term can lead to inconsistency, overparameterisation can lead to inefficiency, even in large samples. This may at first sight seem surprising, since for a fixed regressor model, carrying out estimation on the assumption that the disturbances follow, say, an

AR(1) process when they are really white noise has no effect on the asymptotic efficiency of the ML estimators of β. Once again it is the structure of the information matrix which differentiates a lagged dependent variable model from one with fixed regressors.

Consider (2.1), and suppose that the disturbance term is white noise. If the model is estimated by ML under the assumption that $u_t \sim AR(1)$, the large sample covariance matrix of $(\tilde{\alpha}, \tilde{\beta}, \tilde{\phi})$ is given by

$$\text{Avar}(\tilde{\alpha}, \tilde{\beta}, \tilde{\phi}) = \frac{\sigma^2}{T} \begin{bmatrix} q_y^2 & q_{xy} & \sigma^2 \\ . & q_x^2 & 0 \\ . & . & \sigma^2 \end{bmatrix}^{-1} \tag{2.29}$$

This expression is obtained by setting ϕ equal to zero in (2.15).

For $\tilde{\alpha}$,

$$\text{Avar}(\tilde{\alpha}) = \frac{\sigma^2}{T} \cdot \frac{q_x^2}{q_y^2 q_x^2 q_{xy}^2 - \sigma^{-2} q_x^2} \tag{2.30}$$

but if $u_t \sim NID(0, \sigma^2)$, the OLS estimator is asymptotically efficient. The large sample variance of the OLS estimator of α is given by the top left hand element of (2.2). Using the notation of (2.15) this becomes

$$\text{Avar}(a) = \frac{\sigma^2}{T} \cdot \frac{q_x^2}{q_x^2 q_y^2 - q_{xy}^2} \tag{2.31}$$

On comparing (2.30) with (2.31) it will be seen that

$$\text{Avar}(\tilde{\alpha}) \geqslant \text{Avar}(a) \tag{2.32}$$

The equality only holds if $q_x^2 = 0$, in which case x_t would be a constant term.

3. Testing for Serial Correlation

In a static regression model, tests of randomness appropriate for an observable time series may be applied directly to the OLS residuals. As a rule these tests will only have asymptotic validity, although in certain special cases the exact distribution of a test statistic can be derived. The same tests may be used when a transfer function model is estimated under the assumption that the disturbances are normally distributed white noise. However, the nonlinear nature of the estimation problem means that small sample results are unavailable. Thus, a valid test for AR(1) disturbances may be carried out with the Durbin–Watson d-statistic, but only on the basis of its asymptotic distribution. This is equivalent to calculating the first order sample autocorrelation coefficient from the residuals and taking it to be normally distributed with a mean of zero and a variance of $1/T$.

Given the above results, it might seem reasonable to suppose that the tests described in Sections 6.3 and 6.6 are applicable when a stochastic difference equation is estimated by OLS. However, in general, this turns out not to be the case. The lagged dependent variable affects the test statistic in such a way that the usual distribution theory fails to hold, even in large samples. The net result is that the tests against serial correlation are not, in general, as powerful as they might be.

The Durbin h-Test

Although model (2.1) contains only one lagged value of the dependent variable, it encompasses all the difficulties described above. Suppose the hypothesis to be tested is that the disturbances follow an AR(1) process. As in the classical model, rejection of this hypothesis may be an indication of a more complicated disturbance process or even of a more fundamental equation misspecification. However, the point at issue here is to derive a test which is valid in the sense that the distribution of the test statistic is known, at least asymptotically, under the null hypothesis.

By suitably modifying the d-statistic to allow for the influence of the lagged dependent variable, Durbin (1970) was able to obtain a suitable test procedure. This is based on the 'h-statistic',

$$h = (1 - \tfrac{1}{2}d)\sqrt{\frac{T}{1 - T \cdot \mathrm{avar}(a)}} \qquad (3.1)$$

where $\mathrm{avar}(a)$ is the estimated variance of the OLS estimator of α. Under the null hypothesis, $h \sim AN(0, 1)$, and a test for positive serial correlation at the five percent level would take $h > 1.645$ as the critical region.

Since $d \simeq 2(1 - r_1)$, where r_1 is the first order autocorrelation of the OLS residuals, the h-statistic may be expressed as

$$h \simeq r_1 \sqrt{\frac{T}{1 - T \cdot \mathrm{avar}(a)}} \qquad (3.2)$$

This implies that r_1 is asymptotically normal with a mean of zero and a variance of $\{1 - T \cdot \mathrm{Avar}(a)\}/T$. In the static regression and transfer function models, on the other hand, $r_1 \sim AN(0, 1/T)$. The distribution of r_1 is therefore more concentrated when a lagged dependent variable is present, at least in large samples. This implies that taking r_1 to have a variance of $1/T$ will result in a test in which the actual size is smaller than the nominal size. Hence we find the potential loss in power.

Derivation of the h-Statistic and Some Alternatives

The rationale behind the h-statistic becomes clear once it is recognised that it can be derived using the Lagrange Multiplier principle. The log–likelihood function for (2.1) with AR(1) disturbances is

$$\log L(\alpha, \beta, \phi) = -\frac{T}{2} \log 2\pi - \frac{T}{2} \log \sigma^2 - \frac{1}{2\sigma^2} S(\alpha, \beta, \phi) \quad (3.3)$$

where $S(\alpha, \beta, \phi)$ is defined in (2.4). The theory surrounding formula (5.5.10) is directly appropriate, with $\psi^{(1)} = (\alpha, \beta)'$ and $\psi^{(2)} = \phi$. Thus, using (2.7),

$$\frac{\partial \log L}{\partial \phi} = D_2 \log L(\psi) = -\sigma^{-2} \sum \frac{\partial \epsilon_t}{\partial \phi} \cdot \epsilon_t$$

$$= \sigma^{-2} \sum u_{t-1}(u_t - \phi u_{t-1}) \quad (3.4)$$

where $u_t = y_t - \alpha y_{t-1} - \beta x_t$. Under the null hypothesis, $H_0 : \phi = 0$, expression (3.4) is evaluated as

$$D_2 \log L(\tilde{\psi}_0) = \hat{\sigma}^{-2} \sum \hat{u}_{t-1} \hat{u}_t \quad (3.5)$$

where $\hat{u}_t$ denotes the tth OLS residual from the regression of y_t on y_{t-1} and x_t, and $\hat{\sigma}^2 = T^{-1} \sum \hat{u}_t^2$ is the ML estimator of σ^2.

The information matrix for the unrestricted model is given in (2.15). Partitioning in accordance with (5.5.9) and evaluating at $\phi = 0$ yields

$$I_{22}^1 - I_{21}' I_{11}^{-1} I_{12} = T - T^2 \, \text{avar}(a) \quad (3.6)$$

The vector $I_{12} = (T, 0)'$ effectively picks out the top left-hand element of I_{11}^{-1}. This is a consistent estimator of the asymptotic variance of the OLS estimator of α when the null hypothesis is true; cf. (2.2).

Substituting (3.5) and (3.6) into the expression for the LM statistic given in (5.5.10) gives

$$\text{LM} = \frac{(\hat{\sigma}^{-2} \sum \hat{u}_t \hat{u}_{t-1})^2}{T - T^2 \, \text{avar}(a)} = r_1^2 \left\{ \frac{T}{1 - T \cdot \text{avar}(a)} \right\} \quad (3.7)$$

Under H_0, LM is asymptotically χ_1^2. However, the square root is identical to (3.2), thereby justifying a test procedure in which the critical region for the h-statistic is based on a standardised normal distribution.

Since maximising the log–likelihood function, (3.3), is equivalent to minimising the sum of squares function, (2.4), the LM test statistic can also be expressed in the $T \cdot R^2$ form of (5.5.16). The derivatives of ϵ_t are given by (2.5) to (2.7). These may be evaluated under the restriction that $\phi = 0$. In terms of the notation of Section 5.5 this yields $z_{t1} = y_{t-1}, z_{t2} = x_t$ and $z_{t3} = \hat{u}_{t-1}$. An LM test of $\phi = 0$ is therefore obtained by regressing u_t on y_{t-1}, x_t and $\hat{u}_{t-1}$, and treating the statistic $T \cdot R^2$ as χ_1^2. Alternatively, a test of significance may be carried out on the coefficient of $\hat{u}_{t-1}$. This last approach was suggested by Durbin (1970), although within the framework of Section 5.5 it will be recognised as a modified LM test. The most convenient way to carry out the test is with reference to the t-distribution, although the test is, of course, only valid asymptotically.

The modified LM test can be used with a one-sided or a two-sided alternative. The former will usually be most appropriate, with the null hypothesis being rejected against the alternative of positive serial correlation when the t-ratio is greater than a particular critical value.

The Monte Carlo results reported in Spencer (1975) suggest that the actual size of the modified LM test will be close to the nominal size even for a sample as small as twenty. However, this does not appear to be the case for the h-statistic. In his experiments Spencer found that in only three cases out of twenty-seven did the expected proportion of rejections *not* differ significantly from the proportion expected. Furthermore, the h-test suffers from an important practical disadvantage. If $T \cdot \mathrm{avar}(a) \geqslant 1$, the h-statistic cannot be computed and the test breaks down. Spencer found that the proportion of such cases was relatively high, particularly for small values of T and α.

A General Test of Misspecification

The portmanteau test, described in Section 6.6, provides a useful check on the goodness of fit in a classical regression model. It is also valid for a transfer function model with white noise disturbances. However, for the reasons already given, it is not appropriate to use it with the residuals from a stochastic difference equation.

The portmanteau test may be rationalised in terms of an LM test against an $\mathrm{AR}(P)$ or $\mathrm{MA}(p)$ disturbance term. Given that an effective test against $\mathrm{AR}(1)$ disturbances has just been derived using the LM principle, it seems natural to employ the same approach in developing an overall test for goodness of fit.

Suppose that model (2.1) has been estimated by OLS, and that it is required to test the assumption of white noise disturbances against the alternative, $u_t \sim MA(P)$. The unrestricted model is ARMAX$(1, 0, P)$, with a residual function

$$\epsilon_t = y_t - \alpha y_{t-1} - \beta x_t - \theta_1 \epsilon_{t-1} - \cdots - \theta_P \epsilon_{t-P} \qquad (3.8)$$

The derivatives for the ARMAX$(1, 0, 1)$ model are given in (2.26) to (2.28) and it is not difficult to see that the derivatives for (3.8) with respect to α, β and $\theta_1, \ldots, \theta_P$ have a similar form. Evaluating these derivatives under the null hypothesis that $\theta_1 = \cdots = \theta_P = 0$ gives

$$\frac{\partial \epsilon_t}{\partial \alpha} = -y_{t-1}, \quad \frac{\partial \epsilon_t}{\partial \beta} = -x_t \quad \text{and} \quad \frac{\partial \epsilon_t}{\partial \theta_j} = -\hat{u}_{t-j}, \qquad j = 1, \ldots, P$$

where $\hat{u}_t$ is the tth OLS residual in the regression of y_t on y_{t-1} and x_t. The modified LM procedure therefore consists of regressing $\hat{u}_t$ on y_{t-1}, x_t and $\hat{u}_{t-1}, \ldots, \hat{u}_{t-P}$, and testing the joint significance of $\hat{u}_{t-1}$ to $\hat{u}_{t-P}$. The LM test is based on $T \cdot R^2$ from the same regression, and is, of course, asymptotically equivalent. Exactly the same test procedure is obtained if the alternative hypothesis is taken to be $u_t \sim AR(P)$.

In the general case, the underlying model is of the form (1.3). Extending the model to include an MA(P) disturbance term then leads to the ARMAX model, (1.4), and a test analogous to the one described in the previous paragraph may be summarised as follows. *Add P lagged OLS residuals, $\hat{u}_{t-1}, \ldots, \hat{u}_{t-P}$, to the set of regressors and carry out an F-test of joint significance.* Although the formal justification for this procedure is as a test for an MA(P), or equivalently AR(P), disturbance term, its main use is likely to be in the context of diagnostic checking. As such it will be referred to as the *goodness of fit F-test*, and denoted by $Q^\dagger(P, T^\dagger)$, where $T^\dagger$ denotes the sample size after making an appropriate correction for the loss in degrees of freedom.

Given the above derivation, the form of the modified LM test for any higher order AR or MA process should be clear. For example, a test for fourth-order autocorrelation may be obtained by adding $\hat{u}_{t-4}$ to the set of regressors and testing it for significance. As in the first-order case a one-sided critical region will often be appropriate.

An Estimated ARMA Disturbance Term

The LM approach also enables valid tests to be constructed when a stochastic difference equation is estimated with an ARMA(p, q)

disturbance term. The general principle may be stated as follows: write down an expression for the ϵ_t in the unrestricted model; find expressions for the derivatives of ϵ_t with respect to all the parameters of the model; evaluate ϵ_t and its derivatives under the null hypothesis; regress ϵ_t on all its derivatives and carry out an F-test of joint significance on the coefficients associated with the derivatives of the parameters which are set equal to zero under H_0. Test statistics of this kind can easily be brought into a computer program constructed within an ML framework. However, it is interesting to examine one or two special cases.

Example 1   Suppose that model (2.1) is estimated with an AR(1) disturbance term and it is required to test the hypothesis that $u_t \sim$ AR(2). The residual function in the unrestricted model is given by

$$\epsilon_t = y_t - \alpha y_{t-1} - \beta x_t - \phi_1 (y_{t-1} - \alpha y_{t-2} - \beta x_{t-1})$$
$$- \phi_2 (y_{t-2} - \alpha y_{t-3} - \beta x_{t-2}) \tag{3.9}$$

The derivatives of ϵ_t, evaluated under the restriction $\phi_2 = 0$, are

$$\frac{\partial \epsilon_t}{\partial \alpha} = -(y_{t-1} - \tilde{\phi}_1 y_{t-2})$$

$$\frac{\partial \epsilon_t}{\partial \beta} = -(x_t - \tilde{\phi}_1 x_{t-1})$$

$$\frac{\partial \epsilon_t}{\partial \phi_1} = -(y_{t-1} - \tilde{\alpha} y_{t-2} - \tilde{\beta} x_{t-1}) = -\tilde{u}_{t-1}$$

$$\frac{\partial \epsilon_t}{\partial \phi_2} = -\tilde{u}_{t-2} \tag{3.10}$$

where $\tilde{\alpha}$, $\tilde{\beta}$ and $\tilde{\phi}_1$ are the ML estimators. The modified LM test of the hypothesis $H_0 : \phi_2 = 0$ is then carried out by regressing $\tilde{u}_t$ on the four derivatives given in (3.10), and testing $\partial \epsilon_t / \partial \phi_2$ for significance. If $\tilde{\alpha}$, $\tilde{\beta}$ and $\tilde{\phi}_1$ are calculated by the Gauss–Newton scheme described in Section 2, the first three derivatives in (3.10) are immediately available.

Example 2   Consider the ARMAX model, (2.33) and let the alternative hypothesis be that the disturbance term follows an ARMA(1, 1) process. The residual function is

$$\epsilon_t = y_t - \alpha y_{t-1} - \beta x_t - \phi(y_{t-1} - \alpha y_{t-2} - \beta x_{t-1}) - \theta \epsilon_{t-1}$$
$$\tag{3.11}$$

Under the null hypothesis that $\phi = 0$, the derivatives are evaluated as

$$\frac{\partial \epsilon_t}{\partial \alpha} = -y_{t-1} - \tilde{\theta} \frac{\partial \epsilon_{t-1}}{\partial \theta}$$

$$\frac{\partial \epsilon_t}{\partial \beta} = -x_t - \tilde{\theta} \frac{\partial \epsilon_{t-1}}{\partial \theta}$$

$$\frac{\partial \epsilon_t}{\partial \theta} = -\epsilon_{t-1} - \tilde{\theta} \frac{\partial \epsilon_{t-1}}{\partial \theta}$$

and

$$\frac{\partial \epsilon_t}{\partial \phi} = -\tilde{u}_{t-1} \tag{3.12}$$

while

$$\epsilon_t = \tilde{u}_t - \tilde{\theta} \epsilon_{t-1} \tag{3.13}$$

On comparing the first three derivatives of (3.12) with (2.26) to (2.28), it will be observed that they are automatically available as a by-product of the ML computations for the restricted model. Similarly, (3.13) is given directly as the dependent variable in the final Gauss–Newton iteration. Thus all that is entailed in the modified LM test is the addition of $\tilde{u}_{t-1}$ to the set of regressors, followed by a test of significance on its estimated coefficient.

A goodness of fit test, analogous to the one described in the previous sub-section may be obtained by considering a more general model in which the order of *either* the AR *or* the MA component is increased by a P. However, except for the case when the fitted model is assumed to have white noise disturbances, the test statistics will be different.

4. Model Selection

The methodology which has been built up for determining the specification of stochastic difference equations is somewhat different from that which surrounds transfer functions. In modelling transfer functions, a suitable specification is achieved by the cycle of identification, estimation and diagnostic checking. The most appropriate methodology for specifying a stochastic difference equation, on the other hand, is to begin with a very general specification and then 'test down' until a more specific model is obtained. However, any model which is seriously entertained must satisfy various diagnostic checking procedures.

The main feature of the above methodology is the abandonment of any attempt to identify a suitably parsimonious model at the outset. Instead, there is deliberate 'overfitting'. One of the reasons for this stems primarily from the fact that stochastic difference equations typically contain several explanatory variables. In these circumstances it becomes difficult to identify a suitable model by any of the techniques described in Section 5 of the previous chapter. Furthermore, the decision as to whether or not a particular explanatory variable should be included in the model can only really be made if it is included in the model in the first place.

There is, however, a second reason why overfitting is more popular in stochastic difference equations than in transfer functions, and in a sense this is more fundamental than the reason already given. Consider a transfer function model in which $r = s = 1$, i.e.

$$y_t = \frac{(\beta_0 + \beta_1 L)x_t}{(1 - \alpha L)} + \epsilon_t \tag{4.1}$$

A specification of this kind can lead to computational problems arising from common factors. This happens irrespective of the form of the disturbance term, and so for simplicity it has been assumed to be white noise. In the development that follows it will be convenient to reparameterise the numerator of the rational lag in (4.1) by defining γ and β such that $\beta(1 - \gamma L) = \beta_0 + \beta_1 L$. Maximising the likelihood function of (4.1) is equivalent to minimising a sum of squares function in which the residuals are computed recursively from

$$\epsilon_t = \alpha \epsilon_{t-1} + y_t - \alpha y_{t-1} - \beta x_t + \beta \gamma x_{t-1}, \qquad t = 2, \ldots, T \tag{4.2}$$

with $\epsilon_1 = 0$. The derivatives of (4.2) with respect to α and γ are

$$\frac{\partial \epsilon_t}{\partial \alpha} = \epsilon_{t-1} - y_{t-1} + \alpha \frac{\partial \epsilon_{t-1}}{\partial \alpha} \tag{4.3a}$$

and

$$\frac{\partial \epsilon_t}{\partial \gamma} = \beta x_{t-1} + \alpha \frac{\partial \epsilon_{t-1}}{\partial \gamma} \tag{4.3b}$$

Now suppose that there is a common factor in (4.1), i.e. $\gamma = \alpha$. Imposing this restriction in (4.2) implies that

$$\epsilon_t = y_t - \beta x_t \tag{4.4}$$

This is also apparent from (4.1). Using (4.4), it will be seen that when $\alpha = \gamma$, $\partial \epsilon_t / \partial \alpha$ and $-\partial \epsilon_t / \partial \gamma$ are identical. A Gauss—Newton iterative scheme will therefore break down due to multicollinearity. Furthermore, any other iterative scheme will probably also run into difficulties since $\alpha = \gamma$ implies singularity of the information matrix.

The above argument may be summarised by saying that when the lag structure of a transfer function is overparameterised, problems are likely to arise due to lack of identifiability. In a stochastic difference equation, on the other hand, the existence of common factors in the lag structure will not generally imply that the model is underidentified. Thus starting with a very general model will not usually lead to the kind of problem described in the previous paragraph. The movement to a more specific formulation then proceeds by a search for common factors in the lag structure. Once a set of common factors has been extracted, further simplification will be possible if it turns out that some of the factors have a root of zero.

The common factor approach to dynamic specification has been developed by Sargan (1964, 1980a), and applied by Hendry and Mizon (1978) and Mizon and Hendry (1980). The full procedure is described in some detail in the sub-section below, while the final sub-section considers the ways in which it may be used in an overall model selection strategy. In order to avoid any possible confusion with the multivariate technique of 'factor analysis', the common factor method is often referred to as '*COMFAC analysis*'.

COMFAC Analysis

The essential point behind COMFAC analysis is the integration of systematic and disturbance dynamics. Thus serial correlation in the disturbances is taken as a reflection of a particular structure imposed on the systematic dynamics of the equation. This is best explained by means of a simple example.

Consider the model

$$y_t = \alpha y_{t-1} + \beta_0 x_t + \beta_1 x_{t-1} + \epsilon_t, \qquad t = 2, \ldots, T \qquad (4.5)$$

with $|\alpha| < 1$. This may be rewritten as

$$(1 - \alpha L) y_t = (\beta_0 + \beta_1 L) x_t + \epsilon_t \qquad (4.6)$$

and a common factor will be present if $\beta_0(1 - \alpha L) = \beta_0 + \beta_1 L$. The existence of such a common factor implies that the parameters in (4.5) are subject to the restriction

$$\alpha \beta_0 + \beta_1 = 0 \qquad (4.7)$$

However, in contrast to the transfer function, (4.1), which has exactly the same systematic dynamics as (4.5), no computational problems arise when (4.7) holds. Estimation of (4.5) may still be carried out by OLS, and the only consequence of ignoring the restriction in (4.7) is a loss in efficiency.

If the common factor restriction is imposed on the model, it may be rearranged by dividing both sides of (4.6) by $(1 - \alpha L)$. This gives

$$y_t = \beta_0 x_t + u_t \tag{4.8}$$

where $u_t = \epsilon_t/(1 - \alpha L)$. The original dynamic model is therefore reduced to a static one in which the disturbance term follows a stationary AR(1) process,

$$u_t = \alpha u_{t-1} + \epsilon_t \tag{4.9}$$

Thus although serial correlation may be regarded as a 'nuisance' in static regression, it may actually represent a simplification when viewed in the wider context of dynamic regression.

By starting from (4.5) it is possible to test two hypotheses, one of which is nested within the other. The first hypothesis concerns the validity of the common factor restriction, (4.7). If this restriction is accepted, the hypothesis of a zero root, i.e. $\alpha = 0$, may be tested. Failure to reject this second hypothesis leads to a static regression model with white noise disturbances. A model with three parameters is therefore reduced to a model with only one. This is an example of a *simplification search*, and as was argued in Section 5.8, an approach of this kind is generally preferable to one in which a very restricted model is taken as the starting point.

The common factor framework may be generalised as follows. In the stochastic difference equation, (1.3), the orders of the polynomials $A(L), B_1(L), \ldots, B_{k-1}(L)$ and $B_k(L)$ are $r, s_1, \ldots, s_{k-1}$ and s_k respectively. These polynomials can therefore have at most l common roots, where $l = \min(r, s_1, \ldots, s_k)$. If there are actually $p \leqslant l$ common roots, there exists a polynomial $\phi^*(L)$ of order p which is common to $A(L), B_1(L), \ldots, B_k(L)$. The polynomials may therefore be factorised as $A(L) = \phi(L)A^*(L)$, $B_1(L) = \phi(L)B_1^*(L), \ldots,$ $B_k(L) = \phi(L)B_k^*(L)$, and the model becomes

$$A^*(L)y_t = \sum_{i=1}^{k} B_i^*(L)x_{ti} + \phi^{-1}(L)\epsilon_t \tag{4.10}$$

The net result is that the number of parameters in the model is reduced by kp, while the disturbance term becomes an AR(p) process.

The test procedure implemented in Sargan's COMFAC algorithm

$$W = \{r(\tilde{\psi})' [\tilde{R}' I^{*-1}(\tilde{\psi}) \tilde{R}]^{-1} \{r(\tilde{\psi})\}$$

unrestricted estimates

$\tilde{R}$ has $\partial r(\tilde{\psi})$ on jth column

is based on a series of Wald tests. The reason for this is that estimation of the unrestricted model is relatively easy. If $\epsilon_t \sim NID(0, \sigma^2)$, the ML estimates are given by OLS, and these estimators are used as the basis for a series of common factor tests. Unfortunately, the form of the Wald statistics is rather complicated, in general, and a full discussion is beyond the scope of this book. However, in the case of the simple model introduced in (4.5), the derivation of the common factor test is straightforward, and this provides a useful illustration of the technique underlying the whole approach.

See p8 165

A common factor is present in (4.5) when the constraint (4.7) holds. In terms of the notation of Section 5.4, this may be written

$$r(\psi) = \alpha\beta_0 + \beta_1 = 0 \tag{4.11}$$

where $\psi = (\alpha, \beta_0, \beta_1)'$. The derivatives of $r(\psi)$ with respect to α_1, β_0 and β_1 are $\partial r(\psi)/\partial\alpha = \beta_0$, $\partial r(\psi)/\partial\beta_0 = \alpha$ and $\partial r(\psi)/\partial\beta_1 = 1$. The complete vector of derivatives may therefore be written as

$$\partial r(\psi)/\partial\psi = (\beta_0, \alpha, 1)' \tag{4.12}$$

Expressions (4.11) and (4.12) may be evaluated using the OLS estimates in (4.5) and can then be used to construct the Wald statistic, (5.4.8). The matrix $I^*(\tilde{\psi})$ can be set equal to $s^2(\sum z_t z_t')^{-1}$ where $z_t = (y_{t-1}, x_t, x_{t-1})'$. Under $H_0 : r(\psi) = 0$, W has a χ_1^2 distribution in large samples, but when no common factor is present it tends to become large.

Example 1 Godfrey (1973) applied the above test procedure using the data published in Durbin and Watson (1951). The model considered by Durbin and Watson was

$$y_t = \delta + \beta_1 x_{1t} + \beta_2 x_{2t} + u_t \tag{4.13}$$

where y is the logarithm of consumption of spirits per head, x_1 is the logarithm of real income per head and x_2 is a relative price index. The data refer to the UK over the period 1870–1938, and when (4.13) was estimated by OLS the DW statistic obtained was 0.249. This is obviously highly significant, and it could be taken to imply that the appropriate course of action is to estimate the model with an AR(1) disturbance term. However, a more general dynamic specification is a model of the form (4.5). Estimating such a model by OLS in deviation from the mean form gave

$$\hat{y}_t = 0.947 y_{t-1} + 0.717 x_{1t} - 0.744 x_{1, t-1} - 0.892 x_{2t}$$
$$+ 0.824 x_{2, t-1} \tag{4.14}$$

The question now is whether the restrictions implied by fitting

(4.13) with an AR(1) disturbance term are consistent with the estimates of the unrestricted model displayed in (4.14).

Since the model contains two explanatory variables there are two restrictions of the form (4.11). These are evaluated as:

$$r_1(\bar{\psi}) = (0.717)(0.947) - 0.744 = -0.065$$

$$r_2(\bar{\psi}) = (-0.898)(0.947) - 0.824 = -0.026$$

Differentiating $R(\psi) = \{r_1(\psi), r_2(\psi)\}'$ with respect to the parameters in ψ, in the order in which they appear in (4.14) yields

$$\frac{\partial R(\psi)}{\partial \psi} = \begin{bmatrix} 0.717 & 9.47 & 1.0 & 0 & 0 \\ -0.898 & 0 & 0 & 9.47 & 1.0 \end{bmatrix}$$

when evaluated at the estimates in (4.14). This matrix together with the estimated co-variance matrix in (4.14) gave

$$W = (-0.065, -0.026) \begin{bmatrix} 0.302 \times 10^{-2} & -0.158 \times 10^{-4} \\ -0.158 \times 10^{-4} & 0.536 \times 10^{-3} \end{bmatrix}$$

$$\times \begin{pmatrix} -0.065 \\ -0.026 \end{pmatrix} = 2.702$$

Under the null hypothesis that the restrictions are valid $W \sim \chi_2^2$ in large samples. Since the critical value for a χ_2^2 variate at the 10% level of significance is 4.61, the test supports the notion that estimating the static regression model, (4.13), with an AR(1) disturbance term is an appropriate course of action.

The last aspect of the COMFAC procedure is the way in which a series of common factor tests are carried out. Having decided on the most general dynamic specification, a series of nested hypotheses are mapped out. The overall test procedure is then executed in the way described in Section 5.8. The first test is for one common factor. If this is not rejected, a test for two common factors is carried out, and so on. The procedure terminates when an hypothesis is rejected.

Example 2   The study by Mizon and Hendry (1980) is concerned with the specification of a model for Canada relating expenditure on consumer durables (y) to real income (x_1) and relative price (x_2). The observations are quarterly and so a reasonable specification for a general model is to take four period lags on all variables, i.e.

$$\log y_t = \delta_0 + \sum_{j=1}^{3} \delta_j Q_{jt} + \sum_{j=1}^{4} \alpha_j \log y_{t-j} + \sum_{j=0}^{4} \beta_{1j} \log x_{1,t-j}$$

$$+ \sum_{j=0}^{4} \beta_{2j} \log x_{2,t-j} + \epsilon_t \qquad (4.15)$$

The model also includes a constant and three seasonal dummies.

The results of COMFAC are shown in table 8.1. It is the incremental values of the Wald statistics which are used in the test procedure, since these are independent of each other. Thus, for example, a test for two common factors is based on

$$W_2^* = W_2 - W_1$$

i.e.

$$W_2^* = 1.24 - 0.47 = 0.77$$

If each of these tests is carried out at the 1% level of significance, the overall probability of a Type I error is $1 - (0.99)^4 = 0.0394$. Since $Pr(\chi_2^2 > 9.21) = 0.01$, it is clear from the table that the hypothesis of three common factors cannot be rejected, whereas four can be. The above results suggest that a convenient simplification of (4.15) might be to include c, y and p with one period lags, and to allow the disturbance term to follow an AR(3) process. However, this was not the model which Mizon and Hendry finally adopted. On examining the estimates in the unrestricted model, (4.15), it was found that the coefficients of y_{t-2} and x_{t-2} were far larger than the corresponding coefficients at $t-1$. Furthermore, the restricted model yielded a value of s markedly larger than that in the unrestricted model, and a likelihood ratio test indicated a rejection of the common factor restrictions which COMFAC had suggested were data admissible. These findings point very strongly to the view that the results of a COMFAC analysis should not be accepted uncritically, insofar as they suggest a particular specification. However, there is no doubt that COMFAC has a valuable role to play in *rejecting* models which impose invalid common factor restrictions.

Table 8.1   COMFAC Analysis for Model (4.15)

	Direct Tests			Incremental Tests	
No. of common factors (i)	Degrees of freedom	Value of test statistic (W_i)		Degrees of freedom	Value of test statistic (W_i^*)
1	2	0.47		2	0.47
2	4	1.24		2	0.77
3	6	6.09		2	4.85
4	8	32.01		2	25.92

Source: G. E. Mizon and D. F. Hendry (1980)

A Strategy for Model Selection

Any specification which is being seriously entertained must be subject to diagnostic checks. The modified LM test developed in Section 4 is particularly appropriate for this purpose. However, although a test based on $Q(m, T^\dagger)$ is formally a test against an AR(m) or an MA(m) disturbance term, a rejection of the null hypothesis should not be interpreted as evidence *for* such a specification. It was emphasised in earlier chapters that serial correlation in the residuals need not be symptomatic of serial correlation in the disturbances. A more fundamental problem, such as an omitted variable, may lie at the heart of the problem. This is still the case in dynamic regression, although the matter is now further complicated by the consideration that serial correlation in the residuals may also be a reflection of an inappropriate specification of the systematic dynamics. As an example, suppose that the true model is (4.5). The residuals which emerge from regressing y_t on x_t will tend to be serially correlated, but unless the restriction in (4.7) is valid, extending the model to include an AR(1) disturbance term will not be an appropriate course of action.

When the disturbance term in a stochastic difference equation is modelled as an ARMA(p, q) process, the LM test against an ARMA($p + m$, q) process provides the basis for a convenient diagnostic check. Of course, all the points made in the previous paragraph are still relevant.

The residual sum of squares, or alternatively s^2, provides a useful measure of *goodness of fit*. However, it is worth noting that any stochastic difference equation in which the dependent variable is in levels may be re-arranged so that the dependent variable is in first differences. Thus, subtracting y_{t-1} from both sides of (1.2) gives

$$\Delta y_t = (\alpha - 1)y_{t-1} + \sum_{i=1}^{k} \beta_i x_{it} + \epsilon_t \qquad (4.16)$$

and reformulating the model in this way makes no difference to the parameter estimates or to s^2. However, the coefficient of multiple correlation now gives the proportion of the variance of first differences explained by the model and such a measure is arguably more appropriate for time series regression.

Once a model has been selected, it should be validated against post-sample observations. This can be done using the $\xi(l)$ statistic defined in (5.7.4). However, in an autoregressive distributed lag model, OLS is the appropriate estimation technique and it therefore seems reasonable to adopt some of the procedures proposed in Section 5.7

for classical regression. In particular, $\xi(l)$ may be replaced by the
Chow statistic, $\xi^*(l)$. A test based on the F-distribution will only
have asymptotic validity, but because the prediction errors in $\xi^*(l)$
are 'standardised' — see (5.7.7) — the size of the test should be
reasonably close to the nominal significance level.

Bearing the above points in mind, a general strategy for dynamic
model selection is to proceed in the following way.

(i) Estimate a general model in which the dynamics are of a
relatively high order. Examine the coefficients in order to gain some
idea of which lags are likely to be important, and retain the value
of s^2. This provides a yardstick against which the goodness of fit of
more restricted models may be assessed. Could use LM test for diagnostics

(ii) Formulate a more restricted model than the one just estimated,
using any prior economic theory to suggest a suitable specification
for the dynamics. (The role played by economic theory in this
context is discussed in the next section.) The model so specified $\}$ t-tests
should, however, be consistent with the estimated general model. COMFAC
Thus a large coefficient in the general model would not be arbitrarily
set equal to zero.

In the absence of any prior economic theory, COMFAC analysis
may be used to develop a more parsimonious model.

(iii) Check the estimated restricted model for consistency with
the general model. Large discrepancies between estimated parameters
and/or a substantial increase in s^2 imply that some change in specification might be appropriate. (LR test)

At the same time, diagnostic checks based on LM tests may be
carried out. These procedures may be supplemented by the less
formal graphical procedures described in Section 5.2.

(iv) A particular specification may be regarded as acceptable if:
(a) it satisfies any prior constraints imposed by economic theory;
(b) it is consistent with the most general model estimated; (c) it
passes the diagnostic checks; (d) it compares favourably with other
models in terms of goodness of fit; and (e) it satisfies the post-sample
predictive tests of Section 5.7.

5. Economic Theory and Dynamic Specification

Even if the variables entering an equation are known, specification of
the dynamic structure remains a problem. In the previous section a
strategy for model selection was outlined, and it was pointed out that
economic theory can play a vital role in limiting the class of models
to be placed under serious consideration. In this section, it is shown

how the theory surrounding *steady-state* solutions can be incorporated in the model selection process.

See Davidson et al Econ J, 1978

The Steady-State Solution

Consider the model

$$\log y_t = \alpha \log y_{t-1} + \delta + \beta_0 \log x_t + \beta_1 \log x_{t-1} + \epsilon_t \qquad (5.1)$$

where δ is a constant term and $|\alpha| \leq 1$. If x_t is taken to be constant and equal to $\bar{x}$, an *equilibrium* solution for (5.1) is given by suppressing the disturbance term. This takes the form

$$\log \bar{y} = \frac{\delta}{1-\alpha} + \frac{\beta_0 + \beta_1}{1-\alpha} \log \bar{x} \qquad (5.2)$$

cf. (7.2.2). However, the concept of a static equilibrium is limited, and a more general long-run solution for a model of the type (5.1) is obtained by allowing x to grow at a constant rate, g_x. In the *steady-state* it is assumed that the relationship between x and y is of the form $y = Kx^{\nu}$, i.e.

$$\log y = \log K + \nu \log x \qquad (5.3)$$

where K and ν are parameters. If this steady-state solution is to be maintained in the long run, y will have to grow at the rate $g_y = \nu g_x$.

The parameter ν is the *long-run elasticity* of y with respect to x, and economic considerations will often suggest a value, or plausible range of values, for this parameter. Hence the motivation for considering the steady-state solution implied by a particular dynamic specification. In the case of (5.1), the first step is to re-arrange the equation as follows:

$$\Delta \log y_t = (\alpha - 1)\log y_{t-1} + \delta + \beta_0 \Delta \log x_t$$
$$+ (\beta_0 + \beta_1)\log x_{t-1} + \epsilon_t \qquad (5.4)$$

On further re-arrangement this becomes

error correction mechanism

$$\Delta \log y_t = \delta + \beta_0 \Delta \log x_t$$
$$+ (\alpha - 1)\left[\log y_{t-1} - \frac{(\beta_0 + \beta_1)}{1-\alpha} \log x_{t-1} \right] + \epsilon_t \qquad (5.5)$$

In the steady-state there is a linear relationship between log y and log x. This is reflected in the term in square brackets in (5.5). A comparison of this term with expression (5.3) gives

$$\nu = (\beta_0 + \beta_1)/(1-\alpha) \qquad \textit{long run} \qquad (5.6)$$

thereby making explicit the relationship between the parameters of the model and the implied long run elasticity.

Substituting from (5.3) into (5.5), suppressing the disturbance term, and setting $\Delta \log x_t = g_x$ and $\Delta \log y_t = g_y = v g_x$ yields

$$v g_x = \delta + \beta_0 g_x + (\alpha - 1) \log K$$

Re-arranging this equation then leads to the result

$$K = \exp[\{\delta + (\beta_0 - v) g_x\}/(1 - \alpha)] \tag{5.7}$$

The value taken by the constant term in (5.3) therefore depends on the growth rate of x. When $g_x = 0$, the equilibrium solution, (5.2), is obtained.

Example 1 In the model for consumer durables in Canada referred to in Section 4, Mizon and Hendry (1980) estimated the general model (4.15) as a first step in their model selection procedure. In steady-state growth, the long-run solution for the model is of the form

$$y = K_j x_1^{v(1)} x_2^{v(2)}, \qquad j = 0, 1, 2, 3 \tag{5.8}$$

where

$$v(i) = \frac{\beta_{i0} + \beta_{i1} + \beta_{i2} + \beta_{i3} + \beta_{i4}}{1 - \alpha_1 - \alpha_2 - \alpha_3 - \alpha_4}, \qquad i = 1, 2$$

and $\log K_j$ takes a different value in each quarter depending on the estimated coefficient of the corresponding seasonal dummy. In other words the steady-state equation corresponding to (5.3) contains three seasonal dummy variables in addition to the constant term. From the results in table 8.1,

$$v(1) = \frac{0.93}{0.71} = 1.3 \quad \text{and} \quad v(2) = \frac{-0.39}{0.71} = -0.55$$

Thus the long-run income elasticity of demand for consumer durables is 1.3, while the price elasticity of demand is -0.55. These values are reasonable on the basis of prior knowledge: the share of consumer durables in total expenditure tends to go up with rising income, while a negative price elasticity is consistent with a downward sloping demand curve.

In the above example, prior economic knowledge is helpful insofar as it suggests a reasonable range of values for the long-run parameters. However, in some cases, economic theory is able to place much stronger constraints on the model. Suppose that (5.1) is a simple consumption function model in which y and x are consumers'

d impose $\beta_0 + \beta_1 = 1 - \alpha \Rightarrow$

$\Delta \log y_t = \delta + \beta_0 \Delta \log x_t + (\alpha - 1) \log\left(\frac{y_{t-1}}{x_{t-1}}\right) + \varepsilon_t$

Error correction mechanism

expenditure and personal disposable income respectively, both measured in real terms. In the long run, the elasticity of y with respect to x must be unity. If it were not, the proportion of disposable income spent would change over time. However, for nearly all countries for which data are available, the proportion of income spent remains remarkably constant.

Imposing the constraint $\nu = 1$ on (5.4) leads to the model

$$\Delta \log y_t = \delta + \beta_0 \Delta \log x_t + (\alpha - 1)\log(y_{t-1}/x_{t-1}) + \epsilon_t \qquad (5.9)$$

The number of parameters in the model is therefore reduced by one.

Levels, Differences and the Error Correction Mechanism

Suppose that the observations lie very close to the steady-state growth path. In these circumstances there will be strong multi-collinearity between $\log x_{t-1}$ and $\log y_{t-1}$, resulting in very imprecise estimates of the coefficients in (5.4). In (5.5) the term in square brackets will be approximately equal to $\log K$. The model

$$\Delta \log y_t = \delta' + \beta_0 \Delta \log x_t + \epsilon_t \qquad (5.10)$$

where

$$\delta' \simeq \delta + (\alpha - 1)\log K = (1 - \beta_0)g_x$$

is therefore a parsimonious representation of the *short-run* relationship between x and y.

Within this framework, a model cast in first differences may be regarded as an approximation to a more general dynamic model. However, the approximation is only reasonable when x_t and y_t stay close to this path. If the observations begin to deviate from the steady-state growth path, the behaviour of the system can be adequately captured only by the more general model (5.4), and (5.10) becomes inadequate from the point of view of both estimation and forecasting. The term in square brackets in (5.5) now plays a crucial role in the working of the model. Suppose that y begins to grow at a faster rate than is consistent with the steady-state solution. This could arise because of a series of abnormally large random disturbances, or because of the systematic effect of a third variable which does not appear in the long-run solution. The net effect of this increase in the growth rate of y is that ξ, the term in square brackets in (5.5), becomes positive, since y_{t-1} has drifted above the steady-state growth path. However, because the coefficient $\alpha - 1$ is negative, the effect of ξ being positive is to *reduce* the growth rate of y, and drive y_t back towards its long-run growth path.

For this reason the last term in equation (5.5) is known as the *error correction mechanism*.

When the long-run elasticity in (5.3) is equal to unity, the error correction mechanism takes a particularly simple form. This can be seen in equation (5.9) where the error correction term is the variable $\log(y_{t-1}/x_{t-1})$. Models of this kind have been found to be appropriate in a variety of applications, including wage–price models (Sargan 1964), the demand for money (Hendry 1980) and the consumption function (Davidson *et al.*, 1978).

The error correction model of the consumption function is particularly interesting in that it provides a reconciliation of short-run and long-run behaviour. The empirical evidence available for most countries shows that over a prolonged period of time, the relationship between consumption (y) and personal disposable income is of the form:

$$y = Kx \tag{5.11}$$

This implies that the proportion of income consumed (the 'average propensity to consume') remains constant at a value of K. However, a key feature of the Keynesian consumption function is that the marginal propensity to consume (the increase in consumption induced by an additional unit of income) should be less than the average propensity to consume, whereas in (5.11), they are the same. The error correction model takes account of this apparent contradiction by having a steady-state solution of the form (5.11), coupled with a marginal propensity to consume (MPC) which is considerably less than K in the short run.

Example 2   In their study of the UK consumption function, Davidson *et al.* (1978, p. 684) estimated the model

$$\Delta_4 \widehat{\log} \, y_t = 0.49 \, \Delta_4 \log x_t - 0.17 \, \Delta_1 \Delta_4 \log x_t$$
$$\qquad\quad (0.04) \qquad\qquad\quad (0.05)$$

$$\qquad - 0.06 \log(y_{t-4}/x_{t-4}) + 0.01 \, \Delta_4 D_t^0 \tag{5.12}$$
$$\qquad (0.01) \qquad\qquad\qquad (0.004)$$

$$R^2 = 0.71, \quad s = 0.0067, \quad DW = 1.6, \quad \xi(20) = 80.7$$

where the observations on consumption (y) and income (x) are quarterly, seasonally unadjusted and measured in £ million at 1970 prices. The dummy variable, D_t^0, allows for the effect of the 1968 budget and the introduction of VAT in 1973.

A number of features differentiate this model from (5.9). In the first place, first differences are replaced by fourth differences, and the variables in the error correction term appear with a lag of four

periods rather than one. However, this is simply a reflection of the
fact that the observations are quarterly. The annual growth rates
are now measured by the fourth differences of logarithms and the
theory goes through exactly as before. In particular, the error
correction term, $\log(y_{t-4}/x_{t-4})$, comes in to alter the short-run
consumption decision whenever the proportion of income
consumed strays from the steady-state value, K, implied by (5.11).

A second feature of (5.12) is the introduction of the regressor
$\Delta_1\Delta_4 \log x_t$. Ignoring the dummy variable, the model implies that
consumers plan to spend the same as they did in the corresponding
quarter of the previous year, modified by a proportion of their
annual change in income $(+0.49\Delta_4 \log x_t)$, and by whether the
change itself is increasing or decreasing $(-0.17\Delta_1\Delta_4 \log x_t)$.
These factors determine the short-run consumption decision.
Expressing them in levels implies that $\log y_t$ depends partly on the
linear combination

$$(0.49 - 0.17)\log x_t - 0.17(-\log x_{t-1}) = 0.32 \log x_t$$
$$+ 0.17 \log x_{t-1}$$

The immediate effect of a unit change in $\log x$ (the impact
elasticity) is 0.32. This rises to 0.49 after one quarter but remains
constant for the next two quarters. In the short run, therefore, the
marginal propensity to consume (MPC) is around one half. This
contrasts with a long run MPC of between 0.8 and 0.9, obtained
from a direct examination of the data.

The third feature of (5.12) which differentiates it from (5.9) is
the absence of a constant term. This was specifically excluded
from the estimated model, since to have left it in would have
resulted in relatively imprecise estimates due to multicollinearity;
this arises because the error correction term shows very little
variation when the observations all lie close to the steady-state
growth path. However, in a later paper, Hendry and von Ungern-
Sternberg (1980) found that $\log(y_{t-4}/x_{t-4})$ retained its significance if seasonal dummies were added to the model.

The last point to note about (5.12) is that the post-sample
predictive test statistic, $\xi(20)$ is significant at the 1% level. This
suggests that some change in the specification is necessary, and in
a later part of their paper Davidson *et al.* extend their model to
allow for inflation effects. However, the basic dynamic structure
of (5.12) remains unchanged. Hendry and von Ungern-Sternberg
(1980) take the development one step further by allowing for
changes in liquid assets.

6. Systems of Equations

Systems of regression equations were analysed in Sections 2.9 and 6.9. However, these models were static. A dynamic system is characterised by the appearance of lagged dependent variables among the regressors in some or all of the equations. As might be expected, this raises similar issues to those involved in the single equation case.

Following the approach adopted in Section 6.9, the full model may be written as

$$y_t = Z_t \gamma + u_t, \qquad t = 1, \ldots, T \tag{6.1}$$

where y_t and u_t are $N \times 1$ vectors as in (2.9.15). The matrix Z_t is defined in the same way as X_t in (2.9.16), the difference in notation serving to indicate that some of the regressors are lagged dependent variables. For the same reason the symbol γ, rather than β, is used to denote the parameter vector.

Example 1   Consider the two equation system,

$$\left. \begin{array}{l} y_{1t} = \alpha_1 y_{1, t-1} + \beta_1 x_{1t} + u_{1t} \\ y_{2t} = \alpha_2 y_{2, t-1} + \alpha_2^* y_{1, t-2} + \beta_2 x_{2t} + u_{2t} \end{array} \right\} \quad t = 1, \ldots, T \tag{6.2}$$

where x_{1t} and x_{2t} are exogenous. The matrix Z_t for (6.2) is defined by

$$Z_t = \begin{bmatrix} y_{1, t-1} & x_{1t} & 0 & 0 & 0 \\ 0 & 0 & y_{2, t-1} & y_{1, t-2} & x_{2t} \end{bmatrix} \tag{6.3}$$

while the corresponding vector of parameters is

$$\gamma = (\alpha_1 \beta_1 \alpha_2 \alpha_2^* \beta_2)' \tag{6.4}$$

An alternative way of writing models of this kind is as follows:

$$y_t = A_1 y_{t-1} + \cdots + A_r y_{t-r} + B_0 x_t + \cdots + B_s x_{t-s} + u_t,$$
$$t = 1, \ldots, T \tag{6.5}$$

This is a generalisation of (2.10.1) with x_t denoting a $K \times 1$ vector of exogenous variables, $A_1, \ldots, A_r$ denoting $N \times N$ matrices of parameters, and $B_0, \ldots, B_s$ denoting $N \times K$ matrices of parameters. Model (6.2) may be expressed in the notation of (6.5) by writing it as:

$$\begin{pmatrix} y_{1,t} \\ y_{2,t} \end{pmatrix} = \begin{bmatrix} \alpha_1 & 0 \\ 0 & \alpha_2 \end{bmatrix} \begin{pmatrix} y_{1,t-1} \\ y_{2,t-1} \end{pmatrix} + \begin{bmatrix} 0 & 0 \\ \alpha_2^* & 0 \end{bmatrix} \begin{pmatrix} y_{1,t-2} \\ y_{2,t-2} \end{pmatrix}$$

$$+ \begin{bmatrix} \beta_1 & 0 \\ 0 & \beta_2 \end{bmatrix} \begin{pmatrix} x_{1t} \\ x_{2t} \end{pmatrix} + \begin{pmatrix} u_{1t} \\ u_{2t} \end{pmatrix} \qquad t = 1, \ldots, T \qquad (6.6)$$

Note that the parameter matrices in (6.6) are subject to restrictions, in that certain variables are excluded from each of the equations. This will often be the case with a model developed in the form (6.5). In fact (6.2) is rather extreme, since the equations contain no explanatory variables in common, except insofar as y_1 appears in each with a different lag.

In the discussion below, estimation and testing procedures will be developed using (6.1). Other aspects of dynamic systems, are best considered in terms of (6.5). It is therefore useful to be able to work with both representations.

Systematic Dynamics

The dynamic properties of the system, (6.5), are derived by generalising the results of Section 7.2. Consider first the case when $r = s = 1$, i.e.

$$y_t = A y_{t-1} + B x_t + u_t \qquad (6.7)$$

This model reduces to (7.1.13) when $N = 1$, and it is instructive to compare the results below with the corresponding results obtained in the single equation case.

Repeated substitution for lagged values of y_t in (6.7) gives

$$y_t = \sum_{j=0}^{\infty} A^j B x_{t-j} + \sum_{j=0}^{\infty} A^j u_{t-j} \qquad (6.8)$$

This representation is valid only if $\lim A^j = 0$ as j tends to infinity. A necessary and sufficient condition for this to happen is that the roots of the determinantal equation,

$$| A - \lambda_i I | = 0 \qquad (6.9)$$

are less than one in absolute value. A similar assumption is needed if (6.7) is to have an equilibrium solution. If x_t is constant at $\bar{x}$,

$$E(y_t) = \bar{y} = \left[\sum_{j=0}^{\infty} A^j \right] B \bar{x} \qquad (6.10)$$

Writing (6.10) as an equilibrium solution pre-supposes that the infinite sum of A^j's converges. In fact it can be shown that

$$\sum_{j=0}^{\infty} A^j = (I - A)^{-1} \qquad (6.11)$$

This is a matrix generalisation of the standard result for summing an infinite geometric progression.

The matrix $(I - A)^{-1}B$ is generally referred to as the *matrix of total multipliers*, although the term *dynamic multipliers* is also used by some writers. This gives the total effect on the mean path of y_t of a unit change in the level of all the variables in x. On the other hand, the immediate effect of a unit change in x is given by B, the *matrix of impact* multipliers. These concepts are completely analogous to those developed for the single equation case. Finally, matrices of interim multipliers may be defined. The operation in (6.8) enables the mean path of y to be expressed as an infinite distributed lag in x. Thus

$$\bar{y}_t = E(y_t) = \sum_{j=0}^{\infty} D_j x_{t-j}$$

where

$$D_j = A^j B_j, \qquad j = 0, 1, \ldots \qquad (6.12)$$

By analogy with (7.2.6), the *matrix of Jth interim multipliers* is defined as

$$D_J^* = \sum_{j=0}^{J} D_j \qquad (6.13)$$

Thus, for example, after two time periods the mean path of y will have changed by $(I + A + A^2)B$ in response to a unit change in all the elements of x.

In the general case, (6.5), these formulae are best developed using the lag operator. Defining the matrix polynomials in the lag operator,

$$A(L) = I - A_1 L - \cdots - A_r L^r \qquad (6.14)$$

and

$$B(L) = B_0 + B_1 L + \cdots + B_s L^s \qquad (6.15)$$

(6.5) may be expressed as

$$A(L)y_t = B(L)x_t + u_t \qquad (6.16)$$

The system will be *stable* if the roots of the determinantal polynomial $|A(L)|$ lie outside the unit circle; cf. TSM (Section 2.6). If this condition holds, the polynomial matrix $A(L)$ is non-singular, and the system may be expressed in the form

$$y_t = D(L)x_t + u_t \tag{6.17}$$

where $D(L)$ is an infinite matrix polynomial in the lag operator defined by

$$D(L) = A^{-1}(L)B(L) \tag{6.18}$$

cf. (6.12). The interim multipliers, (6.13), can be obtained by solving (6.18).

The equilibrium solution of the system is given by

$$\bar{y} = D(1)\bar{x}$$

where $D(1)$, the matrix of total multipliers, may be evaluated from the expression

$$D(1) = A^{-1}(1)B(1)$$

Example 2 In (6.6)

$$
D(1) = \begin{bmatrix} 1 - \alpha_1 & 0 \\ -\alpha_2^* & 1 - \alpha_2 \end{bmatrix}^{-1} \begin{bmatrix} \beta_1 & 0 \\ 0 & \beta_2 \end{bmatrix}
$$

$$
= \frac{1}{(1 - \alpha_1)(1 - \alpha_2)} \begin{bmatrix} \beta_1(1 - \alpha_2) & 0 \\ \beta_1 \alpha_2^* & \beta_2(1 - \alpha_1) \end{bmatrix} \tag{6.20}
$$

A change in x_2 will not therefore affect the mean path of y_1. On the other hand, a unit increase in x_1, with x_2 held constant, will eventually raise the level of y_2 by $\beta_1 \alpha_2^*/(1 - \alpha_1)(1 - \alpha_2)$.

Estimation*

If $u_t \sim NID(0, \Omega)$, the estimation of (6.1) reduces to a straightforward application of the SURE technique. The fact that the model is dynamic has no implications for the asymptotic efficiency of any of the estimators already considered. Furthermore, if each equation in the system contains the same set of explanatory variables, OLS applied to each equation in turn will be fully efficient. A situation of this kind arises if no restrictions are placed on the matrices of parameters in (6.5).

The introduction of serial correlation into the disturbances complicates matters. The issues which arise are essentially the same as those noted for the single equation case. However, provided estimation is approached within an ML framework, it is not difficult to ensure asymptotic efficiency.

In the single equation model, the main points concerning estimation were established with respect to an AR(1) disturbance term. For the system (6.1), a vector AR(1) process will be assumed for u_t. Once this case has been sorted out, the theory of estimating models with more general ARMA disturbance terms can be deduced by drawing on the material presented in Section 6.8.

Maximising the likelihood function for (6.1) with an AR(1) disturbance term is equivalent to minimising the sum of squares function

$$S(\Phi, \gamma) = \sum_t \epsilon_t' \Omega^{-1} \epsilon_t \qquad (6.21)$$

The residuals are defined by

$$\epsilon_t = u_t - \Phi u_{t-1} = y_t - Z_t \gamma - \Phi(y_{t-1} - Z_{t-1}\gamma) \qquad (6.22)$$

and a suitable way of minimising (6.21) is by the multivariate Gauss—Newton algorithm. The relevant vectors of derivatives are

$$\frac{\partial \epsilon_t}{\partial \gamma'} = -(Z_t - \Phi Z_{t-1}) \qquad (6.23)$$

and

$$\frac{\partial \epsilon_t}{\partial \phi'} = -u_{t-1} \otimes I_N \qquad (6.24)$$

where $\phi = \text{vec}(\Phi)$.

A step-wise optimisation procedure can also be used to compute the ML estimator. Such a procedure was suggested in connection with the fixed regressor system of Section 6.8, and it was pointed out that this was a natural generalisation of the Cochrane—Orcutt scheme for a single equation. However, while an iterative stepwise procedure is a valid method of computing ML estimates in the dynamic model, a recognition of the fact that the information matrix is no longer block diagonal with respect to Φ and γ may lead to a preference for multivariate Gauss—Newton. (The discussion in Hendry and Tremayne (1976) is relevant here.) A more important point is that a two-step estimator of γ based on the generalisation of Cochrane—Orcutt will not, in general, be asymptotically efficient in the dynamic case. This situation exactly parallels that in the single equation model and the solution is essentially the same: given initial consistent estimates of γ, Φ and Ω, one iteration of multivariate Gauss—Newton will yield asymptotically efficient estimators of Φ and γ.

The *two-step estimation* procedure may therefore be summarised as follows:

(i) Compute a consistent estimator of γ, $\hat{\gamma}$, by applying the method of instrumental variables to each equation in turn. Compute estimates, $\hat{\Phi}$ and $\hat{\Omega}$, from the residual vectors, $\hat{u}_1, \ldots, \hat{u}_T$.

(ii) Carry out a SURE regression, (2.9.17), of $\hat{u}_t - \hat{\Phi}\hat{u}_{t-1}$ on $Z_t - \hat{\Phi}Z_{t-1}$ and $\hat{u}_{t-1} \otimes I_N$. Let the estimated coefficients in this regression be denoted by $\hat{\hat{\gamma}}$ and $\hat{\hat{\Phi}}$. The two-step estimators are then given by $\gamma^* = \hat{\gamma} + \hat{\hat{\gamma}}$ and $\Phi^* = \hat{\Phi} + \hat{\hat{\Phi}}$. Alternatively, take $y_t - \hat{\Phi}y_{t-1}$ as the vector of dependent variables, in which case γ^* is given directly by the vector of regression coefficients.

Some Monte Carlo evidence on the small sample performance of the two-step estimator is given in Spencer (1979). He compares this estimator with a number of other procedures which are known not to be fully efficient in large samples. Taken as a whole, his results are very favourable to the two-step procedure.

Testing for Serial Correlation*

The LM principle may be used to derive an asymptotically valid test for serial correlation when the system of equations is estimated under the assumption that $u_t \sim NID(0, \Omega)$. If the alternative hypothesis is that u_t follows a vector AR(1) process, the resulting test may be regarded as a generalisation of the procedures developed by Durbin for the single equation case; see Section 3.

The unrestricted model — i.e. with a vector AR(1) disturbance term — is the same as that considered in the previous section. The log–likelihood function is

$$\log L = \frac{-TN}{2} \log 2\pi - \frac{T}{2} \log |\Omega| - \tfrac{1}{2}\sum \epsilon_t' \Omega^{-1} \epsilon_t$$

Let $\psi = (\gamma', \phi')'$, where $\phi = \text{vec}(\Phi)$. Because the information matrix is block diagonal with respect to ψ and Ω, the LM statistic for testing any hypothesis regarding the elements of ψ is of the form (5.5.4). However, a test of the null hypothesis $H_0: \Phi = 0$ against the alternative $H_1: \Phi \neq 0$ is a standard example of a subset test. In terms of the notation of (5.5.10)

$$D_2 \log L(\tilde{\psi}_0) = \frac{\partial \log L}{\partial \phi} = -\sum \frac{\partial \epsilon_t'}{\partial \phi} \Omega^{-1} \epsilon_t \tag{6.25}$$

Under the null hypothesis, $\tilde{\psi}_0 = (\hat{\gamma}', 0')'$ where $\hat{\gamma}$ is the SURE estimator of γ. Using (6.22) and (6.23), expression (6.25) becomes

$$D_2 \log L(\bar{\psi}_0) = -\sum Z_t' \Omega^{-1}(y_t - Z_t \hat{\gamma})$$

$$= -\sum Z_t' \Omega^{-1} \hat{u}_t \qquad (6.26)$$

The matrix $I_{22}^{-1} - I_{21} I_{11}^{-1} I_{12}$ must also be computed in order to evaluate the LM statistic, (5.5.10). This matrix is the asymptotic covariance matrix of the ML estimator of Φ evaluated at $\Phi = 0$ and $\gamma = \hat{\gamma}$. It can be estimated as follows. In large samples the information matrix for ψ is given by

$$I^*(\psi) = \sum \frac{\partial \epsilon_t'}{\partial \psi} \Omega^{-1} \frac{\partial \epsilon_t}{\partial \psi'} \qquad (6.27)$$

where the elements of $\partial \epsilon_t / \partial \psi'$ are given by (6.23) and (6.24). Inverting $I^*(\psi)$, evaluating at $\Phi = 0$, $\gamma = \hat{\gamma}$ and extracting the $N^2 \times N^2$ sub-matrix in the lower right hand corner gives a suitable estimator of the matrix in question. Under H_0 the LM statistic will be asymptotically χ^2 with N^2 degrees of freedom. If the alternative hypothesis is modified to Φ being a diagonal matrix, the calculations involved will be simplified somewhat, and the resulting LM statistic will be tested as a χ_N^2 variate. The reduction in degrees of freedom may make this test more appealing, particularly if it is used in a diagnostic checking capacity.

The LM test can, in principle, be extended to situations where the alternative hypothesis is a higher order AR or MA process. Furthermore, when the model is estimated with a vector ARMA disturbance term, tests of goodness of fit may be developed using a similar methodology to that employed in the single equation case.

Identification*

When a dynamic system is estimated with a vector ARMA disturbance term, problems of identification can arise due to common factors. However, with pure AR disturbances, difficulties of this kind are unlikely. In the vector AR(1) case, Sargan (1972) has shown that a sufficient condition for the model to be globally identified is that the system should contain at least N exogenous variables which do not become redundant when lagged (i.e. constant terms and certain dummy variables are not included). This condition is not particularly stringent. The reasoning underlying Sargan's condition can be sketched out for a simple model as follows. Consider the model (6.7), with the disturbances generated by a vector AR(1) process. Subtracting Φy_{t-1} from both sides of the model yields

$$y_t = (\Phi + A)y_{t-1} - \Phi A y_{t-2} + Bx_t - \Phi B x_{t-1} + \epsilon_t \qquad (6.28)$$

If the model contained no exogenous variables, the values of Φ and A would be interchangeable. This is precisely the situation highlighted in (3.6.7). In that case the introduction of an exogenous variable is sufficient to ensure the global identifiability of all the parameters in the model.

Model (6.28) may be written as

$$y_t = A_1^* y_{t-1} + A_2^* y_{t-2} + B_0^* x_t + B_1^* x_{t-1} + \epsilon_t \tag{6.29}$$

The restrictions in (6.28) imply that $B_1^* = -\Phi B$ and $B_0^* = B_0$. Taken together these imply:

$$B_1^* + \Phi B_0^* = 0$$

These equations can be solved for Φ if B_0^* has full column rank. Since $B_0^* = B_0$, this implies that B_0 should have full column rank, and unless B_0 is subject to restrictions, the condition that $k \geqslant N$ is sufficient to ensure that this is the case. Once Φ is identifiable, it follows that A is identifiable also.

7. Causality

The concept of cause and effect is fundamental in any science. However, when it is not possible to conduct a controlled experiment, it becomes very difficult to produce convincing evidence that a cause and effect relationship actually exists. This is almost invariably the case in economics, and as a result the traditional approach in econometrics has been to set up a model on the basis of prior economic theory. Not only is a cause and effect relationship assumed to hold but, furthermore, the direction of causality is also taken to be known.

Suppose, for simplicity, that a system contains only two variables, x and y. Given that a cause and effect relationship can be in only one direction, say from x to y, the only question that remains is whether such a relationship actually exists. The most common approach to answering this question is to regress y_t on x_t and test the coefficient of x_t for 'significance'. However, a high correlation between two sets of non-experimental observations does not constitute evidence for a relationship between the underlying variables. Furthermore, the problem of spurious correlation remains even if some dynamic structure is imposed on the model by taking y to depend on current and past values of x. Thus even though it is assumed at the outset that y cannot cause x, there is no way in which a causal relationship from x to y can be established on the basis of the data alone.

Given the limitations imposed by non-experimental data, it is not unreasonable to take the view that fitting a regression model is primarily an exercise in measurement. That a relationship exists is not actually questioned, but is taken for granted on the basis of economic theory. In these circumstances, tests of significance are no longer used to judge whether or not a relationship exists between two variables. What is of most importance is the measured effect of one variable on another. While this approach underlies a good deal of econometric model building, it suffers from the disadvantage that an inordinate amount of faith is placed in prior knowledge from economic theory. Furthermore, it means that certain untestable features are being brought into a model, thereby violating an important principle of scientific research.

The need to examine the assumptions underlying an econometric model estimated from non-experimental time series data has led to the development of the concept of *causality*. However, it must be stressed at the outset that this notion of causality is a purely statistical one, and it does *not* correspond to any acceptable definition of cause and effect in the philosophical sense. Instead, it refers to the more limited concept of *predictability*, and although 'causality' is clearly too strong a word in this context, its use in the literature is so well established that a change in terminology would serve no useful purpose.

The idea of causality dates back to Wiener, but was first formalised by Granger (1969). Two basic rules are taken to apply. The first is that the future cannot predict the past. Thus strict causality can only occur with the past causing the present or future. Secondly, it is assumed that it is only meaningful to discuss causality for a group of stochastic variables. A variable x is then said to 'cause' a variable y if taking account of past values of x enables better predictions to be made for y, all other things being equal.

Granger's Definition of Causality

The essence of Granger's concept of causality is that x causes y if taking account of past values of x leads to improved predictions for y. However, in order to make this concept operational, it is necessary to be more precise about the way in which predictions are generated, and the way in which their accuracy is measured. In his definition, Granger restricts attention to unbiased least squares predictions, and measures the accuracy of such predictions by the variance of the one-step ahead prediction error. The full definition, which is given below, may therefore be characterised by the phrase: 'reduction in forecasting variance with respect to a given information set'.

Definition   Let U be an information set including all past and present information, and let $\bar{U}$ denote the same set, but excluding present information. Similarly, let X denote all past and present information on the variable x, i.e. $X = (x_t, \tau \leqslant t)$, and let $\bar{X}$ be the past information alone, i.e. $\bar{X} = (x_\tau, \tau < t)$. The variable x is then said to *cause* y if the one step ahead predictor of y, $\tilde{y}$, based on all past information has a smaller mean square error than the predictor of y based on all past information excluding x. More precisely, x causes y if

$$\mathrm{MSE}(\tilde{y}/\bar{U}) < \mathrm{MSE}(\tilde{y}/\bar{U} - \bar{X}) \tag{7.1}$$

Similarly, x causes y *instantaneously* if

$$\mathrm{MSE}(\tilde{y}/U) < \mathrm{MSE}(\tilde{y}/U - X) \tag{7.2}$$

However, as Granger and Newbold (1977, p. 255) observe, the occurrence of instantaneous causality may partly be a function of the data, insofar as a time lag would exist if observations could be taken sufficiently close together in time.

As it stands, Granger's definition of causality is non-operational, insofar as U represents *all* available information. In dealing with this problem, Granger suggests replacing all the information in the universe with the concept of 'all relevant information'. Once this is done, however, the notion that causality may be determined on purely statistical grounds is no longer valid since *a priori* theory must play a key role in deciding what constitutes relevant information. Nevertheless, these requirements may be regarded as being fairly minimal.

Given the above definitions, a complete classification of the patterns of causality has three dimensions: (a) whether x causes y; (b) whether y causes x; and (c) whether instantaneous causality exists. This leads to $2^3 = 8$ basic outcomes. These are listed in table 8.2, together with a scheme of notation. Thus, for example, $x \Rightarrow y$ indicates that x causes y instantaneously, with no feedback.

Table 8.2   Patterns of Causality

Description	*Notation*
(1) Instantaneous causality only	$(x - y)$
(2) x causes y only and not instantaneously	$(x \rightarrow y)$
(3) . . . . instantaneously	$(x \Rightarrow y)$
(4) y causes x only and not instantaneously	$(x \leftarrow y)$
(5) . . . . instantaneously	$(x \Leftarrow y)$
(6) Feedback, not instantaneously	$(x \leftrightarrow y)$
(7) Feedback and instantaneous causality	$(x \Leftrightarrow y)$

In addition to these symbols, the statement 'does not cause' may be represented by '$\nrightarrow$'.

Tests of Causality

In order to simplify matters it will be assumed that the relevant information set, U, consists only of information on the two variables x and y. A statistical framework for characterising causality may then be developed by assuming that (x_t, y_t) is a bivariate time series which is linear, covariance stationary and purely non-deterministic; cf. (1.5.25).

The autoregressive representation of such a process is of the form

$$\Pi(L)\begin{bmatrix} y_t \\ x_t \end{bmatrix} = \begin{bmatrix} \epsilon_t \\ \eta_t \end{bmatrix} \tag{7.3}$$

where $(\epsilon_t, \eta_t)'$ is a bivariate white noise process with a mean of zero and covariance matrix, Ω, and $\Pi(L)$ is a 2 × 2 matrix polynomial in the lag operator, i.e.

$$\Pi(L) = \Pi_0 + \Pi_1 L + \cdots + \Pi_r L^r + \cdots \tag{7.4}$$

Setting Π_0, the leading term in (7.4), equal to the identity matrix implies no loss in generality.

Expression (7.3) may be re-written in terms of scalar polynomials as

$$\begin{bmatrix} \pi_{11}(L) & \pi_{21}(L) \\ \pi_{12}(L) & \pi_{22}(L) \end{bmatrix}\begin{pmatrix} y_t \\ x_t \end{pmatrix} = \begin{pmatrix} \epsilon_t \\ \eta_t \end{pmatrix} \tag{7.5}$$

When $\pi_{21}(L) = 0$, the system depicted by (7.5) is characterised by *unidirectional causality* since, to the extent that a relationship exists, the direction of causality can only be from x to y.

Stated formally, the tests set out below are tests of the hypothesis that y does not cause x, irrespective of whether or not there is instantaneous causality or causality from x to y. If there is causality from x to y, and the hypothesis that y does not cause x is rejected, the relationship may be characterised by 'feedback'. Tests of unidirectional causality are sometimes referred to as tests of exogeneity (of x). However, this concept of exogeneity does not correspond precisely to the definition given towards the end of Section 1.2. The use of the term exogenous in this context is therefore best avoided; see Engle *et al.* (1979).

(i) *The 'Direct' Test*   The second line of (7.5) may be written out as

$$\pi_{21}(L)y_t + \pi_{22}(L)x_t = \eta_t \tag{7.6}$$

Bearing in mind that setting $\Pi_0 = I$ implies that the leading terms in $\pi_{21}(L)$ and $\pi_{22}(L)$ are zero and one respectively, expression (7.6) may be re-arranged in such a way that x_t is equal to a linear combination of past values of x_t and y_t. The hypothesis that y causes x can then be examined by regressing x_t on $y_{t-1}, \ldots, y_{t-m}$, $x_{t-1}, \ldots, x_{t-n}$ and testing the joint significance of the lagged values of y. If $\pi_{21}(L) = 0$, it can be seen from (7.6) that x_t will be determined solely by η_t and its own past values.

In adopting this approach, it is important to ensure that the specification is adequate so that the disturbance term in the regression of x_t on lagged values of x_t and y_t is close to being white noise. This means choosing suitable values of m and n. The value of n is particularly important, since omitting relevant lagged values of x could inflate the coefficients of the lagged y's.

(ii) *Sims' Test*   Sims' test is based on the following theorem. *When $(y, x)'$ has an autoregressive representation, y can be expressed as a distributed lag function of current and past x with a residual which is not correlated with any values of x, past or future, if, and only if, y does not cause x in Granger's sense.* A practical test procedure may therefore be based on a regression of y_t on past and future values of x_t. The regression model is effectively

$$y_t = \sum_{\tau = n}^{-m} \gamma_\tau x_{t-\tau} + w_t \tag{7.7}$$

with n and m sufficiently large to exclude any expected non-negligible coefficients. A test that y does not cause x is then a test of the null hypothesis, $H_0 : \gamma_{-1} = \gamma_{-2} = \cdots = \gamma_{-m} = 0$. However, in implementing this procedure, it is important to take account of any serial correlation in the disturbance term, w_t, since otherwise the results of the F-test could be misleading. One possibility is to fit a suitable ARMA model to the disturbances and apply generalised least squares. In his 1972 paper, however, Sims preferred to simplify matters by assuming a particular structure for the disturbances *a priori*. He effectively took w_t to be generated by an AR(2) process with $\phi_1 = 1.5$ and $\phi_2 = -0.5625$, and so GLS was carried out by multiplying each variable in the regression by $1 - 1.5L + 0.5625L^2$. Such an approach is often referred to as *ad hoc filtering*. A second way of avoiding the need to build a model for the disturbance term, is to fit an autoregressive disturbance term of an arbitrarily high order. This is effectively a non-parametric approach, and it would seem preferable to *ad hoc* filtering since in some cases the pre-chosen filter may not result in the transformed disturbances being even approximately white noise.

(iii) *Geweke's Test* One difficulty with Sims' procedure is the need to treat serial correlation in the disturbances. It was suggested above that one possibility is to estimate the regression by a non-parametric procedure. Either a time domain or a spectral approach could be adopted, but assuming that w_t can be approximated by an $AR(P)$ process leads to a convenient modification of Sims' test.

By defining suitable polynomials in the lag operator, (7.7) may be expressed as

$$y_t = \gamma(L)x_t + \phi^{-1}(L)\zeta_t \tag{7.8}$$

where ζ_t is assumed to be white noise. Multiplying (7.8) through by the AR polynomial, $\phi(L)$, gives

$$\phi(L)y_t = \phi(L)\gamma(L)x_t + \zeta_t \tag{7.9}$$

This suggests regressing y_t on $y_{t-1}, \ldots, y_{t-p}$ together with m future values of x_t and $n+p$ past values. The hypothesis that y does not cause x can then be tested by computing the F-statistic for the future values of x_t.

Tests for the Direction of Causality

The test procedures described in the previous sub-section can be used in a more general framework to test the *direction* of causality. In Sims (1972) the relationship between GNP and the money supplied is examined. The regressions were carried out using quarterly US data from 1947 to 1969 with both GNP and the money supply measured in logarithms. All variables were subject to the *ad hoc* filter referred to in the description of Sims' test above. The values of n and m in (7.7) were set equal to eight and four respectively.

Two definitions of money supply were used by Sims: MB — currency plus reserves adjusted for changes in reserve requirements; and MI — currency plus demand deposits. However, as can be seen from the results in table 8.3, the way in which the money supply is defined makes little difference to the results. When GNP(y) is regressed on the money supply (x), the future values of x are clearly insignificant. On the other hand, regressing money supply on GNP leads to F-statistics which are significant at the 5% level. Sims (1972, p. 540) summarises his conclusion as follows. 'The main empirical finding is that the hypothesis that causality is unidirectional from money to income agrees with post-war US data, whereas the

Table 8.3   *F*-Tests on Four Future Quarters Coefficients

Regression Equation	F
GNP on MI	0.36
GNP on MB	0.39
MI on GNP	4.29*
MB on GNP	5.89*

*Significant at 5% level for an *F*-distribution with (4, 60)
degrees of freedom.
Source: Sims (1972, p. 547).

hypothesis that causality is unidirectional from income to money is rejected.'

An alternative approach to assessing the pattern of causality between two variables is to *pre-whiten* both series. Given that $(y_t, x_t)'$ has the vector autoregressive representation, (7.3), it can be shown that x_t and y_t each have representations as univariate ARMA processes. The disturbances in these processes constitute the pre-whitened series, and the cross-correlations between them can be expected to yield information on the causality patterns between x and y since they are the components of x and y which cannot be predicted from their own past.

If x_t^* and y_t^* denote the pre-whitened series, the theoretical cross-correlations are defined by

$$\rho_\tau(x^*, y^*) = E[(y_t^* - \mu_y^*)(x_{t-\tau}^* - \mu_x^*)]/(\sigma_{x^*}\sigma_{y^*}),$$

$$\tau = 0, \pm 1, \pm 2, \dots \quad (7.10)$$

These cross-correlations characterise the various causality patterns. The essential features, corresponding to the three dimensional classification noted earlier, are as follows:

(a) $\rho_\tau(x^*, y^*) = 0$ for some $\tau > 0$ indicates $x \to y$
(b) $\rho_\tau(x^*, y^*) = 0$ for some $\tau < 0$ indicates $y \to x$
(c) $\rho_0(x^*, y^*) \neq 0$ indicates instantaneous causality.

The sample cross-correlations, $r_\tau(x^*, y^*)$, are calculated from the residuals obtained by fitting suitable ARMA models to x_t and y_t. The patterns of causality can then be assessed by examining the sample cross-correlation function. Under the null hypothesis that x and y are not causally related, the sample cross-correlations of the residuals are independently and normally distributed with mean zero and variance $1/T$ when the sample size is large.

A test that there is some sort of causal relationship between x and y can be based on the statistic

$$U = T \sum_{\tau=-n}^{m} r_\tau^2(x^*, y^*) \tag{7.11}$$

where m and n are chosen so as to include any expected non-negligible cross-correlations. Under the null hypothesis that x and y are not causally related, U is asymptotically χ^2 with $m + n + 1$ degrees of freedom. This test procedure is used in Pierce (1977) in order to assess the evidence for causal relationships between certain economic time series. By and large Pierce concluded that the evidence for causal relationships was not very strong, a result which has led to some debate as to whether the pre-whitening approach is really appropriate.

Notes

Section 3   Breusch and Pagan (1980), Godfrey (1978), Kenward (1976).
Section 6   Pagan and Byron (1977), Hatanaka (1976).

Exercises

1. (a) If $\alpha = 0.6$ and $\beta = 2$ in (2.1) find the total multiplier, the mean lag and the median lag.
   (b) In (7.1.16), $B(L) = \beta_0 + \beta_1 L$ and $A(L) = 1 - \alpha L$. Find an expression for the coefficients in the polynomial $D(L) = B(L)/A(L)$.
   (c) If the disturbance term in (7.1.16) follows an AR(1) process, explain how you would compute ML estimates of the parameters.
   (d) Calculate the total multiplier and the mean lag if $\beta_0 = 2$, $\beta_1 = 3$ and $\alpha = 0.5$.

2. Determine the mean lag in (1.6).

3. Suppose that (2.1) is estimated with an AR(1) disturbance term. Construct a general test of misspecification analogous to the one described in Section 4.

4. Suppose that (1.7) is estimated by OLS. How would you test this specification against the alternative (7.4.1)?

5. Show that Durbin's estimator of ϕ computed from (6.1.10) is: (i) consistent, (ii) asymptotically efficient.

6. Show that the h-statistic in (3.1) is the LM statistic for testing against an MA(1) disturbance.

7. Generalise (3.1) to test for a fourth-order seasonal AR process, $u_t = \phi u_{t-4} + \epsilon_t$, in (2.1).

8. Using the estimates in table 8.1, show that K_1 in (5.8) is equal to $\exp(-2.17 - 2.21 g_1 - 1.55 g_2)$ where g_1 and g_2 are the annual growth rates of x_1 and x_2 respectively.

9. Derive an expression for the mean lag in equation (5.9).

10. Suppose that a consumption function is estimated on the basis of equation (5.9). Show that an increase in the growth rate of income implies a

decrease in the long run average propensity to consume. Is this what you would expect?

11. How would you test the hypothesis that the long-run elasticity in a model of the form (5.1) is unity?

12. Consider the system

$$y_t = Ay_{t-1} + B_0 x_t + B_1 x_{t-1} + u_t$$

Find an expression for (i) the matrix of total multipliers; (ii) the matrix of interim multipliers after two time period.

13. Calculate the interim multipliers in (5.12) up to a lag of 8 and comment on the pattern. Find the total multiplier, the mean lag and the median lag.

14. Under what circumstances would the ARMAX model, (1.4), be under-identified?

9
Simultaneous Equation Models

1. Introduction

In the dynamic model introduced towards the end of the previous chapter, the explanatory variables in each equation consisted of observations on exogenous variables together with observations on *lagged* values of the endogenous variables. However, in econometric problems, it often makes sense to suppose that an endogenous variable is also affected by *current* values of the other endogenous variables in the system. Although instantaneous feedback as such is unlikely to occur, the time interval between observations means that the system must effectively be treated as a simultaneous one.

The model of Section 8.6 may be extended to allow for simultaneity by the introduction of an $N \times N$ matrix, Γ. Thus (8.6.5) becomes

$$\Gamma y_t = A_1 y_{t-1} + \cdots + A_r y_{t-r} + B_0 x_t + \cdots + B_s x_{t-s} + u_t \quad (1.1)$$

Apart from Γ, the notation is the same as in (8.6.5) with u_t a disturbance vector with mean zero and covariance matrix Ω. It will be assumed throughout that u_t has a multivariate normal distribution.

An interesting special case arises when the equations can be arranged in such a way that Γ is triangular. Consider a three equation system in which

$$\Gamma = \begin{bmatrix} 1 & 0 & 0 \\ \gamma_{21} & 1 & 0 \\ \gamma_{31} & \gamma_{32} & 1 \end{bmatrix} \quad (1.2)$$

The first endogenous variable y_1 depends only on the exogenous variables and past values of the endogenous variables. The second variable, y_2, depends on the current value of y_1 but there is no instantaneous feedback from y_2 to y_1. Similarly, y_3 depends on the current values of both y_2 and y_1, but there is no feedback from y_3 to y_1 and y_2.

A system in which Γ is triangular and Ω is diagonal is said to be *recursive*. Such a model is to be contrasted with an *interdependent* system. Suppose the matrix, (1.2), is modified to

$$\Gamma = \begin{bmatrix} 1 & 0 & \gamma_{13} \\ \gamma_{21} & 1 & 0 \\ \gamma_{31} & \gamma_{32} & 1 \end{bmatrix} \tag{1.3}$$

This model exhibits instantaneous feedback, since the current value of y_3 now influences y_1 directly through γ_{13}. Furthermore since y_{2t} depends on y_{1t}, the current value of y_3 also affects the current value of y_2. Hence y_{1t}, y_{2t} and y_{3t} are determined simultaneously.

This chapter is primarily concerned with systems which are inter-dependent, although some attention is paid to recursive models insofar as they represent special features. In particular it is shown that the assumption of recursiveness can simplify estimation considerably.

The Reduced Form

The equations in (1.1) give the *structural form* of the model. Pre-multiplying both sides of (1.1) by Γ^{-1} leads to the *reduced form*:

$$y_t = \Gamma^{-1} A_1 y_{t-1} + \cdots + \Gamma^{-1} A_r y_{t-r} + \Gamma^{-1} B_0 x_t + \cdots$$
$$+ \Gamma^{-1} B_s x_{t-s} + \Gamma^{-1} u_t$$
$$= \Pi_1^+ y_{t-1} + \cdots + \Pi_r^+ y_{t-r} + \Pi_0 x_t + \cdots + \Pi_s x_{t-s} + v_t \tag{1.4}$$

The $N \times N$ matrices Π_i^+, $i = 1, \ldots, r$ and the $N \times K$ matrices Π_j, $j = 1, \ldots, s$ contain the reduced form parameters, while v_t is the reduced form disturbance. Given the properties of the structural disturbance term, u_t, it follows that

$$E(v_t) = E(\Gamma^{-1} u_t) = \Gamma^{-1} E(u_t) = 0$$

and

$$E(v_t v_t') = \Gamma^{-1} E(u_t u_t') \Gamma^{-1\prime} = \Gamma^{-1} \Omega \Gamma^{-1\prime}$$

Furthermore the linear transformation from u_t to v_t means that a normal distribution for the structural disturbances implies normality for the reduced form disturbances.

The reduced form represents a solution to the simultaneity of the structural equations. Given a particular realisation of the disturbances, together with a set of values for the exogenous and lagged endogenous variables, (1.4) solves for the values of the current endogenous variables. However, on comparing (1.4) with (8.6.5) it will be seen that the reduced form is exactly the kind of model considered in Section 8.6. Thus all the theory developed there is directly applicable to (1.4). In particular, the systematic dynamics may be derived, and the predictions of future values obtained, using material which has already been presented. Furthermore, since the reduced form is just the solved version of the structural form, the systematic dynamics and predictions derived for (1.4) are also directly applicable to (1.1). Suppose the total multipliers for (1.1) are required. The matrix of total multipliers for (1.4) is

$$(I - \Pi_1^+ - \cdots - \Pi_1^+)^{-1}(\Pi_0 + \cdots + \Pi_s)$$

Substituting structural for reduced form parameters gives the matrix of total multipliers in terms of (1.1), i.e.

$$(I - \Gamma^{-1}A_1 - \cdots - \Gamma^{-1}A_r)^{-1}(\Gamma^{-1}B_0 + \cdots + \Gamma^{-1}B_s)$$

$$- (\Gamma - A_1 \quad \cdots \quad A_r)^{-1}(B_0 + \cdots + B_s)$$

More generally the systematic dynamic properties may be explored by converting the reduced form to the *final form*, in which y_t is expressed in terms of current and lagged values of the exogenous variables only; cf. (8.6.17).

The reduced form coefficients may be estimated directly. If we ignore the way in which it was derived, the system in (1.4) appears simply as a linear dynamic model in which the disturbances are normally distributed with mean vector zero and an unknown covariance matrix. Since the explanatory variables in each equation are the same, the ML estimators are given by applying OLS to each equation in turn. Given a set of estimates, predictions for future values of y_t may be computed straightforwardly. Furthermore this whole approach generalises to models in which v_t is assumed to be generated by a multivariate ARMA process.

The fact that the reduced form may be estimated directly, immediately raises the question: why do we need to consider methods of estimating the structural form? There are two reasons, although they are closely related. The first is that economic theory is framed in terms of the structural equations. The numerical values of the structural parameters are therefore of interest in their own right, and indeed part of the motivation for estimating a model is to test economic hypotheses. Furthermore, economic theory will impose *a priori* restrictions on the structural form. Such restrictions play a vital role in setting up a simultaneous equation model, since without them it becomes impossible to obtain meaningful estimates of the structural parameters. However, the restrictions also have implications for prediction. This leads to the second reason for being interested in the structural form. If sufficient restrictions are placed on the structural form, the reduced form will be subject to restrictions as well. Ignoring these restrictions will then lead to inefficient estimators of the reduced form parameters, and hence inefficient predictions. Thus although it may be convenient to base prediction on the *unrestricted reduced form* (URF), this will not be desirable insofar as it leads to a loss in efficiency. However, this argument presupposes that any restrictions placed on the structural form are valid. Fortunately, the validity of the restrictions is a testable proposition, the appropriate methodology being described in Section 5.

Identifiability

In the previous paragraph it was observed that restrictions must be placed on the structural equations in order to obtain meaningful estimates. This is an issue of identifiability, a concept which was touched on in the first chapter, and then explored more thoroughly within a statistical framework in Section 3.6. The question of identifiability is central in any consideration of a simultaneous equation system, and Section 2 gives a discussion of the problem in the context of a static model,

$$\Gamma y_t = B x_t + u_t, \qquad \iota = 1, \ldots, T \tag{1.5}$$

The reduced form of (1.5) is

$$y_t = \Pi x_t + v_t, \qquad t = 1, \ldots, T \tag{1.6}$$

where $\Pi = \Gamma^{-1} B$ and $v_t = \Gamma^{-1} u_t$. The distribution of the endogenous variables is determined by the reduced form, the expected value of y_t depending on the NK parameters in Π, and its variance depending on the $N(N+1)/2$ parameters in the covariance matrix of v_t. The covariance matrix of the disturbance term in the structural form likewise contains $N(N+1)/2$ distinct parameters. However, the matrices B and Γ can contain up to $N^2 + NK$ parameters between them. Therefore, unless restrictions are placed on some of the elements in B, Γ and Ω, the structural form will contain N^2 more parameters than the reduced form. In Section 2 it is shown that at least N^2 restrictions must be placed on the structural form if every equation is to be identifiable. If more than N^2 restrictions are imposed on the structural form, the reduced form will also be subject to restrictions, and it is in these circumstances that there is likely to be a gain in basing predictions on estimates of the reduced form parameters derived from the structural parameters.

Most of the restrictions on the structural form are obtained from *a priori* theory. However, N restrictions are given simply as a result of *normalisation*. One endogenous variable in each structural equation is usually regarded as the dependent variable, and so its coefficient may be arbitrarily set at unity. Thus the matrix Γ must have a one in each of its N columns. Since the endogenous variables in the reduced form have already been implicitly normalised, the minimum number of restrictions which must be provided by economic theory is not N^2, but $N^2 - N$. In general, these restrictions are imposed by setting certain parameters in B and Γ equal to zero, and it is this form of restriction which is emphasised in the text. Constraining a parameter to be zero is equivalent to excluding the corresponding variable from the equation in question.

Estimation

The complete set of equations in the structural form may be estimated jointly by a *full information* procedure. Alternatively attention may be focused on a single equation within the system and a *limited information* approach employed. A limited information estimator takes no account of any *a priori* restrictions on the other equations within the system. It will therefore be less efficient than the corresponding full information estimator, except in certain special cases. However, if there is some uncertainty surrounding the specification of the model as a whole, a limited information estimator may be more attractive. Its failure to incorporate all the prior information will probably be a virtue if that 'information' is of dubious validity. Thus a contrast between full information and limited information methods essentially involves a trade-off between robustness and efficiency.

Maximum likelihood estimation is considered in Section 3. The method appropriate for the model as a whole is known as *full information maximum likelihood* (FIML). Although it is straightforward to set up the likelihood function, the likelihood equations are highly nonlinear. This is also true for the *limited information maximum likelihood* (LIML) estimator. Although calculating these estimators does not present an insurmountable problem with modern computing facilities, there is still a good case for employing estimators which are easier to handle computationally. The *two-stage least squares* (2SLS) and *three-stage least squares* (3SLS) estimators are both relatively straightforward to compute, and under fairly

general conditions they can be shown to have the same large sample
properties as the LIML and FIML estimators respectively. Both
2SLS and 3SLS can be regarded as instrumental variable procedures,
and they are derived within this framework in Section 4.

In developing both the 2SLS and 3SLS estimators it is convenient
to have a notation for individual equations. Bearing in mind that
certain variables will be excluded from each equation for reasons of
identifiability, the ith equation in (1.5) may be written as:

$$y_{it} = z_{it}' \delta_i + u_{it}, \qquad \begin{array}{l} t = 1, \ldots, T, \\ i = 1, \ldots, N \end{array} \tag{1.7}$$

The vector z_i contains only the exogenous and endogenous
variables, other than the dependent variable, which have not been
excluded from the equations. Let k_i denote the number of included
exogenous variables, and let n_i denote the number of included
endogenous variables, including the dependent variable. Define
$m_i = n_i + k_i - 1$. Then z_i and the corresponding vector of
parameters, δ_i, are both of length m_i for $i = 1, \ldots, N$. The number
of variables excluded is $K + N - k_i - n_i, i = 1, \ldots, N$.

Outline of Remainder of Chapter

As already indicated, Sections 2 to 5 concentrate on developing
particular features of (1.1) which arise purely because of the
simultaneity aspect. Sections 6 and 7 bring these results together
with those of earlier chapters to examine the dynamic case. This
model, (1.1), is the most general considered in the book, and it
forms the basis for modern econometric model building. In the last
section, it is contrasted with three other approaches to model building. These approaches are based on estimation of an unrestricted
reduced form, a multivariate ARMA model, and a set of univariate
ARMA models. The amount of prior information incorporated in
each model diminishes in moving from (1.1) to the univariate ARMA
models, and the main purpose of Section 8 is to examine the implications of these different approaches for prediction and control.

2. Identifiability

This section considers the problem of identifiability within the
context of a static simultaneous equation model, (1.5). A heuristic
introduction is first given in terms of a simple demand and supply
model. The discussion is then widened and rules for identification

are derived within the general framework established in Section 3.6. A consideration of the dynamic model, (1.1), is delayed until Section 6.

Demand and Supply

A demand and supply model was introduced in Chapter 1. Assuming that the market always clears, the model reduces to a system of just two behavioural equations as in (1.2.7). Adding a disturbance term to each equation yields the stochastic specification

$$S : q_t = \gamma_{11} p_t + \beta_{11} + \beta_{12} x_{2t} + \epsilon_{1t}, \qquad (2.1a)$$
$$t = 1, \ldots, T$$
$$D : q_t = \gamma_{21} p_t + \beta_{21} \qquad\qquad + \epsilon_{2t}, \qquad (2.1b)$$

where $(\epsilon_{1t} \epsilon_{2t})' \sim NID(0, \Omega)$. No restrictions are placed on the covariance matrix, Ω, and an equation can only be identified if restrictions are placed on the coefficients of the variables themselves.

As it stands, (2.1) is simply a stochastic version of the deterministic specification in (1.2.7). In figure 1.1 it was shown that assigning different values to x_2 effectively traces out the demand curve. The introduction of a disturbance term into (2.1) means that this is no longer strictly the case, although as figure 9.1 shows the

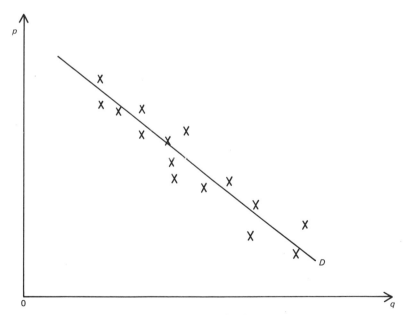

Figure 9.1   *Path Generated by Model (2.1)*

observations will be close to the underlying demand curve. This suggests that the demand curve can still be sensibly estimated, although as we shall see later, an OLS regression of q_t on p_t will not be an appropriate way of achieving this objective.

The crucial feature of (2.1) is the exclusion of x_2 from the demand equation. No restrictions have been placed on the supply equation, however, and the observations themselves are unable to provide sufficient information to enable meaningful estimates of the parameters to be constructed. Of course if x_2 were excluded from the supply equation, but included in the demand equation, the situation would be reversed. This leaves the question of what happens if x_2 is excluded from neither equation. In a sense the answer is obvious. Since both the supply and demand equations have exactly the same specification, there is no way in which it is possible to discriminate between them on the basis of the sample information. Hence neither equation is identifiable.

It is instructive to relate these findings to the reduced form. Treating the supply and demand equations as a pair of simultaneous equations in p and q, we can solve out to obtain expressions for p and q in terms of the exogenous variables only, i.e.

$$q_t = \pi_{11} + \pi_{12} x_{2t} + v_{1t} \tag{2.2a}$$

$$p_t = \pi_{21} + \pi_{22} x_{2t} + v_{2t}, \qquad t = 1, \ldots, T \tag{2.2b}$$

In the matrix notation of (1.5),

$$\Gamma = \begin{bmatrix} 1 & -\gamma_{11} \\ 1 & -\gamma_{21} \end{bmatrix} \quad \text{and} \quad B = \begin{bmatrix} \beta_{11} & \beta_{12} \\ \beta_{21} & \beta_{22} \end{bmatrix} \tag{2.3}$$

so that

$$\Pi = \begin{bmatrix} \pi_{11} & \pi_{12} \\ \pi_{21} & \pi_{22} \end{bmatrix} = \begin{bmatrix} 1 & -\gamma_{11} \\ 1 & -\gamma_{21} \end{bmatrix}^{-1} \begin{bmatrix} \beta_{11} & \beta_{12} \\ \beta_{21} & \beta_{22} \end{bmatrix}$$

Thus,

$$\pi_{11} = \frac{\gamma_{11}\beta_{21} - \gamma_{21}\beta_{11}}{\gamma_{11} - \gamma_{21}} \tag{2.4a}$$

$$\pi_{21} = \frac{\beta_{21} - \beta_{11}}{\gamma_{11} - \gamma_{21}} \tag{2.4b}$$

$$\pi_{12} = \frac{\gamma_{11}\beta_{22} - \gamma_{21}\beta_{12}}{\gamma_{11} - \gamma_{21}} \tag{2.4c}$$

$$\pi_{22} = \frac{\beta_{22} - \beta_{12}}{\gamma_{11} - \gamma_{21}} \tag{2.4d}$$

The joint distribution of p_t and q_t is determined through the reduced form equations (2.2). However, while there are only $KN = 4$ reduced form parameters, the number of structural parameters in B and Γ totals six, i.e. $KN + N(N-1)$. It can be seen from (2.4) that an infinite number of structural form parameters will be consistent with a particular set of values for the π_{ij}'s.

In model (2.1), a restriction is introduced as β_{22} is set equal to zero. On examining (2.4c) and (2.4d) it will be seen that this implies

$$\pi_{12}/\pi_{22} = \gamma_{21} \tag{2.5}$$

Thus there is only one value of γ_{21} which is consistent with the values of π_{12} and π_{22}. Similarly, only one value of β_{21} is consistent with the reduced form parameters, since from (2.4a) and (2.4b),

$$\pi_{11} - \gamma_{21}\pi_{21} = \beta_{21}$$

and from (2.5)

$$\beta_{21} = \pi_{11} - \pi_{12}\pi_{21}/\pi_{22} \tag{2.6}$$

However, it is not difficult to see that there are still an infinite number of values of γ_{11}, β_{11} and β_{12} consistent with any set of values for the reduced form parameters.

We can now move on to bring out the implications of identifiability for estimation. The reduced form parameters can be estimated consistently by applying OLS to each equation in (2.2). Expressions (2.5) and (2.6) suggest that consistent estimators of γ_{21} and β_{21} can be constructed from the reduced form estimators. However, because an infinite set of values of the supply equation parameters can be associated with any set of values for the reduced form parameters, it is not possible to obtain consistent estimators of these parameters from the reduced form estimators. In terms of figure 9.1, this means that for a given value of x_2 we can sketch in a supply curve with any slope and intercept we care to choose, and there will be nothing in the data to contradict our particular choice.

The discussion of identifiability in Section 1 implied that for all the equations in the structural form to be identifiable, at least $N(N-1)$ restrictions would have to be found. For (2.1), $N = 2$ and so if the model as a whole is to be identifiable at least one more restriction must be imposed. One possibility is to set $\beta_{12} = 0$. However, if this were done x_2 would not appear anywhere in the system and so neither equation would be identifiable. An alternative possibility is to set $\beta_{11} = 0$, so that the model becomes

$$S: q_t = \gamma_{11}p_t + \beta_{12}x_{2t} + \epsilon_{1t} \tag{2.7a}$$

$$D: q_t = \gamma_{21}p_t + \beta_{21} \quad\quad + \epsilon_{2t}, \quad\quad t = 1, \dots, T \tag{2.7b}$$

Equations (2.4a) and (2.4b) may then be re-arranged to give

$$\gamma_{11} = \pi_{11}/\pi_{21} \tag{2.8}$$

while β_{11} may similarly be expressed in terms of the reduced form parameters only; cf. (2.5) and (2.6). Thus only one set of values of the supply equation parameters will be consistent with a given set of reduced form parameters and so the equation is identifiable.

The restriction that $\beta_{11} = 0$ implies that the supply curve passes through the origin. This may not be an unreasonable assumption although it is not very common for *a priori* restrictions to be imposed on the constant term. Figure 9.2 shows why a restriction of this kind is effective in identifying an equation. In order to illustrate the point in two dimensions the structural form is taken to be

$$S: q_t = \gamma_{11}p_t \quad\quad + \epsilon_{1t} \tag{2.9a}$$

$$D: q_t = \gamma_{21}p_t + \beta_{21} + \epsilon_{2t} \tag{2.9b}$$

The observations are clustered around a single point and while the

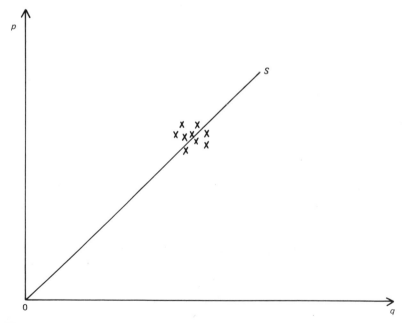

Figure 9.2   *Data Generated by Model (2.9)*

demand curve is no longer identifiable, an estimate of γ_{11} is given by the slope of the line marked S.

The motivation behind the introduction of (2.7) was the desire to have a structural form with the same numbers of parameters as the reduced form. Thus $N(N-1) = 2$ parameters were set equal to zero. However, had the two parameters set equal to zero been in the same equation, it would no longer have been the case that the model as a whole was identifiable. Thus consider the model

$$S: q_t = \gamma_{11} p_t \qquad\qquad\qquad + \epsilon_{1t} \qquad\qquad (2.10a)$$

$$D: q_t = \gamma_{21} p_t + \beta_{21} + \beta_{22} x_{2t} + \epsilon_{2t} \qquad\qquad (2.10b)$$

Although there are two restrictions on the structural form they both serve to identify the same equation, namely the supply curve. On comparing (2.10) with (2.9) it will be apparent that one restriction is quite sufficient for this purpose. The supply curve in (2.10) is therefore said to be *over-identified*, whereas in (2.9) it is *exactly identified*. The fact that the supply curve is over-identified in (2.10), however, is of no consolation to the demand curve: this is still *under-identified*.

Condition for Identifiability

The concept of identifiability was defined quite precisely in Section 3.6 in terms of the joint density function of the endogenous variables; a structure of a model is identifiable if and only if no other structure is observationally equivalent. When the observations are generated by a simultaneous equation model, their distribution is determined by the reduced form. If the disturbances in (1.5) are normally distributed and independent the model may be written as

$$\Gamma y_t = B x_t + \epsilon_t \qquad\qquad (2.11)$$

where $\epsilon_t \sim NID(0, \Omega)$. It follows that

$$y_t \sim NID(\Gamma^{-1} B x_t, \Gamma^{-1} \Omega \Gamma^{-1\prime})$$

Thus any two structures which have reduced forms in which both the expected value of y_t and its covariance matrix are identical will be observationally equivalent.

A class of observationally equivalent structures may be defined by pre-multiplying (2.11) by a nonsingular matrix F. This yields

$$F\Gamma y_t = FB x_t + F\epsilon_t \qquad\qquad (2.12)$$

The reduced form of (2.12) is

$$y_t = (F\Gamma)^{-1}FBx_t + (F\Gamma)^{-1}F\epsilon_t$$
$$= \Gamma^{-1}Bx_t + \Gamma^{-1}\epsilon_t$$

This is precisely the same as the reduced form of (2.11).

It was argued earlier that identifiability of a structural form can only be achieved if restrictions are placed on the parameters of the model. Thus although (2.12) is observationally equivalent to (2.11) it will only be a *model admissible* structure if it satisfies the same *a priori* restrictions as does (2.11). If, for example, B has a zero in a certain position, FB must also have a zero in exactly the same position. If it does not, it cannot be considered as a structure of the model with which we are concerned. A particular structure of (2.11) will therefore be identifiable if the constraints on the model imply that (2.12) can only be model admissible if F is equal to the identity matrix. Since this argument holds for all structures of (2.11), this condition is also a necessary and sufficient one for the identifiability of the model itself. What is therefore needed is a set of rules which tell us the conditions under which the only transformation matrix consistent with an admissible structure is the identity matrix. As will be shown below these rules can be established for one equation at a time, and the identifiability of the system as a whole determined from the identifiability of each individual equation. However, in a limited information approach, attention is focused on a single equation, and the identifiability of the other equations need never be considered.

Throughout this sub-section it will be assumed that the only way in which restrictions can be imposed on the model is by setting certain coefficients in B and Γ equal to zero. More generally, it is possible to impose restrictions by considering linear combinations of the parameters in an equation. Thus in the context of model (2.1) it can be shown that the restriction $\gamma_{11} = \beta_{12}$ is sufficient to ensure the identifiability of the supply equation. Restrictions of this kind will not be considered explicitly here. Nor will a great deal of attention be paid to restrictions on the Ω matrix, although in the next sub-section it is shown that the identifiability of a recursive model is guaranteed by taking Ω to be diagonal. However, in the remainder of the section it is supposed that Ω is unrestricted.

Consider a single equation in the model (2.11). Without any loss of generality, this can be taken to be the first equation, and it may be written as

$$\sum_{i=1}^{N} \gamma_{1i}y_{it} = \sum_{j=1}^{K} \beta_{1j}x_{jt} + \epsilon_{1t} \tag{2.13}$$

The normalisation rule implies that one of the γ_{1i}'s must be set equal to unity. In order for the equation to be identifiable a certain number of variables must be excluded from (2.13). When this has been done, the equation will contain $n_1 \leqslant N$ endogenous variables and $k_1 \leqslant K$ exogenous variables. Another specification of this equation will only be model admissible if it includes the same set of exogenous and endogenous variables.

Let the first row of the matrix F be denoted by $(1 \ f')$, where f is an $(N-1) \times 1$ vector. Setting the first element equal to unity is done simply to comply with the normalisation requirement. The first rows of the matrices $F\Gamma$ and FB in (2.12) are given by $(1 \ f')\Gamma$ and $(1 \ f')B$ respectively. The vector, f, will be deemed admissible if $(1 \ f')\Gamma$ and $(1 \ f')B$ obey the same restrictions as the first rows of Γ and B. If the only admissible f is the null vector, the equation is identifiable. In other words what is being asked is whether it is possible to reproduce the characteristics of the first equation in the model by a linear combination of the other equations.

Some examples should make these concepts clearer as well as suggesting some possible rules for identification.

Example 1   In (2.9) the Γ matrix is subject to no restrictions, apart from normalisation, but B is

$$B = \begin{bmatrix} 0 \\ \beta_{21} \end{bmatrix}$$

Since $N = 2$, f is a scalar and so

$$(1 \ f)B = f\beta_{21} \tag{2.14}$$

The model for the supply equation (2.9a) contains the *a priori* restriction that $\beta_{11} = 0$. The transformation in (2.14) is therefore only admissible if $f = 0$. Hence (2.9a) is identifiable.

Example 2   Now consider a three equation system:

$$
\begin{aligned}
y_{1t} + \gamma_{12}y_{2t} + \gamma_{13}y_{3t} &= \beta_{11}x_{1t} + \quad\quad\quad + \beta_{13}x_{3t} + \epsilon_{1t} \\
y_{2t} + \gamma_{23}y_{3t} &= \quad\quad \beta_{22}x_{2t} + \beta_{23}x_{3t} + \epsilon_{2t} \\
\gamma_{11}y_{1t} \quad\quad + \quad y_{3t} &= \quad\quad \beta_{32}x_{2t} \quad\quad\quad + \epsilon_{3t}
\end{aligned}
\tag{2.15}
$$

In examining the identifiability of the first equation, all we need to consider is the second element in $(1 \ f')B$. If there exists an $f \neq 0$ such that this element is zero, no structure of the equation will be identifiable, since it will always be possible to find an observationally equivalent structure. Let g and h denote the two elements in f, i.e. $f' = (g \quad h)$. The question is whether it is possible to satisfy the equation

$$g\beta_{22} + h\beta_{32} = 0 \qquad (2.16)$$

with either g or h, or both, taking a non-zero value. The answer is obviously in the affirmative, since all that needs to be done is to set $g/h = -\beta_{32}/\beta_{22}$.

Excluding one variable from an equation is obviously not sufficient to ensure identifiability in this case. However, suppose that x_3 is excluded as well, i.e. $\beta_{13} = 0$. The matrix FB will only satisfy the same restrictions as B if g and h can be chosen such that both (2.16) and the equation

$$g\beta_{23} = 0 \qquad (2.17)$$

are satisfied. However, unless $\beta_{23} = 0$, this implies that g must be zero and unless $\beta_{32} = 0$ it follows from (2.16) that $h = 0$. Therefore the only admissible f vector is $f = 0$ and so the equation is now identifiable. Specifying that β_{23} and β_{32} be non-zero is obviously rather pedantic insofar as the specification of the model states that the corresponding variables are indeed included in the relevant equations.

These examples suggest a rule for identifiability, namely that at least $N - 1$ variables must be excluded from the equation in question. Indeed this ties in with some of the earlier discussion, since if $N - 1$ variables are excluded from each equation, the total number of restrictions in the structural form is $N(N - 1)$. The structural form therefore contains the same number of parameters as the reduced form. However, while the requirement that $N - 1$ variables be excluded from an equation will usually ensure identifiability, there are cases where it is not enough.

Example 3   Suppose a minor amendment is made to the three equation model in Example 2. Including x_3 in the third equation means that the model is now written as

$$
\begin{aligned}
y_{1t} + \gamma_{12} y_{2t} + \gamma_{13} y_{3t} &= \beta_{11} x_{1t} + && + \epsilon_{1t} \\
y_{2t} + \gamma_{23} y_{3t} &= && \beta_{22} x_{2t} + \beta_{23} x_{3t} + \epsilon_{2t} \\
\gamma_{11} y_{1t} + y_{3t} &= && \beta_{32} x_{2t} + \beta_{33} x_{3t} + \epsilon_{3t}
\end{aligned}
$$

$$(2.18)$$

Equation (2.17) now becomes

$$g\beta_{23} + h\beta_{33} = 0 \qquad (2.17)'$$

In general, equations (2.16) and (2.17)' will only both be satisfied by setting $g = h = 0$. However, if

$$\beta_{22}\beta_{33} - \beta_{32}\beta_{23} = 0$$

it is possible to find non-zero values of g and h which solve the equations provided that

$$g/h = -\beta_{32}/\beta_{22} = -\beta_{23}/\beta_{33}$$

Such possibilities must therefore be specifically excluded in stating conditions which are both necessary and sufficient to ensure the identifiability of an equation.

Now consider the general case of (2.12). The $N \times (N + K)$ matrix $[\Gamma : B]$ may be partitioned as

$$[\Gamma : B] = \left[\begin{array}{c|c|c|c} 1, -\gamma_1' & 0' & \beta_1' & 0' \\ \hline \bar{\Gamma}_1 & \Gamma_1 & \bar{B}_1 & B_1 \end{array}\right] \begin{array}{l} \} \ 1 \\ \} \ N-1 \end{array} \qquad (2.19)$$

$$\underbrace{}_{n_1} \quad \underbrace{}_{N - n_1} \quad \underbrace{}_{k_1} \quad \underbrace{}_{K - k_1}$$

The first row in (2.19) corresponds to the first equation in the model. The variables have been arranged in such a way that the coefficients corresponding to the n_1 endogenous variables included in the first equation appear in the first n_1 columns. The next set of $N - n_1$ columns contains the coefficients of the endogenous variables excluded from the first equation. The remaining K columns are arranged on a similar basis.

The first row in (2.19) shows the coefficients for the first equation in the model, the notation (γ_1, β_1) being the same as that employed later in the chapter in (4.3). However, the important point about the first row of (2.19) in the present context is that it contains two null row vectors. These correspond to the coefficients of the excluded variables. If the row vector $(1 \ f')[\Gamma : B]$ is to be subject to the same restrictions as the first row of (2.19), the vector f must satisfy the two sets of equations, $f'\Gamma_1 = 0$ and $f'B_1 = 0$. These may be written jointly as

$$f'[\Gamma_1 : B_1] = 0 \qquad (2.20)$$

The two equations (2.16) and (2.17)' represent a special case of (2.20).

The matrix $[\Gamma_1 : B_1]$ is of order $(N - 1) \times (N - n_1 + K - k_1)$. Unless $N - n_1 + K - k_1 \geqslant N - 1$, it will always be possible to find an $f \neq 0$ such that (2.20) is satisfied. This argument holds for any equation in the model and so the condition

$$N - n_i + K - k_i \geqslant N - 1, \qquad i = 1, \ldots, N \qquad (2.21)$$

is *necessary* for the ith equation to be identifiable. This is usually referred to as the *order condition*.

Although the order condition is necessary for identification it is not sufficient. It is only possible to be certain that $f = 0$ is a unique solution to (2.20) if $[\Gamma_1 : B_1]$ includes an $(N - 1) \times (N - 1)$ sub-matrix which is nonsingular. This is equivalent to demanding that $[\Gamma_1 : B_1]$ has full column rank. If this condition — the *rank condition* — is met the equation must be identified. Hence it is both necessary and sufficient.

To check the identifiability of an equation, the order condition is first examined. The inequality (2.21) may be interpreted as saying that the number of excluded variables must be at least as great as $N - 1$. Alternatively a re-arrangement of (2.21) yields

$$K \geqslant n_i + k_i - 1 = m_i, \qquad i = 1, \ldots, N \qquad (2.22)$$

This has the interpretation that the number of exogenous variables in the system must be greater than or equal to the number of explanatory variables in the ith equation. If the order condition is not satisfied the equation is under-identified, and there is no need even to consider the rank condition. If it is satisfied, the rank condition may be examined. The first step is to construct the $(N - 1) \times (N - n_i + K - k_i)$ matrix $[\Gamma_i : B_i]$, which is made up of parameters in Γ and B corresponding to variables *not* included in the ith equation. The equation is then identified if at least one of the $(N - 1) \times (N - 1)$ submatrices of $[\Gamma_i : B_i]$ is nonsingular. In (2.18)

$$[\Gamma_1 : B_1] = \begin{bmatrix} \beta_{22} & \beta_{23} \\ \beta_{32} & \beta_{33} \end{bmatrix}$$

and so the rank condition for the first equation is

$$\beta_{22}\beta_{33} - \beta_{23}\beta_{32} \neq 0 \qquad (2.23)$$

This is exactly what emerged in the discussion below (2.17)'. A similar condition may be derived for the second and third equations. In model (2.15), however, the third equation is over-identified and

$$[\Gamma_3 : B_3] = \begin{bmatrix} \gamma_{12} & \beta_{11} & \beta_{13} \\ 1 & 0 & \beta_{23} \end{bmatrix}$$

Any one of the following three conditions is therefore sufficient to ensure identifiability:

$$\begin{vmatrix} \gamma_{12} & \beta_{11} \\ 1 & 0 \end{vmatrix} = -\beta_{11} \neq 0 \qquad\qquad (2.24a)$$

$$\begin{vmatrix} \gamma_{12} & \beta_{13} \\ 1 & \beta_{23} \end{vmatrix} = \beta_{23}\gamma_{12} - \beta_{13} \neq 0 \qquad\qquad (2.24b)$$

or

$$\begin{vmatrix} \beta_{11} & \beta_{13} \\ 0 & \beta_{23} \end{vmatrix} = \beta_{11}\beta_{23} \neq 0 \qquad\qquad (2.24c)$$

The difficulty in applying the rank condition in practice is that it may require more *a priori* knowledge than is normally available. This point is highlighted by (2.23). There may be economic reasons to suggest that $\beta_{22}/\beta_{23} = \beta_{22}/\beta_{33}$, and if this is the case the first equation is clearly not identifiable. On the other hand if economic theory has nothing to say about the relationship between these parameters, identifiability can only be established by default. If the equation is not, in fact, identifiable this may or may not become apparent in the process of estimating the model. For example, in computing the FIML estimator the algorithm may fail to converge due to the singularity of the information matrix at the true parameter values.

A further point about the rank condition is that it can only be established if we are prepared to specify the complete model. In limited information estimation the researcher may be unable, or unwilling, to fulfil this requirement. Fortunately, the order condition is usually sufficient to ensure identifiability, and although it is important to be aware of the rank condition, a failure to verify it will rarely result in disaster.

Restrictions on Disturbances*

Throughout the discussion of the previous sub-section Ω was taken to be unrestricted. However, placing constraints on the form of Ω can aid identifiability in much the same way as can constraints on B and Γ.

If Ω is specified to be diagonal, the observationally equivalent structures given by (2.12) will only be model admissible if $F\Omega F'$ is also diagonal. However, constraining Ω to be diagonal is not sufficient to ensure identifiability by itself. This is basically because Ω is symmetric and so contains only $N(N-1)/2$ distinct off-diagonal terms. More formally, if the model, (2.12), is normalised by setting the N diagonal terms in Ω equal to unity, the diagonality restriction implies $\Omega = I$. If the matrix F is then chosen as any orthogonal matrix, this restriction will be preserved. However, the structure of the model is clearly altered since $F\Gamma$ and FB need not be the same as Γ and B.

If Ω is specified to be diagonal at least $N(N-1)/2$ further restrictions are needed to ensure identifiability of the model as a whole. In a recursive model — cf. (1.2) — the assumption of a triangular Γ matrix imposes this number of constraints and every equation in the model can be shown to be identifiable.

3. Maximum Likelihood Estimation

The joint distribution of the observations $y_1, \ldots, y_T$ in the reduced form of (2.11), is given by

$$\log L = -\frac{TN}{2}\log 2\pi - \frac{T}{2}\log |\Gamma^{-1}\Omega\Gamma^{-1\prime}|$$

$$-\frac{1}{2}\sum_{t=1}^{T}(y_t - \Pi x_t)'\Gamma'\Omega^{-1}\Gamma(y_t - \Pi x_t) \qquad (3.1)$$

as $E(v_t v_t') = \Gamma^{-1}\Omega\Gamma^{-1\prime}$. Since $\pi = \Gamma^{-1}B$, the model may be expressed entirely in terms of the structural parameters. Within the last term in (3.1)

$$(y_t - \pi x_t)'\Gamma' - \Gamma y_t - \Gamma\pi x_t = \Gamma y_t - B x_t$$

while

$$-\frac{T}{2}\log |\Gamma^{-1}\Omega\Gamma^{-1\prime}| = T\log |\Gamma| - \frac{T}{2}\log |\Omega|$$

This follows from a standard result on determinants. Thus (3.1) becomes

$$\log L = -\frac{TN}{2}\log 2\pi + T\log |\Gamma| - \frac{T}{2}\log |\Omega|$$

$$-\frac{1}{2}\sum_{t=1}^{T}(\Gamma y_t - B x_t)'\Omega^{-1}(\Gamma y_t - B x_t) \qquad (3.2)$$

If $\Gamma = I$, the system collapses to the multivariate regression model, (2.10.1). However, if some elements of B are specified to be zero the model is best handled within a SURE framework. The complications in ML estimation of a simultaneous system (i.e. $\Gamma \neq I$) arise because of the inclusion of the determinant of Γ in (3.2). The derivatives of $\log |\Gamma|$ with respect to the non-zero elements in Γ depend on elements in Γ^{-1}, and so although differentiating the last term in (3.2) with respect to Γ is straightforward, the likelihood equations will be highly nonlinear.

There are a number of approaches to computing the full information maximum likelihood (FIML) estimator in the simultaneous equation model. They will not be described in detail here. However, two points are worth noting. Firstly, the covariance matrix, Ω, may be concentrated out of the likelihood function in the same way as in the SURE model; thus

$$\tilde{\Omega}(\Gamma, B) = T^{-1} \sum_{t=1}^{T} (\Gamma y_t - B x_t)(\Gamma y_t - B x_t)'$$

$$= T^{-1} \sum_{t=1}^{T} \epsilon_t \epsilon_t' \tag{3.3}$$

and substituting for Ω in (3.2) yields the concentrated log–likelihood function

$$\log L_c(\Gamma, B) = -\frac{TN}{2} \log(2\pi + 1) + T \log |\Gamma| - \frac{T}{2} \log |\tilde{\Omega}| \tag{3.4}$$

cf. (4.4.25). Secondly, analytic expressions for the derivatives of $\log L$ with respect to B and Γ may be obtained. A derivation of the relevant formulae is beyond the scope of this book, but the interested reader may refer to Phillips and Wickens (1978), Hendry (1976) or Rothenberg and Leenders (1964) for further details. However, one interesting point to note is that deriving expressions for first and second derivatives enables a *two-step estimator* to be constructed. The initial consistent estimators needed to ensure asymptotic efficiency can be obtained relatively easily using the two stage least squares method outlined in the next section.

The ML estimators of B and Γ can be shown to have the usual properties under regularity conditions. An expression for the covariance matrix of the asymptotic distribution of $\tilde{\Gamma}$ and $\tilde{B}$ will not be given here. However, in the next section the three stage least squares (3SLS) estimator will be derived. Unless Ω is subject to *a priori* restrictions, the 3SLS estimator has the same large sample distribution as the ML estimator, and so the properties established for that estimator are directly applicable for FIML.

Recursive Systems

Recursive systems were introduced in Section 1. Although it is arguable whether such systems should really be regarded as simultaneous equation models, they represent an interesting special case within the context of ML estimation.

A recursive system has two characteristics. Firstly the matrix Γ can be arranged in such a way that it is triangular, and so the determinant of Γ is equal to the product of the elements on the (main) diagonal. In view of the normalisation rule all these elements are unity and so $|\Gamma| = 1$. The second term in the log–likelihood function, (3.2), is therefore zero.

The second characteristic of a recursive system is the diagonality of Ω. This simplifies the third term in (3.2). More importantly, it enables the last term in (3.2) to be split up into separate terms for each equation. In the notation of (1.7) the ith element in the vector $\Gamma y_t - Bx_t$ is given by $y_{it} - z_{it}'\delta_i$, and so (3.2) may be written

$$\log L = -\frac{TN}{2}\log 2\pi - \frac{T}{2}\sum_{i=1}^{N}\log \omega_{ii}$$

$$-\frac{1}{2}\sum_{i=1}^{N}\sum_{t=1}^{T}\omega_{ii}^{-1}(y_{it} - z_{it}'\delta)^2$$

$$= \sum_{i=1}^{N}\left\{-\frac{T}{2}\log 2\pi - \frac{T}{2}\log \omega_{ii} - \frac{1}{2\omega_{ii}}\sum_{t=1}^{T}(y_{it} - z_{it}'\delta_i)^2\right\} \quad (3.5)$$

where ω_{ii} is the ith diagonal element in Ω. The log–likelihood function is the sum of N log–likelihood functions corresponding to the individual structural equations. The full ML estimators of $\delta_1,\ldots,\delta_N$ are therefore given by applying OLS separately to each equation in the system.

Limited Information Maximum Likelihood

The limited information maximum likelihood (LIML) procedure is appropriate for estimating a single equation in a system. Data is assumed to be available for all exogenous variables in the system, but no *a priori* information on the specification of other equations in the system is employed.

Unfortunately the theory underlying the derivation of the LIML estimator is relatively complicated. The two-stage least squares estimator, which is described in the next section, has the same asymptotic properties as LIML while being far easier to compute. A description of LIML will not therefore be given here, and the interested reader should consult a text such as Theil (1971) for further details.

4. Two-Stage and Three-Stage Least Squares

In developing the two-stage and three-stage least squares estimation procedures it is convenient to write the model, (2.12) in the form

$$y_{it} = z'_{it}\delta_i + \epsilon_{it}, \qquad i = 1, \ldots, N, \quad t = 1, \ldots, T \qquad (4.1)$$

where z_i is an $m_i \times 1$ vector consisting of the $n_i - 1$ endogenous variables and k_i exogenous variables considered to be explanatory variables in the ith equation. The $m_i \times 1$ vector δ_i consists of the corresponding parameters. Thus it is supposed that identifiability restrictions are imposed by excluding variables from each equation. Any equation which is to be estimated will be taken as satisfying the order condition, $K \geqslant m_i$. In discussing distributional theory it will be implicitly assumed that any relevant rank condition is also fulfilled.

The equations in (4.1) may be written in matrix form as

$$y_i = Z_i\delta_i + \epsilon_i, \qquad i = 1, \ldots, N \qquad (4.2)$$

where y_i and ϵ_i are $T \times 1$ vectors and Z_i is a $T \times m_i$ matrix for $i = 1, \ldots, N$. The relationship between (4.2) and (4.1) is the same as the relationship between the vector and matrix formulations of the classical regression model, and so an explicit definition of y_i, Z_i and ϵ_i is unnecessary; cf. (2.1.3) and (2.1.4).

When it is necessary to differentiate between endogenous and exogenous regression variables, (4.2) will be written as

$$y_i = Y_i\gamma_i + X_i\beta_i + \epsilon_i, \qquad i = 1, \ldots, N \qquad (4.3)$$

where Y_i is a $T \times (n_i - 1)$ matrix of observations on the endogenous variables and X_i is a $T \times k_i$ matrix of observations on the exogenous variables. In terms of the notation of (4.1), $Z_i = [Y_i : X_i]$ and $\delta_i = (\gamma'_i, \beta'_i)'$ for $i = 1, \ldots, N$.

The purpose of this section is to obtain estimators which are simpler than LIML and FIML while retaining desirable statistical properties. One obvious candidate is OLS. However, although FIML collapses to a set of OLS regressions in the special case of a recursive system, OLS will not be an appropriate estimation technique for an interdependent system. When an equation contains endogenous variables one of the critical assumptions of the classical regression model is violated, as these variables will, in general, be correlated with the disturbance term. As a result OLS will not even be

consistent. This is precisely the situation highlighted at the beginning of Section 2.11. The IV estimator was introduced there as an
appropriate method of handling the problem and it turns out that
this technique can be effectively employed in the present context.
However, before considering these matters the OLS estimator is
placed in perspective by examining its performance in a specific
case.

Ordinary Least Squares

In discussing the identifiability of model (2.1) it was remarked, in
passing, that simply regressing q_t on p_t would not be an appropriate
way of estimating the demand curve. Even though the demand
curve is identifiable, OLS is inconsistent as can be seen by evaluating
its probability limit for γ_{21}, the slope of the demand curve.

The OLS estimator of γ_{21} is

$$c_{21} = \frac{\Sigma(p_t - \bar{p})(q_t - \bar{q})}{\Sigma(p_t - \bar{p})^2}$$

$$= \gamma_{21} + \frac{\Sigma(p_t - \bar{p})(\epsilon_{2t} - \bar{\epsilon}_2)}{\Sigma(p_t - \bar{p})^2}$$

Hence

$$\text{plim } c_{21} = \gamma_{21} + \frac{\text{Cov}(p_t, \epsilon_{2t})}{\text{Var}(p_t)}$$

As p is an endogenous variable, determined by the market, random
shifts in the demand curve, due to ϵ_{2t}, affect p_t and imply that
$\text{Cov}(p_t, \epsilon_{2t})$ is non-zero.

In order to evaluate $\text{Cov}(p_t, \epsilon_{2t})$ consider the reduced form
equation (2.2b). Since x_2 is exogenous it follows immediately that

$$\text{Cov}(p_t, \epsilon_{2t}) = \text{Cov}(v_{2t}, \epsilon_{2t})$$

where

$$v_{2t} = (\epsilon_{2t} - \epsilon_{1t})/(\gamma_{11} - \gamma_{21})$$

$$\therefore \quad \text{Cov}(p_t, \epsilon_{2t}) = (\omega_{22} - \omega_{12})/(\gamma_{11} - \gamma_{21})$$

Meanwhile,

$$\text{Var}(p_t) = \text{Var}(v_{2t}) = (\omega_{11} + \omega_{22} - 2\omega_{12})/(\gamma_{11} - \gamma_{21})^2$$

Therefore

$$\text{plim } c_{21} - \gamma_{21} = (\gamma_{11} - \gamma_{21}) \cdot \frac{(\omega_{22} - \omega_{12})}{\omega_{11} + \omega_{22} - 2\omega_{12}} \neq 0 \qquad (4.4)$$

This quantity measures the 'inconsistency' of c_{21}. It is sometimes referred to as simultaneous equation bias.

If $\omega_{12} = 0$ the disturbances are independent and (4.4) simplifies to

$$\text{plim } c_{21} - \gamma_{21} = (\gamma_{11} - \gamma_{21}) \cdot \frac{\omega_{22}}{\omega_{11} + \omega_{22}} \qquad (4.5)$$

On the basis of *a priori* economic theory $\gamma_{11} > 0$ and $\gamma_{21} < 0$. Therefore the large sample bias in c_{21} is unambiguously positive, implying that the estimated curve will be too flat.

Two-Stage Least Squares

The difficulty with applying OLS to (4.1) arises because of the endogenous variables appearing on the right hand side of the equation. The idea underlying 2SLS is to first 'purge' the endogenous variables of their stochastic component, and then to regress y_{it} on z_{it}. The purged observations are obtained by applying OLS to the reduced form equations: hence the name two-stage least squares.

The notation of (4.3) is most appropriate for describing the procedure. The $n_i - 1$ reduced form equations corresponding to the endogenous variables on the right hand side of (4.3) may be written

$$Y_i = X\Pi_i' + v_i, \qquad i = 1, \ldots, N \qquad (4.6)$$

where Π_i' is a $K \times (n_i - 1)$ matrix of reduced form coefficients, X is a $T \times K$ matrix of observations on all the exogenous variables, and v_i is a $T \times (n_i - 1)$ matrix of reduced form disturbances. These equations are a standard example of a multivariate regression model, and as shown in Section 2.10, OLS applied to each equation in turn is fully efficient. The matrix of OLS estimators may be written as

$$P_i' = (X'X)^{-1}X'Y_i \qquad (4.7)$$

The predicted values of the observations in Y_i are therefore given by

$$\hat{Y}_i = XP_i' = X(X'X)^{-1}X'Y_i \qquad (4.8)$$

The 2SLS estimator of δ_i is given by regressing y_i on $\hat{Y}_i$ and X_i. An explicit expression for the estimator is:

$$\tilde{d} = \begin{bmatrix} \tilde{c} \\ \tilde{d} \end{bmatrix} = \begin{bmatrix} \hat{Y}_i'\hat{Y}_i & \hat{Y}_i'X_i \\ X_i'\hat{Y}_i & X_i'X_i \end{bmatrix}^{-1} \begin{bmatrix} \hat{Y}_i'y_i \\ X_i'y_i \end{bmatrix} \qquad (4.9)$$

On substituting for $\hat{Y}_i$ from (4.8) this becomes

$$\tilde{d} = \begin{bmatrix} \tilde{c} \\ \tilde{b} \end{bmatrix} = \begin{bmatrix} Y_i'X(X'X)^{-1}X'Y_i & Y_i'X(X'X)^{-1}X'X_i \\ X_i'X(X'X)^{-1}X'Y_i & X_i'X_i \end{bmatrix}^{-1}$$

$$\times \begin{bmatrix} Y_i'X(X'X)^{-1}X'y_i \\ X_i'y_i \end{bmatrix} \qquad (4.10)$$

The 2SLS estimator can also be derived by the method of instrumental variables. This is perhaps a more natural approach to adopt since it places the estimator within a more general framework. This is important, since the IV approach provides a more fruitful way of extending 2SLS to more complicated situations.

A set of instruments are needed for the endogenous variables in (4.3). However, the problem falls more easily into the framework established in Section 2.11, if we look for a set of instruments for all the variables in the equation. The argument will therefore be conducted in terms of (4.1); cf. (2.11.1).

The exogenous variables in the system as a whole provide suitable candidates for the instrumental variables. They are, by definition, uncorrelated with the structural disturbances. On the other hand, they will be highly correlated with the endogenous variables through the reduced form. Sufficient instruments will exist provided that $K \geqslant m_i$, but, of course, this is just the usual order condition for identifiability. When the equation in question is exactly identified $K = m_i$, and the IV estimator is simply

$$\tilde{d}_i = (X'Z_i)^{-1}X'y_i, \qquad i = 1, \dots, N \qquad (4.11)$$

In the more general case where the equation is overidentified an optimal set of m_i instruments must be constructed. The GIVE procedure, (2.11.20), suggests the following instruments:

$$W_i = X(X'X)^{-1}X'Z_i, \qquad i = 1, \dots, N \qquad (4.12)$$

The GIVE estimator of δ_i is

$$\tilde{d}_i = (W_i'Z_i)^{-1}W_i'y_i \qquad (4.13)$$

On substituting for W_i from (4.12) this becomes

$$\tilde{d}_i(Z_i'X(X'X)^{-1}X'Z_i)^{-1}Z_i'X(X'X)^{-1}X'y_i, \qquad i = 1, \dots, N \qquad (4.14)$$

As the notation suggests, (4.14) is identical to the 2SLS estimator, (4.10). This can be shown by observing that the matrix of instruments in (4.12) decomposes into two parts:

$$W_i = [XP_i' : X_i] = [\hat{Y}_i : X_i] \qquad (4.15)$$

where $\hat{Y}_i$ is the matrix of predicted values of Y_i obtained from the reduced form; cf. (4.8). The second part of W_i simply states that the exogenous variables in the equation are used as their own instruments. Substituting $Z_i = [Y_i : X_i]$ and (4.15) into (4.13) gives the 2SLS estimator (4.10) when $\hat{Y}_i$ is replaced by the expression on the right hand side of (4.8).

Asymptotic Properties of 2SLS

The results given for the large sample properties of GIVE estimators suggest that $\tilde{d}_i$ will be asymptotically normally distributed with a mean of δ_i and a covariance matrix

$$\text{Avar}(\tilde{d}_i) = \sigma_i^2 [Z_i' X(X'X)^{-1} X'Z_i]^{-1} \qquad (4.16)$$

where $\sigma_i^2 = \omega_{ii} = \text{Var}(\epsilon_{it})$ for $i = 1, \ldots, N$. However, in order for this result to hold the conditions (2.11.8) and (2.11.9) must be satisfied. This can be shown to be the case if

$$\text{plim } T^{-1} X'X = Q \qquad (4.17)$$

where Q is a p.d. matrix.

Three-Stage Least Squares

Two-stage least squares is a limited information procedure which can be used to estimate any identifiable equation in isolation from the other equations in the system. Three-stage least squares, on the other hand, is a method for estimating the system as a whole. Hence it is an alternative to FIML.

The N individual equations in (4.2) can be expressed in the form of a single equation by writing

$$\begin{bmatrix} y_1 \\ \vdots \\ y_N \end{bmatrix} = \begin{bmatrix} Z_1 & & 0 \\ & \ddots & \\ 0 & & Z_N \end{bmatrix} \begin{bmatrix} \delta_1 \\ \vdots \\ \delta_N \end{bmatrix} + \begin{bmatrix} \epsilon_1 \\ \vdots \\ \epsilon_N \end{bmatrix} \qquad (4.18)$$

or, in more compact notation,

$$y = Z\delta + \epsilon \qquad (4.19)$$

An IV approach can again be used. However, the covariance matrix of the disturbances in (4.19) is

$$E(\epsilon\epsilon') = \Omega \otimes I_N \qquad (4.20)$$

and so it is first necessary to transform the observations. Let L be a nonsingular $N \times N$ matrix such that

$$L'L = \Omega^{-1} \qquad (4.21)$$

and define $\bar{L} = L \otimes I_N$. The disturbances in the transformed model,

$$\bar{L}y = \bar{L}Z\delta + \bar{L}\epsilon \qquad (4.22)$$

have a scalar covariance matrix since

$$E(\bar{L}\epsilon\epsilon'\bar{L}') = (L \otimes I_N)(\Omega \otimes I_N)(L \otimes I_N)'$$
$$= L\Omega L' \otimes I_N$$
$$= I_{NT}$$

As in 2SLS the exogenous variables can be used to form the instruments in each equation. Let

$$\bar{X} = \begin{bmatrix} X & 0 & \cdots & 0 \\ 0 & X & \cdots & 0 \\ \vdots & \vdots & \ddots & \vdots \\ 0 & 0 & \cdots & X \end{bmatrix} = I \otimes X$$

The optimal set of instruments for the GIVE estimator are constructed from a multivariate regression of $\bar{L}Z$ on $\bar{X}$, i.e.

$$\bar{W} = \bar{X}(\bar{X}'\bar{X})^{-1}\bar{X}'\bar{L}Z$$

The GIVE estimator is then

$$\tilde{d} = (\bar{W}'\bar{Z})^{-1}\bar{W}\bar{y}$$

where $\bar{Z} = \bar{L}Z$ and $\bar{y} = \bar{L}y$. This can be re-written in the same way as (2.11.22), i.e.

$$\tilde{d} = (Z'\bar{L}'\bar{P}\bar{L}Z)^{-1}Z'\bar{L}'\bar{P}\bar{L}y \qquad (4.23)$$

where $\bar{P} = \bar{X}(\bar{X}'\bar{X})^{-1}\bar{X}'$. On substituting for $\bar{X}$ and $\bar{L}$, and rearranging, it will be found that

$$\bar{L}'\bar{P}\bar{L} = Z'\{\Omega^{-1} \otimes X(X'X)^{-1}X'\}Z \qquad (4.24)$$

Since Ω will, in general, be unknown, it must be replaced by an appropriate estimator. The usual procedure is to apply 2SLS to each individual equation in turn and to estimate Ω from the residuals, i.e.

$$\hat{\Omega} = T^{-1} \sum_{t=1}^{T} e_t e_t' \qquad (4.25)$$

where the ith element in the $N \times 1$ vector e_t is $e_{it} = y_{it} - z'_{it}\tilde{d}_i$. Substituting (4.24) into (4.23) and replacing Ω by $\hat{\Omega}$ yields the three-stage least-squares estimator,

$$\tilde{d} = [Z'\{\hat{\Omega}^{-1} \otimes X(X'X)^{-1}X'\}Z]^{-1}Z'\{\hat{\Omega}^{-1} \otimes X(X'X)^{-1}X'\}y$$

(4.26)

The asymptotic properties of $\tilde{d}$ may be established on the basis of the general results for IV estimators. Subject to the relevant conditions, which may be verified in much the same way as for 2SLS, the 3SLS estimator is asymptotically normally distributed with a mean of δ and a covariance matrix,

$$\text{Avar}(\tilde{d}) = [Z'(\Omega^{-1} \otimes X(X'X)^{-1}X')Z]^{-1}$$

(4.27)

cf. (2.11.23). Note that replacing Ω by the consistent estimator, $\hat{\Omega}$, has no effect on the asymptotic distribution of $\tilde{d}$.

There are two special cases where 3SLS collapses to 2SLS applied separately to each equation. These are

(i) when Ω is diagonal; and
(ii) when every equation in the system is exactly identified.

Classes of IV Estimators: LIVE and FIVE *

The instruments defined in (4.12) may be replaced by the instruments

$$W_i^\dagger = [X\hat{\Pi}_i \vdots X_i], \qquad i = 1, \ldots, N$$

(4.28)

where $\hat{\Pi}_i$ is a consistent estimator of Π_i. An IV estimator of δ_i constructed using $W_i^\dagger$ will have the same asymptotic properties as the 2SLS estimator. Estimators which take this form are known as *limited information instrumental variables efficient* (LIVE) estimators. The corresponding *full information instrumental variables efficient* (FIVE) estimator uses a matrix of instrumental variables defined by

$$W^\dagger = (\hat{\Omega}^{-1} \otimes I_T)\bar{W}$$

(4.29)

where $\bar{W}$ is the matrix

$$\bar{W} = \begin{bmatrix} W_1^\dagger & 0 & \cdots & 0 \\ 0 & W_2^\dagger & \cdots & 0 \\ \vdots & \vdots & \ddots & \vdots \\ 0 & 0 & & W_N^\dagger \end{bmatrix}$$

(4.30)

and the $W_i^\dagger$'s are as defined in (4.28). The 3SLS estimator is a special case in which each $W_i^\dagger$ is the 2SLS matrix of instruments defined in (4.12). However, as long as all the reduced form coefficients in $\tilde{W}$ are estimated consistently and Ω is estimated consistently, any FIVE estimator will have the same asymptotic distribution as 3SLS.

The LIVE and FIVE procedures are of particular importance in large systems where the size of K means that the construction and inversion of the matrix $X'X$ becomes unmanageable. In these circumstances 2SLS, and hence 3SLS, is no longer a viable proposition. Further discussion will be found in Brundy and Jorgenson (1971).

FIML as an IV Estimator*

The FIML procedure can be placed within the IV framework; see, for example, Hausman (1975), Hendry (1976) or Phillips and Wickens (1978, pp. 333–335). The estimator has the form of a FIVE estimator, with the estimates of the reduced form parameters computed from estimates of the structural form parameters, i.e. $\hat{\Pi} = \hat{\Gamma}^{-1}\hat{B}$. If this procedure is iterated to convergence, with $\hat{\Omega}$ updated after each iteration, the resulting estimators of B and Γ are FIML estimators.

An obvious way to start off the FIML iteration would be to compute the initial estimates of the Π_i's from the reduced form as in 2SLS, and then to estimate Ω from (4.25). The first FIVE estimator is then 3SLS. Further iteration does not affect asymptotic efficiency since all that is required to ensure this is that the reduced form parameters are estimated consistently. However, if the procedure is iterated to convergence, the reduced form parameters will be perfectly consistent with the structural parameters. In this sense FIML makes use of all the over identifying restrictions in the model, whereas 3SLS uses only some. Since FIML does truly utilise all the *a priori* information in the model it might reasonably be expected to have more desirable properties than 3SLS in small samples.

5. Testing the Validity of Restrictions on the Model

A key feature of the specification of a simultaneous equation model is the imposition of *a priori* restrictions on the parameters. If more than $N(N-1)$ restrictions are placed on the structural form the reduced form must also be subject to restrictions. Hence a gain in predictive efficiency can reasonably be expected from an estimator based on the structural form, as opposed to an estimator based on the unrestricted reduced form.

The validity of any restrictions over and above those needed to exactly identify each equation in the model can be tested. However, the test can be regarded as a test of the validity of the model only in a rather limited sense: it is testing whether it is reasonable to over-identify some of the equations. While a rejection of the null hypothesis can be taken as evidence that some restrictions on the model are invalid, a failure to reject does not constitute evidence that the model as a whole is reasonable.

The material below is divided into two sections, the first dealing with tests on the model as a whole and the second dealing with tests in a single equation.

Tests on the Model

Consider the multivariate regression model,

$$y_t = Bx_t + \epsilon_t, \qquad t = 1, \dots, T \tag{5.1}$$

with $\epsilon_t \sim NID(0, \Omega)$; cf. (2.10.1). If restrictions are placed on the parameters of B, the model must be estimated as a SURE system. However, if the restrictions are valid, taking account of them in the estimation procedure should lead to a gain in predictive efficiency. When the model is estimated both with and without the restrictions, an LR test of the validity of the restrictions may be carried out. The test statistic is of the form (5.3.15).

In the static simultaneous equation model (2.11), the log–likelihood function of the *unrestricted reduced form* (URF) is given by

$$\log L = \frac{-TN}{2} \log 2\pi - \frac{T}{2} \log |\Sigma| - \frac{1}{2} \sum_{t=1}^{T} v_t' \Sigma^{-1} v_t \tag{5.2}$$

where $v_t = y_t - \Pi x_t$ and $\Sigma = E(v_t v_t')$. This has exactly the same form as the likelihood function of (5.1) and, as in that case, the maximised likelihood depends on the estimated covariance matrix, which for (5.2) is denoted by $\tilde{\Sigma}$. The maximised log–likelihood function for the restricted model is obtained by substituting the ML estimator of Ω,

$$\tilde{\Omega} = T^{-1} \sum_{t=1}^{T} (\tilde{\Gamma} y_t - \tilde{B} x_t)(\tilde{\Gamma} y_t - \tilde{B} x_t)' \tag{5.3}$$

into (3.2) to yield

$$\log L = \frac{-TN}{2} \log(2\pi + 1) - \frac{T}{2} \log |\tilde{\Gamma}^{-1}\tilde{\Omega}\tilde{\Gamma}^{-1'}|$$

Hence the LR test statistic is

$$\mathrm{LR} = T \log |\tilde{\Sigma}_0|/|\tilde{\Sigma}| \tag{5.4}$$

where $\tilde{\Sigma}_0 = \tilde{\Gamma}^{-1}\tilde{\Omega}\tilde{\Gamma}^{-1'}$ is the ML estimator of the disturbance covariance matrix in the restricted reduced form. The unrestricted reduced form will only be a more general model if some of the structural equations are overidentified. Thus under the null hypothesis that the over-identifying restrictions on the structural form are valid, (5.4) will be asymptotically χ^2_R where

$$R = \sum_{i=1}^{N} (K - m_i) = NK - \sum_{i=1}^{N} m_i \tag{5.5}$$

Maximum likelihood estimation of the URF may be carried out by N OLS regressions. While this is straightforward enough, FIML should ideally be used to estimate the structural equations. However, the test procedure is still valid if Γ, B and Σ are estimated by 3SLS.

Tests on Single Equations

The over-identifying restrictions on a single equation in the system may be tested within a limited information framework. In some ways this is more attractive than testing all the over-identifying restrictions in the system as a whole. Attention can be focused on one equation, and the results are not conditional on the specification of the rest of the model.

If the equation is estimated by 2SLS, the following test statistic is appropriate:

$$\frac{T-K}{K-m_i} \frac{(y_i - Y_i\tilde{c})'(M_i - M)(y_i - Y_i\tilde{c})}{(y_i - Y_i\tilde{c})'M(y_i - Y_i\tilde{c})} \tag{5.6}$$

where $M_i = I - X_i(X_i'X_i)^{-1}X_i'$ and $M = I - X(X'X)^{-1}X'$. This has the form of the F-statistic for testing for the inclusion of a subset of parameters in a classical linear regression model. Thus

$$(y_i - Y_i\tilde{c})'M(y_i - Y_i\tilde{c})$$

is the residual sum of squares obtained by regressing $y_i - Y_i\tilde{c}$ on X. However, it will be observed that $K - m_i$ is used as a divisor rather than $K - k_i$.

Under the null hypothesis that the model is correctly specified, the distribution of this statistic can be approximated by an F-distribution with $(K - m_i, T - K)$ degrees of freedom; see the Monte Carlo evidence in Basmann (1960). A high value of the statistic will lead to a rejection of the null hypothesis in favour of the alternative that some of the excluded exogenous variables should have been included in the equation.

6. Dynamic Models

The general form of a dynamic linear simultaneous equation model was given in (1.1) as

$$\Gamma y_t = A_1 y_{t-1} + \cdots + A_r y_{t-r} + B_0 x_t + \cdots + B_s x_{t-s} + u_t \quad (6.1)$$

More complicated models, which are nonlinear in parameters, may also be constructed. However, the development here is restricted to (6.1); the additional technical problems raised by the introduction of nonlinearities are not considered explicitly.

The first point to be explored in connection with (6.1) concerns its dynamic properties, and the relationship between a model of this kind and the basic time series models examined in Chapter 1. An explicit link is found to connect these different classes of model, and the implications of this for forecasting are discussed in the final section of the chapter. The other topic dealt with in the final section is control, and the relationship between the forecasting and control functions of the dynamic simultaneous equation model are placed in some kind of perspective.

Estimation, and the question of identifiability, are considered in the later parts of this section. The assumption that the disturbances in (6.1) are serially uncorrelated makes a significant difference from the technical point of view, and this case is considered first.

Dynamic Properties

As pointed out in Section 1, the systematic dynamics of a simultaneous equation model may be derived from the reduced form (1.4). However, the reduced form equations can be re-arranged so that each endogenous variable is expressed in terms of current and lagged values of the exogenous variables only. This is known as the final form. Let $A(L)$ be the matrix polynomial

$$A(L) = \Gamma - A_1 L - \cdots - A_r L^r \quad (6.2)$$

and define $B(L)$ in terms of $B_0, \ldots, B_s$ in a similar way; cf. (8.6.15).

Model (6.1) may then be written as

$$A(L)y_t = B(L)x_t + u_t \tag{6.3}$$

and pre-multiplying by $A^{-1}(L)$ yields the *final form*

$$y_t = A^{-1}(L)B(L)x_t + A^{-1}(L)u_t \tag{6.4}$$

This is directly comparable with the operation in (8.6.18), and the matrices of interim multipliers may be deduced directly from $A^{-1}(L)B(L)$.

Subtracting the mean path, $\bar{y}_t$, from both sides of (6.4) gives

$$y_t - \bar{y}_t = A^{-1}(L)u_t \tag{6.5}$$

The vector $y_t - \bar{y}_t$ gives the deviations of the observations from their mean path and this represents the stochastic part of the model. If u_t is specified to be a particular multivariate ARMA process,

$$\Phi_u(L)u_t = \Theta_u(L)\epsilon_t \tag{6.6}$$

the stochastic properties of the model are available immediately since $y_t - \bar{y}$ is governed by the multivariate ARMA process

$$y_t - \bar{y}_t = A^{-1}(L)\Phi_u^{-1}(L)\Theta_u(L)\epsilon_t \tag{6.7}$$

Further insight into the properties of the model may be obtained by writing

$$A^{-1}(L) = A^*(L)/|A(L)|$$

where $A^*(L)$ is the adjoint matrix of $A(L)$. Substituting in (6.4) gives the *final equations*,

$$y_t = \frac{A^*(L)B(L)}{|A(L)|}x_t + \frac{A^*(L)}{|A(L)|}u_t \tag{6.8}$$

Note that $|A(L)|$ is just a scalar polynomial. Thus the equations in (6.8) may be regarded as a set of N transfer functions. However, they are subject to two special characteristics. Assuming no cancellation of common factors, the denominator polynomials are the same, firstly for every exogenous variable and disturbance term, and secondly for every equation. The first of these constraints is a feature of stochastic difference equations, and it suggests writing the system in (6.8) as

$$|A(L)|y_t = A^*(L)B(L)x_t + A^*(L)u_t \tag{6.9}$$

This is sometimes known as the *autoregressive final form*. Each equation in (6.9) is a stochastic difference equation of the kind considered in Chapter 8. However, in all N equations the coefficients of the lagged values of the dependent variable are the same. This 'remarkable result', as it was called by Kendall, implies a common pattern of dynamic behaviour for the endogenous variable in the system.

Example Consider the dynamic supply and demand model, (1.2.8), with a white noise disturbance term attached to each equation. The system may be expressed in the form (6.3) by writing

$$\begin{bmatrix} 1 & -\gamma_{11}L \\ 1 & -\gamma_{21} \end{bmatrix} \begin{bmatrix} q_t \\ p_t \end{bmatrix} = \begin{bmatrix} \beta_{11} & 0 \\ \beta_{21} & \beta_{22} \end{bmatrix} \begin{bmatrix} 1 \\ x_t \end{bmatrix} + \begin{bmatrix} \epsilon_{1t} \\ \epsilon_{2t} \end{bmatrix} \qquad (6.10)$$

Since $A(L) = -\gamma_{21} + \gamma_{11}L$, the autoregressive final form is:

$$(-\gamma_{21} + \gamma_{11}L) \begin{bmatrix} q_t \\ p_t \end{bmatrix} = \begin{bmatrix} -\gamma_{21}\beta_{11} + \gamma_{11}\beta_{21}L & \gamma_{11}\beta_{22}L \\ \beta_{21} - \beta_{11} & \beta_{22} \end{bmatrix} \begin{bmatrix} 1 \\ x_t \end{bmatrix}$$

$$+ \begin{bmatrix} -\gamma_{21} & \gamma_{11}L \\ -1 & 1 \end{bmatrix} \begin{bmatrix} \epsilon_{1t} \\ \epsilon_{2t} \end{bmatrix} \qquad (6.11)$$

Alternatively,

$$q_t = (\gamma_{11}/\gamma_{21})q_{t-1} + \{\beta_{11} - \gamma_{11}\beta_{21}/\gamma_{21}\}$$
$$- (\gamma_{11}\beta_{22}/\gamma_{21})x_{t-1} + \epsilon_{1t} - (\gamma_{11}/\gamma_{21})\epsilon_{2,t-1} \qquad (6.12a)$$

$$p_t = (\gamma_{11}/\gamma_{21})p_{t-1} + \{(\beta_{11} - \beta_{21})/\gamma_{21}\}$$
$$- (\beta_{22}/\gamma_{21})x_t + (\epsilon_{1t} - \epsilon_{2t})/\gamma_{21} \qquad (6.12b)$$

The coefficients attached to the lagged dependent variables are identical. That they satisfy the dynamic stability condition, $|\gamma_{11}/\gamma_{21}| < 1$, follows immediately from the condition for stability in a cobweb model, (1.2.9).

Stochastic Exogenous Variables

If the exogenous variables are taken to be stochastic, rather than fixed, the random behaviour in the endogenous variables emanates from two sources. Consider the model (6.1), (6.6) and suppose that x_t is generated by a multivariate ARMA process,

$$\Phi_x(L)x_t = \Theta_x(L)\xi_t \qquad (6.13)$$

where $\xi_t \sim WS(0, \Omega_x)$ and $E(\xi_t) = 0$ for all t, s. The complete vector of observations $(y_t'x_t')'$ can now be regarded as being generated by a multivariate ARMA process

$$\begin{bmatrix} A(L) & B(L) \\ 0 & \Phi_x(L) \end{bmatrix} \begin{bmatrix} y_t \\ x_t \end{bmatrix} = \begin{bmatrix} \bar{\Theta}_u(L) & 0 \\ 0 & \Theta_x(L) \end{bmatrix} \begin{bmatrix} \epsilon_t \\ \xi_t \end{bmatrix} \tag{6.14}$$

where $\bar{\Theta}_u(L) = \Phi_u^{-1}(L)\Theta_u(L)$ on the assumption that u_t has an MA representation for all t, s. Substituting (6.1) into the final form, together with (6.6), yields

$$y_t = \frac{B(L)\Theta_x(L)}{A(L)\Phi_x(L)} \xi_t + \frac{\Theta_u(L)}{A(L)\Phi_x(L)} \epsilon_t \tag{6.15}$$

and so y_t is a stochastic process driven by the innovations in the disturbance term and the explanatory variables. It can be shown to have a multivariate ARMA representation, since it is itself the sum of two multivariate ARMA processes.

A multivariate ARMA process can be decomposed into a set of seemingly unrelated univariate ARMA processes. Therefore if (6.13) does hold, it will be possible to represent each endogenous variable in the dynamic simultaneous equation model, (6.1), by a univariate ARMA process. This result is of some importance in understanding the relationship between the forecasts produced by econometric models and those produced by 'naive' time series methods.

Not all exogenous variables can reasonably be regarded as being generated by stochastic processes. For example, policy variables, such as the tax rate or the rate of interest, typically show step changes. Other exogenous variables also behave in such a way that an ARMA representation would be inappropriate. An extreme example is a dummy variable used to measure the effect of some external event such as a change in government. Rather less extreme is the behaviour of a variable like the price of oil, which is subject to rapid changes in a short period of time due to price fixing by the OPEC cartel.

Bearing the above considerations in mind, it is useful to divide the exogenous variables into two groups. At the risk of oversimplification these may be labelled '*deterministic*' and '*stochastic*'. The deterministic variables will be denoted by x_1, and the stochastic variables by x_2. Thus $x_t = (x_{1t}' x_{2t}')'$ and, corresponding to this division, the polynomial matrix $B(L)$ may be partitioned as $[B_1(L) \vdots B_2(L)]$. The final equations, (6.8), therefore become:

$$y_1 = \frac{A^*(L)B_1(L)}{|A(L)|} x_{1t} + \frac{A^*(L)B_2(L)}{|A(L)|} x_{2t} + \frac{A^*(L)}{|A(L)|} u_t \tag{6.16}$$

If x_{2t} is generated by a multivariate ARMA process, it follows from the argument above that y_t may be expressed in the form

$$y_t = \frac{A^*(L)B_1(L)}{|A(L)|} x_{1t} + u_t^\dagger \tag{6.17}$$

where $u_t^\dagger$ is generated by a multivariate ARMA process, the form of which depends on both u_t and x_{2t}. However, as in the case of (6.14) it follows that u_t may be decomposed into a vector of univariate ARMA processes. This implies that each endogenous variable has a transfer function representation in which only the elements of x_1 appear as explanatory variables; cf. (6.8). Alternatively, because of the common factor $|A(L)|$ in (6.16), a set of stochastic difference equations with ARMA disturbance terms might present a better specification.

7. Estimation and Identification of Dynamic Models

If the disturbance term in (6.1) is white noise, the fact that the model is dynamic raises no difficulties beyond those encountered in the static case. The lagged endogenous variables may be classified with the exogenous variables as *predetermined*. All the results of Sections 2 to 5 carry over directly with the predetermined variables playing exactly the same role as the exogenous variables in the static model.

Example Writing the supply and demand model of (6.10) in the format of (6.1) yields:

$$\begin{bmatrix} 1 & 0 \\ 1 & -\gamma_{21} \end{bmatrix} \begin{pmatrix} q_t \\ p_t \end{pmatrix} = \begin{bmatrix} 0 & \gamma_{11} \\ 0 & 0 \end{bmatrix} \begin{pmatrix} q_{t-1} \\ p_{t-1} \end{pmatrix}$$

$$+ \begin{bmatrix} \beta_{11} & 0 \\ \beta_{21} & \beta_{22} \end{bmatrix} \begin{pmatrix} 1 \\ x_t \end{pmatrix} + \begin{bmatrix} \epsilon_{1t} \\ \epsilon_{2t} \end{bmatrix} \tag{7.1}$$

The order condition for a dynamic model is that the number of *predetermined* variables in the system must be greater than, or equal to, the number of explanatory variables included in that equation. The second equation in (7.1) – the demand equation – contains $m_2 = 3$ explanatory variables. This is exactly equal to the number of predetermined variables appearing in the system and so the equation is exactly identified. The rank condition is, of course, trivial in this case since the model contains only two behavioural equations.

The supply equation has no current endogenous variables

appearing as explanatory variables and it is automatically identifiable. The fact that both equations are identifiable is in
accordance with the heuristic reasoning of Section 1.2.

Although the treatment of dynamic simultaneous equation models
is straightforward when the disturbances are serially uncorrelated, a
number of new issues arise once this assumption is relaxed. The
resulting complications can, to a certain extent, be anticipated on
the basis of he material presented in the previous chapter.

Identification

Consider the dynamic model, (6.1), and let $F(L)$ be any non-singular
$N \times N$ polynomial matrix in the lag operator, i.e.

$$F(L) = F_0 + F_1 L + \cdots + F_m L^m \tag{7.2}$$

A class of observationally equivalent structures is then defined by

$$F(L)A(L)y_t = F(L)B(L)x_t + F(L)u_t \tag{7.3}$$

since the final form is given by (6.4) irrespective of the specification
of $F(L)$; cf. (2.12). As in the static case, (7.3) is said to be model
admissible if it satisfies the same *a priori* restrictions as (6.1). If the
only way of obtaining a model admissible structure is by setting
$F(L) = I$, then (6.1) is identifiable.

The assumption that the disturbance terms in the model are
serially uncorrelated effectively constrains the coefficients of the lag
operator in $F(L)$ to be zero. Thus $F(L) = F$, where F is an $N \times N$
matrix of the type considered in Section 2, and so the question of
identifiability can be handled as in the static case. Once the assumption that the disturbances are serially uncorrelated is relaxed, the sort
of identifiability questions raised in the context of multivariate
ARMA models become relevant. As might be expected, MA and
mixed disturbance terms are more difficult to handle than pure AR
disturbances. Indeed for a vector AR(1) specification of u_t, Sargan's
condition, which was stated in Section 8.6, is directly applicable.
Thus, if the system contains at least N exogenous variables which do
not become redundant when lagged, the AR coefficient matrix, Φ, is
globally identifiable. The conventional static criteria may then be
applied to the identifiability of the remaining parameters in the
model. The conditions developed for more complicated cases by
Hannan (1970), Sargan (1972), Deistler (1976), and others will not,
however, be described in this book. Instead, attention will be focused
on the approach adopted by Hatanaka (1975). Here, the problem is
viewed in a rather different way.

Hatanaka argues that the dynamic specification of a model will, in general, be determined empirically. He therefore assumes: (i) that the orders of the lags on the endogenous and exogenous variables are not known *a priori*; and (ii) that the disturbance term is known only to be a stationary process. Within this framework, he shows that it is possible to derive a simple order condition for identifiability in terms of the exclusion of exogenous variables from an equation.

Consider the first equation in (6.1). This may be written as

$$y_t'\alpha(L) = x_t'\beta(L) + u_t \tag{7.4}$$

where $\alpha(L)$ is an $N \times 1$ vector polynomial in the lag operator and $\beta(L)$ is a $k \times 1$ vector polynomial. It is assumed that none of the variables in x_t is redundant when lagged. Thus, for example, x_t must contain no constant terms or seasonal dummies. Variables of this kind will typically be found in an econometric model, but within this context they are simply ignored.

The only relevant *a priori* information in (7.4) is whether or not a variable is excluded from the equation. Excluding a variable in this way means not only that the current values do not appear, but also that lagged values are inadmissible. Thus if a variable in y_t or x_t is excluded from the equation, the corresponding element in $\alpha(L)$ or $\beta(L)$ will be zero. The $N \times (N + K)$ polynomial matrix $[A(L) \vdots B(L)]$ may then be partitioned as

$$[A(L) \vdots B(L)] = \begin{bmatrix} 1, -\alpha_1(L)' & \vdots & 0' & \vdots & \beta_1'(L) & \vdots & 0' \\ \hline A_1(L) & \vdots & A_1(L) & \vdots & B_1(L) & \vdots & B_1(L) \end{bmatrix} \tag{7.5}$$

The notation in (7.5) is analogous to that in (2.19), and the implications for identifiability are similar. The *rank condition* states that the rank of $[A_1(L) \vdots B_1(L)]$ must be $N - 1$. The *order condition*, which is a generalisation of (2.21), is that the number of excluded *exogenous* variables is greater than or equal to, the number of included endogenous variables minus one. It is important to note that with serially uncorrelated disturbances the classical order condition would emerge in terms of excluded predetermined variables, albeit with a different definition of exclusion.

Maximum Likelihood Estimation

The system of equations defined by (6.1) may be written more compactly by defining $w_t = (y_t', y_{t-1}', \ldots, y_{t-r}', x_t', \ldots, x_{t-s}')'$ and $A = [\Gamma \vdots A_1 \vdots \ldots \vdots A_r \vdots -B_0 \vdots \ldots \vdots -B_s]$. Then

$$Aw_t = u_t, \qquad t = \tau, \ldots, T \tag{7.6}$$

where $\tau = \max(r + 1, s + 1)$.

Suppose that the disturbances are normally distributed and follow a vector AR(1) process, i.e.

$$u_t = \Phi u_{t-1} + \epsilon_t \tag{7.7}$$

where $\epsilon_t \sim NID(0, \Omega)$. The equations in (7.6) then become

$$Aw_t - \Phi Aw_{t-1} = \epsilon_t, \qquad t = \tau + 1, \ldots, T \tag{7.8}$$

The logarithm of the joint distribution of $\epsilon_{\tau+1}, \ldots, \epsilon_T$ is given by

$$p(\epsilon_{\tau+1}, \ldots, \epsilon_T) = \frac{-(T-\tau)N}{2} \log 2\pi - \frac{(T-\tau)}{2} \log |\Omega|$$

$$- \frac{1}{2} \sum_{t=\tau+1}^{T} \epsilon_t' \Omega^{-1} \epsilon_t \tag{7.9}$$

Under the assumption that the initial observations, $y_1, \ldots, y_\tau$ are fixed, the log–likelihood function is given by

$$\log L = - \frac{(T-\tau)N}{2} \log 2\pi - (T-\tau)\log |\Gamma| - \frac{(T-\tau)}{2} \log |\Omega|$$

$$- \frac{1}{2} \sum_{t=\tau+1}^{T} (Aw_t - \Phi Aw_{t-1})' \Omega^{-1} (Aw_t - \Phi Aw_{t-1})$$

$$\tag{7.10}$$

since the Jacobian of the transformation from $\epsilon_{\tau+1}, \ldots, \epsilon_T$ to $y_{\tau+1}, \ldots, y_T$ is

$$\frac{\partial(\epsilon_{\tau+1}, \ldots, \epsilon_T)}{\partial(y_{\tau+1}, \ldots, y_T)} = \begin{vmatrix} \Gamma & & & & \\ -\Phi\Gamma & \Gamma & & 0 & \\ & -\Phi\Gamma & \Gamma & & \\ & & \ddots & \ddots & \\ & & & \ddots & \Gamma \\ 0 & & & -\Phi\Gamma & \Gamma \end{vmatrix}$$

$$= |\Gamma|^{T-\tau}$$

The ML estimators of Ω and Φ, conditional on A, are given by expressions analogous to (5.3) and (2.10.2) respectively. The likelihood function may therefore be concentrated with respect to A, and then maximised by a numerical optimisation procedure. As in the static case, analytic derivatives may be derived and these may be incorporated within an appropriate algorithm. Furthermore if initial consistent estimators are available, these derivatives may be used to construct asymptotically efficient two-step estimators.

Testing for Serial Correlation*

Consider a test of the null hypothesis $H_0: \Phi = 0$ in (7.7). The fact that the model is both dynamic and interdependent means that any test statistic constructed on an intuitive basis is almost certain to be invalid. However, a valid test procedure can be based on the LM principle. An expression for the test statistic is given by Breusch (1979). Unfortunately this expression is rather complicated, but Breusch suggests an alternative approach which is likely to be more practical. This is based on the form of the LM statistic given in (5.5.8). Writing the test statistic in this way implies that it can be computed as a Wald statistic based on the estimates obtained by one step of Newton–Raphson on the method of scoring. The initial estimates are computed under the null hypothesis which means that Φ is set equal to 0.

Formulating the LM test in this way fits very neatly into the two-step estimation framework. The derivatives needed to compute a two-step estimator are the same as those needed to compute the LM statistic. However, the starting values are different. If the two-step estimator is to be efficient, the initial estimators of A and Φ must be consistent estimators of the parameters in the unrestricted model.

Estimation of a Single Equation

In Section 4, the 2SLS estimator was derived within an IV framework. This approach may be extended to handle serially correlated disturbances, the simplest assumption being that an AR(1) process is appropriate.

The ith equation in (6.1) may be written as

$$y_{it} = z_{it}'\delta_i + u_{it}, \qquad i = 1, \ldots, N, \quad t = 1, \ldots, T \tag{7.11}$$

The notation is similar to that adopted in (4.1) except that z_i now includes lagged values of both exogenous and endogenous variables. The disturbance term is assumed to follow a stationary AR(1) process,

$$u_{it} = \phi_i u_{i, t-1} + \epsilon_{it} \tag{7.12}$$

and so (7.11) may be transformed to

$$y_{it} - \phi_i y_{i, t-1} = (z_{it} - \phi_i z_{i, t-1})' \delta_i + \epsilon_{it}$$

$$i = 1, \ldots, N, \quad t = 2, \ldots, T \quad (7.13)$$

Given ϕ_i, a suitable set of instruments is obtained by regressing $z_{it} - \phi_i z_{i, t-1}$ on x_{it}^*, $x_{i, t-1}^*$ and x_t, where x_{it}^* is a vector containing the predetermined variables, including lagged values, in the ith equation. In the notation of Section 2.11, $w_t^* = (x_{it}^{*\prime}, x_{it-1}^{*\prime}, x_t')'$, while the matrix L is as defined in (6.1.4) but with the first row deleted. Furthermore, conditional on δ_i, an estimator of ϕ_i may be obtained in the usual way, simply by regressing the residuals on their lagged values. This suggests a stepwise optimisation procedure in which ϕ_i is estimated by regression and δ_i is estimated by instrumental variables. The GIVE estimator is obtained on convergence.

8. Forecasting, Prediction and Control

It is possible to distinguish three motives for building an econometric model. These are: (a) forecasting; (b) prediction and policy evaluation; and (c) control. The forecasting aspect of econometric modelling ties in closely with time series analysis, while the theory surrounding the third function, control, is derived mainly from the engineering literature. Naturally the relative importance of forecasting, prediction and control depends on the particular application of the model. Furthermore, there is considerable overlap between all three. However, by classifying the roles of a model in this way, it is possible to gain some insight, not only into the role of econometrics as such, but also into its role in relation to time series analysis and control engineering. The aim of this final section, therefore, is to develop some kind of overall perspective.

Forecasting from a dynamic simultaneous equation model of the form (1.1) may be viewed as a generalisation of the process described in Section 7.7. Predictions of future values of the exogenous variables are made in some kind of systematic fashion, and unconditional predictions of the endogenous variables are then derived from the model. These are the forecasts of the endogenous variables, and they may be compared directly with forecasts made by any other method. Such methods may range from fitting 'naive' time series models to consulting the stars, although within the context of this book it is clearly the first of these methods which is the more interesting.

Prediction differs from forecasting in that the predictions are made conditional on future values of the exogenous variables. This does not necessarily imply that these future values are known, although from the theoretical point of view it is convenient to evaluate the accuracy of such predictions on this basis. The essential feature of prediction in practice is that it embodies certain assumptions about the future time paths of the exogenous variables. Thus the user of the model is free to bring his own views and prejudices into the forecasting process. However, these views and prejudices are made *explicit* in the declared values of future exogenous variables.

There is a further aspect to prediction. In many situations, the user of the model can actually manipulate some of the exogenous variables. Thus, for example, in the case of a macroeconomic model, the government is able to set a variable such as the rate of indirect taxation equal to whatever value it chooses. A variable which can be manipulated in this way is known as a *control variable* or *instrument*. Assigning a particular set of values to the control variables in the model then represents a *policy*. The effect of the policy on the endogenous variables may then be assessed from the model, conditional on the remaining, uncontrolled, exogenous variables. This exercise may be repeated with different policies and different assumptions about the uncontrolled variables.

Policy evaluation is an important aspect of econometric model building. However, a rather different approach to problems of this kind has been developed by control engineers. They assume at the outset that any policy objectives are well defined. This implies two things: firstly, that there should be *target* values for both endogenous and control variables, and secondly, that penalties can be assigned when these targets are not met and that these penalties can be combined in an overall *loss function*. The *theory of optimal control* is concerned with methods of deriving rules for finding values of the control variables which minimise this loss function.

There is an increasing awareness of the potential of optimal control in econometrics, and this is reflected in a growing literature on economic applications. Chow (1975) gives a review of some of these applications as well as providing an excellent introduction to the subject as a whole. The material presented in the later part of this section is merely intended to give a brief review of the main concepts involved.

The different uses to which a model may be put represents only one aspect of the discussion in this section. A second theme concerns the role of *a priori* assumptions derived from economic theory. This issue was raised right at the beginning of the first chapter. The discussion below draws on the material presented in Section 5 to compare the forecasting performance of simultaneous equation models, single equation models and time series models. By comparing the models within a unified framework the trade off between robustness and efficiency becomes clear.

The assumptions used in the construction of a model also have implications for policy evaluation and control. Again the treatment below attempts to place the various models in perspective, and to explore the relationship between different strands of the literature.

Prediction

Predictions of future values of the endogenous variables in a dynamic simultaneous equation model, (1.1), are made from the reduced form. Consider a simple example in which $r = 1$ and the disturbances are serially uncorrelated, i.e.

$$\Gamma y_t = A y_{t-1} + B x_t + \epsilon_t \tag{8.1}$$

Given that future values of the exogenous variables are known, the value of s has no bearing on the discussion below, and both current and past values of the exogenous variables can conveniently be included in x_t.

The reduced form can then be written

$$y_t = \Pi_1 y_{t-1} + \Pi x_t + v_t \tag{8.2}$$

where $\Pi_1 = \Gamma^{-1} A$ and $\Pi = \Gamma^{-1} B$. The optimal predictions are obtained from the recursive expression:

$$\tilde{y}_{T+l/T} = \Pi_1 \tilde{y}_{T+l-1/T} + \Pi x_{T+l}, \qquad l = 1, 2, \ldots \tag{8.3}$$

where $\tilde{y}_{T/T} = y_T$. This may be solved to yield

$$\tilde{y}_{T+l/T} = \Pi_1^l y_T + \sum_{j=0}^{l-1} \Pi_1^j \Pi x_{T+l-j}, \qquad l = 1, 2, \ldots \tag{8.4}$$

Conditional on the parameters in Π_1 and Π, the MSE of $\tilde{y}_{T+l/T}$ is given by

$$\text{MSE}(\tilde{y}_{T+l/T}) = \sum_{j=0}^{l-1} \Pi_1^j \Omega \Pi_1'^j, \qquad l = 1, 2, \ldots \tag{8.5}$$

In practice, Π_1 and Π will be unknown. They must therefore be estimated, and the usual procedure is to derive the estimators from the structural form. If $\tilde{\Gamma}$, $\tilde{A}$ and $\tilde{B}$ denote the estimators of the structural parameters in (8.1), obtained by an efficient full information procedure such as 3SLS or FIML, then

$$\tilde{\Pi}_1 = \tilde{\Gamma}^{-1} \tilde{A} \quad \text{and} \quad \tilde{\Pi} = \tilde{\Gamma}^{-1} \tilde{B} \tag{8.6}$$

These estimators are used in constructing the optimal predictor of y_{T+l} from (8.3). It is important to stress that $\tilde{\Pi}$ and $\tilde{\Pi}_1$ will not, in general, be the same as the estimators obtained by applying OLS directly to the reduced form. These estimators represent the first stage of 3SLS. However, they are inefficient insofar as they fail to make use of any over-identifying restrictions in the model. As a rule, A, B and Γ contain a large number of zeros and this information can only be incorporated in the estimators of the reduced form parameters if they are derived from the structural form. The exception is when every equation in the system is exactly identified, for in that case the structural and reduced forms contain the same number of parameters; cf. the discussion in Section 5.

Replacing Π and Π_1 by estimates results in an increase in the predictive MSE. Under standard regularity conditions it can be shown that the additional term which must be added to (8.5) is of $0(T^{-1})$. However, the results in Schmidt (1974, 1977) suggest that the contribution made by this term can be relatively large in small samples. Over-identifying restrictions on the structural form will therefore tend to have a significant impact on predictive efficiency. In the static case these gains may be assessed more precisely since a theoretical analysis of the problem is less complex. Reference should be made to Goldberger, Nagar and Odeh (1961).

Forecasting

The analysis of the previous sub-section was based on the assumption that future values of the exogenous variables are known. In forecasting, on the other hand, the concern is with unconditional or *ex ante* prediction. Future values of the exogenous variables must themselves be predicted and so the dynamic structure of the exogenous variables is now critical, whereas in conditional prediction it is irrelevant. This should be apparent from the results cited in Section 7.7. It will be assumed throughout this sub-section that future values of the exogenous variables are forecast by modelling them as a multivariate ARIMA process.

In the discussion of predictive efficiency in the previous sub-section, the main emphasis was on the *a priori* restrictions incorporated in the structural form, but not incorporated in the reduced form. However, in both cases one piece of *a priori* knowledge was taken for granted: the distinction between endogenous and exogenous variables. If this distinction cannot be made, predictions must be made on the basis of a time series model.

Suppose that the exogenous variables are generated by a multivariate ARMA process. Both the endogenous and exogenous variables can then be considered as being generated by a multivariate ARMA process of the form (6.13). However, because of the assumed exogeneity certain assumptions are embodied in (6.13). These are, firstly, that the autoregressive polynomial matrix is block triangular and, secondly, that the disturbance terms, $\bar{\theta}_u(L)\epsilon_t$ and $\Theta_x(L)\xi_t$, are uncorrelated. If the assumption of exogenous x's were dropped, the complete vector $(y_t', x_t')'$ could still be modelled as a multivariate ARMA process but without the restrictions. Alternatively, attention might be restricted to modelling only y_t as a multivariate ARMA process as suggested by (6.15). In the latter case a generalisation of the analysis in Section 7.7 would imply that, in general, this will involve a loss in predictive efficiency when compared with predictions made from the reduced form with the exogenous variables forecast from a multivariate ARMA model. However, this assumes that the variables in x_t really are exogenous. If this is not the case, the time series model may perform better. Again there is a trade-off between robustness and efficiency.

A multivariate time series model still embodies some *a priori* assumptions insofar as it treats the variables as being jointly determined. A univariate model, however, considers each variable in isolation, thereby invoking no *a priori* knowledge whatsoever. As was pointed out in Section 6, a set of variables generated by a multivariate ARMA process can be modelled individually. Predictions can then be constructed from these models, although if the multivariate model is correctly specified this will normally involve a loss in efficiency.

If the exogenous variables do indeed follow a multivariate ARMA process, modelling the variables in y_t as univariate ARMA processes is quite consistent with the dynamic econometric model (1.1). Forecasting on this basis comes bottom of the following hierarchy:

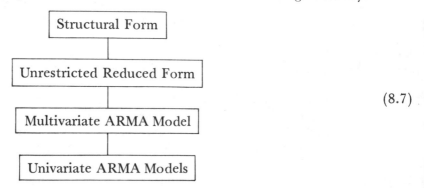

(8.7)

The extent to which *a priori* information is incorporated in the model specification decreases as one moves down the hierarchy. If the *a priori* information is valid, the forecasts of the endogenous

variables derived from the structural form of an econometric model
will normally be the best available. Conversely, forecasts obtained
from fitting individual ARMA processes will be the least efficient.
However, if the economic theory underlying the specification of the
structural form is based on invalid premises, it is conceivable that a
forecast from a univariate ARMA model might yield better results.
The crucial point, though, is that were this to happen, it would be an
indictment of the economic theory rather than the econometric
methodology.

One qualification must be made to the above statement. Dynamic
economic theory is not well-developed and is rarely definitive. The
dynamic specification of an econometric model is therefore largely
an empirical matter. In this respect, econometric methodology is of
critical importance. Even if the economic theory relating to the
inclusion and exclusion of variables is basically correct, an inappropriate specification of the dynamic structure may lead to a very poor
forecasting performance. In these circumstances it is possible that a
'naive' time series model could outperform an econometric model.
However, the rapid development of the dynamic aspects of econometric model building over the last ten years has made this occurrence less likely. A comparison of the predictive performance of a
dynamic econometric model with a set of univariate time series
models will be found in Prothero and Wallis (1976). On p. 485 they
summarise their results as follows:

Overall, the econometric approach, which makes the inter-relationships
between the variables quite explicit, provides a better-fitting representation than
the univariate time series approach, . . . [and] we conclude that to take account
of the specific interdependence of economic variables, whether or not these
variables can be individually represented as ARIMA processes, provides a more
fruitful approach to model building.

Control

Let the vector of exogenous variables in an econometric model be
partitioned as $x_t = (x'_{1t} x'_{2t})'$, where x_{1t} denotes the control variables
and x_{2t} denotes the uncontrolled variables. Both the endogenous and
the control variables are assumed to have target values over the
period $t = 1, \ldots, T$. These targets will be denoted by $y^\dagger_t$ and $x^\dagger_{1t}$,
$t = 1, \ldots, T$. In order to assess the impact of deviations from these
targets a loss function must be constructed. Mathematically, the
most convenient specification for such a function is as the expectation of a quadratic form, i.e.

$$W = E \sum_{t=1}^{T} (\dot{y}'_t \dot{x}'_{1t}) Q_t (\dot{y}'_t \dot{x}'_{1t})' \tag{8.8}$$

where Q_t is a matrix of weights, and a dot indicates a deviation from the target value, i.e. $\dot{y}_t = y_t - y_t^\dagger$ and $\dot{x}_t = x_t - x_t^\dagger$. If the expectation were not taken in (8.8) the stochastic nature of y_t would mean that the loss function was a random variable. Note that if the model assumed to be generating the endogenous variables is dynamic, as in (1.1), the expectation in (8.8) is conditional on the initial values $y_0, \ldots, y_{1-r}$.

The optimal control problem is to minimise the loss function, (8.8), subject to the econometric model (1.1). This means finding a rule for determining the optimal values of the control variables, $x_{11}^0, x_{12}^0, \ldots, x_{1T}^0$. If the econometric model were deterministic, i.e. $u_t \equiv 0$ for all t, the time path of x_1 could be determined at the outset. However, in the usual case, where y_t is stochastic, a feedback rule is needed. This incorporates the observations currently available and so the optimal values of the control variables evolve over time. There are a number of ways of deriving the *feedback equation*. These will not be described here, but the book by Chow (1975) provides an excellent reference. The solution takes the form

$$x_{1t}^0 = G_t z_{t-1} + g_t \qquad\qquad (8.9)$$

where $z_t = (y_t, y_{t-1}, \ldots, y_{t-r+1}, x_t, x_{t-1}, \ldots, x_{t-s+1})'$ and G_t is a matrix, the elements of which depend on the structural parameters of the econometric model and the weights in the loss function. The vector g_t also depends on these quantities, together with the target values of y_t and x_{1t}, and the actual values of x_{2t}.

Although taking the loss function to be a quadratic form permits an elegant solution to the problems of optimal control, it does pose a number of conceptual problems. In the first place, a positive deviation from a target is assigned the same cost as a negative deviation. Secondly, the function is additive so that an expectation is taken for the *sum* of functions of variables in different periods. Both of these properties may be restrictive in certain circumstances.

In a macroeconomic application the variables entering the loss function will be indicators of economic performance. These indicators are traded off against each other in a way that reflects the preferences of the policy makers. Since the economic indicators may include variables such as the level of unemployment and the rate of increase of prices, it is clear that these preferences will be determined by political, as well as economic, considerations. Within this framework the loss function is sometimes referred to as a *welfare function*.

It may not always be easy for the policy makers to formulate their preferences in the precise manner needed for an optimal control calculation to be made. One way of attempting to cope with this uncertainty is to determine the solutions to a number of related loss functions. The desirability of various strategies can then be assessed with regard to their implications for optimal control. Somewhat related to this observation is the point that the recommendations from an optimal control solution may not be the end of an analysis. It may also be desirable to take account of the dynamic behaviour implied by that solution. If the feedback equation settles down into a steady state solution,

$$x_{1t}^0 = \bar{G}z_{t-1} + \bar{g}, \qquad t = 1, \ldots, T \tag{8.10}$$

this is relatively easy to do. Substituting from (8.10) into the reduced form, (1.4), and re-arranging enables the dynamic behaviour of the endogenous variables to be assessed using the methods already developed.

A final point concerns the link with time series analysis. Just as forecasting can be carried out with virtually no *a priori* economic theory, so can control. Suppose that the control variables are identified with the variables labelled as deterministic in (6.16). These variables were denoted by x_1, while the remaining stochastic variables were x_2. Assuming the variables in x_2 to be generated by a multivariate ARMA process, it follows from the argument after (6.17) that each endogenous variable can be modelled by a transfer function. The explanatory variables in each transfer function are control variables. Therefore the *a priori* theory leading to such a specification can be regarded as minimal, since the control variables can be expected to have an effect on the endogenous variables almost by definition.

Notes

Section 6   Wallis (1977), Zellner and Palm (1974).
Section 7   On estimation see Hendry (1971, 1976), Pagan and Byron (1977), Chow and Fair (1973), Reinsel (1979) and Sargan (1961). A large sample test against serial correlation in single equations is given in Godfrey (1978); the test in Harvey and Phillips (1980a) is exact.
Section 8   Econometric and time series models are compared in Nelson (1972), Prothero and Wallis (1976) and McNees (1980).

Exercises

1. What is the relationship between the final equations, (6.8), and the reduced form in the static model, (1.5)?
2. Models (2.7) and (2.10) both incorporate two restrictions. Show that the latter implies restrictions on the reduced form whereas the former does not.

3. Construct an estimator of γ_{11} in (2.9a) from the OLS estimates of the reduced form and show that it is consistent. Show that it is identical to the 2SLS estimator. Repeat the exercise for (2.10a) and comment.

4. Prove (8.5).

5. Show that 3SLS collapses to 2SLS applied to each equation in turn when (i) Ω is diagonal or (ii) every equation is exactly identified.

6. Derive the autoregressive final form for the Keynesian model, (1.6.1).

7. In his comments on the paper by Prothero and Wallis (1976, pp. 487–490), Dr O. D. Anderson said: 'No one doubts that . . . econometric model building . . . ought to be the obvious choice to back in a bout between an adult and an infant — univariate time series analysis.' He continued: '. . . what econometric modelling will have to compete with very soon is multivariate time series analysis. I do not see how it can hope to cope then.' Comment.

8. Show that 3SLS is a special case of a FIVE estimator.

Appendix on Matrix Algebra

Differentiation

Differentiation of linear functions and quadratic forms play an important role in the derivation of estimators.

Let a and x be $n \times 1$ vectors and consider the linear function, $a'x = \Sigma a_i x_i$. Since $\partial(a'x)/\partial x_i = a_i$, $i = 1, \ldots, n$, differentiating $a'x$ with respect to the complete vector x may be expressed as

$$\frac{\partial(a'x)}{\partial x} = a \qquad\qquad\qquad (A.1)$$

Alternatively, differentiating with respect to the transpose of x yields

$$\frac{\partial(a'x)}{\partial x'} = a' \qquad\qquad\qquad (A.2)$$

For the quadratic form $x'Ax$

$$\frac{\partial(x'Ax)}{\partial x} = (A + A')x \qquad\qquad\qquad (A.3)$$

$$\frac{\partial(x'Ax)}{\partial x \, \partial x'} = A + A' \qquad\qquad\qquad (A.4)$$

while

$$\frac{\partial(x'Ax)}{\partial A} = xx' \qquad\qquad\qquad (A.5)$$

Finally

$$\frac{\partial \log |A|}{\partial A} = (A')^{-1}, \qquad \text{if } |A| > 0 \qquad\qquad\qquad (A.6)$$

Kronecker Products

Consider the $m \times n$ matrix $A = \{a_{ij}\}$ and the $p \times q$ matrix B. The Kronecker (or tensor) product of A and B (in that order) is the $mp \times nq$ matrix.

$$A \otimes B = \begin{bmatrix} a_{11}B & a_{12}B & \cdots & a_{1n}B \\ a_{21}B & a_{22}B & \cdots & a_{2n}B \\ \vdots & \vdots & & \vdots \\ a_{m1}B & a_{m2}B & \cdots & a_{mn}B \end{bmatrix} \qquad (A.7)$$

The basic result on the multiplication of two Kronecker products is:

$$(A \otimes B)(C \otimes D) = AC \otimes BD \qquad\qquad (A.8)$$

Given (A.8) the following results are easily established:

$$(A \otimes B)^{-1} = A^{-1} \otimes B^{-1} \qquad\qquad (A.9)$$

$$(A \otimes B)' = A' \otimes B' \qquad\qquad (A.10)$$

$$A \otimes (B + C) = A \otimes B + A \otimes C \qquad\qquad (A.11)$$

$$A \otimes (B \otimes C) = (A \otimes B) \otimes C \qquad\qquad (A.12)$$

Furthermore, it can be shown that

$$|A \otimes B| = |A|^n |B|^m \qquad\qquad (A.13)$$

and

$$\mathrm{tr}(A \otimes B) = (\mathrm{tr} A)(\mathrm{tr} B) \qquad\qquad (A.14)$$

Table A.1   Lower and Upper Bounds of the 5% Points of the Durbin–Watson Test Statistic[a]

T	d_L	d_U	d_L	d_U	d_L	d_U	d_L	d_U	d_L	d_U
	k = 2		k = 3		k = 4		k = 5		k = 6	
15	1.08	1.36	0.95	1.54	0.82	1.75	0.69	1.97	0.56	2.21
16	1.10	1.37	0.98	1.54	0.86	1.73	0.74	1.93	0.62	2.15
17	1.13	1.38	1.02	1.54	0.90	1.71	0.78	1.90	0.67	2.10
18	1.16	1.39	1.05	1.53	0.93	1.69	0.82	1.87	0.71	2.06
19	1.18	1.40	1.08	1.53	0.97	1.68	0.86	1.85	0.75	2.02
20	1.20	1.41	1.10	1.54	1.00	1.68	0.90	1.83	0.79	1.99
21	1.22	1.42	1.13	1.54	1.03	1.67	0.93	1.81	0.83	1.96
22	1.24	1.43	1.15	1.54	1.05	1.66	0.96	1.80	0.86	1.94
23	1.26	1.44	1.17	1.54	1.08	1.66	0.99	1.79	0.90	1.92
24	1.27	1.45	1.19	1.55	1.10	1.66	1.01	1.78	0.93	1.90
25	1.29	1.45	1.21	1.55	1.12	1.66	1.04	1.77	0.95	1.89
26	1.30	1.46	1.22	1.55	1.14	1.65	1.06	1.76	0.98	1.88
27	1.32	1.47	1.24	1.56	1.16	1.65	1.08	1.76	1.01	1.86
28	1.33	1.48	1.26	1.56	1.18	1.65	1.10	1.75	1.03	1.85
29	1.34	1.48	1.27	1.56	1.20	1.65	1.12	1.74	1.05	1.84
30	1.35	1.49	1.28	1.57	1.21	1.65	1.14	1.74	1.07	1.83
31	1.36	1.50	1.30	1.57	1.23	1.65	1.16	1.74	1.09	1.83
32	1.37	1.50	1.31	1.57	1.24	1.65	1.18	1.73	1.11	1.82
33	1.38	1.51	1.32	1.58	1.26	1.65	1.19	1.73	1.13	1.81
34	1.39	1.51	1.33	1.58	1.27	1.65	1.21	1.73	1.15	1.81
35	1.40	1.52	1.34	1.58	1.28	1.65	1.22	1.73	1.16	1.80
36	1.41	1.52	1.35	1.59	1.29	1.65	1.24	1.73	1.18	1.80
37	1.42	1.53	1.36	1.59	1.31	1.66	1.25	1.72	1.19	1.80
38	1.43	1.54	1.37	1.59	1.32	1.66	1.26	1.72	1.21	1.79
39	1.43	1.54	1.38	1.60	1.33	1.66	1.27	1.72	1.22	1.79
40	1.44	1.54	1.39	1.60	1.34	1.66	1.29	1.72	1.23	1.79
45	1.48	1.57	1.43	1.62	1.38	1.67	1.34	1.72	1.29	1.78
50	1.50	1.59	1.46	1.63	1.42	1.67	1.38	1.72	1.34	1.77
55	1.53	1.60	1.49	1.64	1.45	1.68	1.41	1.72	1.38	1.77
60	1.55	1.62	1.51	1.65	1.48	1.69	1.44	1.73	1.41	1.77
65	1.57	1.63	1.54	1.66	1.50	1.70	1.47	1.73	1.44	1.77
70	1.58	1.64	1.55	1.67	1.52	1.70	1.49	1.74	1.46	1.77
75	1.60	1.65	1.57	1.68	1.54	1.71	1.51	1.74	1.49	1.77
80	1.61	1.66	1.59	1.69	1.56	1.72	1.53	1.74	1.51	1.77
85	1.62	1.67	1.60	1.70	1.57	1.72	1.55	1.75	1.52	1.77
90	1.63	1.68	1.61	1.70	1.59	1.73	1.57	1.75	1.54	1.78
95	1.64	1.69	1.62	1.71	1.60	1.73	1.58	1.75	1.56	1.78
100	1.65	1.69	1.63	1.72	1.61	1.74	1.59	1.76	1.57	1.78

Source: This table is taken from Durbin and Watson (1951) with the kind permission of *Biometrika*, the publisher, and the authors.

[a]The value of k at the head of each pair of columns is the number of elements of the parameter vector, and it is assumed that one of these elements is a constant term. See Section 6.3 for details.

Table A.2   Lower and Upper Bounds of the 1% Points of the Durbin—Watson
Test Statistic

T	k = 2 d_L	d_U	k = 3 d_L	d_U	k = 4 d_L	d_U	k = 5 d_L	d_U	k = 6 d_L	d_U
15	0.81	1.07	0.70	1.25	0.59	1.46	0.49	1.70	0.39	1.96
16	0.84	1.09	0.74	1.25	0.63	1.44	0.53	1.66	0.44	1.90
17	0.87	1.10	0.77	1.25	0.67	1.43	0.57	1.63	0.48	1.85
18	0.90	1.12	0.80	1.26	0.71	1.42	0.61	1.60	0.52	1.80
19	0.93	1.13	0.83	1.26	0.74	1.41	0.65	1.58	0.56	1.77
20	0.95	1.15	0.86	1.27	0.77	1.41	0.68	1.57	0.60	1.74
21	0.97	1.16	0.89	1.27	0.80	1.41	0.72	1.55	0.63	1.71
22	1.00	1.17	0.91	1.28	0.83	1.40	0.75	1.54	0.66	1.69
23	1.02	1.19	0.94	1.29	0.86	1.40	0.77	1.53	0.70	1.67
24	1.04	1.20	0.96	1.30	0.88	1.41	0.80	1.53	0.72	1.66
25	1.05	1.21	0.98	1.30	0.90	1.41	0.83	1.52	0.75	1.65
26	1.07	1.22	1.00	1.31	0.93	1.41	0.85	1.52	0.78	1.64
27	1.09	1.23	1.02	1.32	0.95	1.41	0.88	1.51	0.81	1.63
28	1.10	1.24	1.04	1.32	0.97	1.41	0.90	1.51	0.83	1.62
29	1.12	1.25	1.05	1.33	0.99	1.42	0.92	1.51	0.85	1.61
30	1.13	1.26	1.07	1.34	1.01	1.42	0.94	1.51	0.88	1.61
31	1.15	1.27	1.08	1.34	1.02	1.42	0.96	1.51	0.90	1.60
32	1.16	1.28	1.10	1.35	1.04	1.43	0.98	1.51	0.92	1.60
33	1.17	1.29	1.11	1.36	1.05	1.43	1.00	1.51	0.94	1.59
34	1.18	1.30	1.13	1.36	1.07	1.43	1.01	1.51	0.95	1.59
35	1.19	1.31	1.14	1.37	1.08	1.44	1.03	1.51	0.97	1.59
36	1.21	1.32	1.15	1.38	1.10	1.44	1.04	1.51	0.99	1.59
37	1.22	1.32	1.16	1.38	1.11	1.45	1.06	1.51	1.00	1.59
38	1.23	1.33	1.18	1.39	1.12	1.45	1.07	1.52	1.02	1.58
39	1.24	1.34	1.19	1.39	1.14	1.45	1.09	1.52	1.03	1.58
40	1.25	1.34	1.20	1.40	1.15	1.46	1.10	1.52	1.05	1.58
45	1.29	1.38	1.24	1.42	1.20	1.48	1.16	1.53	1.11	1.58
50	1.32	1.40	1.28	1.45	1.24	1.49	1.20	1.54	1.16	1.59
55	1.36	1.43	1.32	1.47	1.28	1.51	1.25	1.55	1.21	1.59
60	1.38	1.45	1.35	1.48	1.32	1.52	1.28	1.56	1.25	1.60
65	1.41	1.47	1.38	1.50	1.35	1.53	1.31	1.57	1.28	1.61
70	1.43	1.49	1.40	1.52	1.37	1.55	1.34	1.58	1.31	1.61
75	1.45	1.50	1.42	1.53	1.39	1.56	1.37	1.59	1.34	1.62
80	1.47	1.52	1.44	1.54	1.42	1.57	1.39	1.60	1.36	1.62
85	1.48	1.53	1.46	1.55	1.43	1.58	1.41	1.60	1.39	1.63
90	1.50	1.54	1.47	1.56	1.45	1.59	1.43	1.61	1.41	1.64
95	1.51	1.55	1.49	1.57	1.47	1.60	1.45	1.62	1.42	1.64
100	1.52	1.56	1.50	1.58	1.48	1.60	1.46	1.63	1.44	1.65

Source: This table is taken from Durbin and Watson (1951) with the kind
permission of *Biometrika*, the publisher, and the authors.

Table B   5%, 1%, and 0.1% Points of the Modified Von Neumann Ratio[a]

Degrees of Freedom	5%	1%	0.1%	5%	1%	0.1%
	One-tailed test against positive autocorrelation			One-tailed test against negative autocorrelation		
2	0.025	0.001	0.000	3.975	3.999	4.000
3	0.252	0.052	0.005	4.142	4.427	4.493
4	0.474	0.170	0.037	3.827	4.295	4.496
5	0.598	0.292	0.095	3.571	4.076	4.378
6	0.701	0.386	0.163	3.413	3.881	4.233
7	0.790	0.464	0.228	3.299	3.731	4.095
8	0.861	0.537	0.285	3.206	3.618	3.973
9	0.922	0.601	0.339	3.131	3.524	3.871
10	0.975	0.657	0.390	3.069	3.445	3.784
11	1.020	0.708	0.438	3.016	3.378	3.710
12	1.060	0.753	0.482	2.970	3.319	3.645
13	1.096	0.795	0.523	2.930	3.268	3.587
14	1.128	0.832	0.561	2.895	3.222	3.535
15	1.157	0.866	0.597	2.863	3.181	3.488
16	1.183	0.898	0.630	2.835	3.144	3.445
17	1.207	0.927	0.661	2.809	3.110	3.406
18	1.228	0.954	0.691	2.785	3.079	3.370
19	1.249	0.979	0.718	2.764	3.051	3.337
20	1.267	1.003	0.744	2.744	3.025	3.306
21	1.285	1.024	0.769	2.725	3.000	3.277
22	1.301	1.045	0.792	2.708	2.978	3.250
23	1.316	1.064	0.814	2.692	2.957	3.225
24	1.330	1.082	0.834	2.677	2.937	3.201
25	1.344	1.100	0.854	2.663	2.918	3.179
26	1.356	1.116	0.873	2.650	2.901	3.157
27	1.368	1.131	0.891	2.638	2.884	3.137
28	1.380	1.146	0.908	2.626	2.868	3.118
29	1.390	1.160	0.925	2.615	2.854	3.100
30	1.400	1.173	0.940	2.605	2.839	3.083

[a]See Section 5.2 for details.

Degrees of Freedom	5%	1%	0.1%	5%	1%	0.1%
	One-tailed test against positive autocorrelation			One-tailed test against negative autocorrelation		
31	1.410	1.186	0.955	2.595	2.826	3.066
32	1.419	1.198	0.970	2.585	2.813	3.051
33	1.428	1.209	0.984	2.576	2.801	3.036
34	1.437	1.221	0.997	2.567	2.789	3.021
35	1.445	1.231	1.010	2.559	2.778	3.007
36	1.452	1.241	1.022	2.351	2.767	2.994
37	1.460	1.251	1.034	2.544	2.757	2.982
38	1.467	1.261	1.045	2.536	2.747	2.969
39	1.474	1.270	1.057	2.529	2.738	2.957
40	1.480	1.279	1.067	2.522	2.729	2.946
41	1.487	1.287	1.078	2.516	2.720	2.935
42	1.493	1.295	1.088	2.510	2.711	2.925
43	1.499	1.303	1.097	2.504	2.703	2.914
44	1.504	1.311	1.107	2.498	2.695	2.904
45	1.510	1.318	1.116	2.492	2.687	2.895
46	1.515	1.325	1.125	2.487	2.680	2.885
47	1.520	1.332	1.133	2.482	2.673	2.876
48	1.525	1.339	1.142	2.477	2.666	2.868
49	1.530	1.346	1.150	2.472	2.659	2.859
50	1.535	1.352	1.158	2.467	2.653	2.851
51	1.540	1.358	1.165	2.462	2.646	2.843
52	1.544	1.364	1.173	2.458	2.640	2.835
53	1.548	1.370	1.180	2.453	2.634	2.828
54	1.552	1.376	1.187	2.449	2.628	2.820
55	1.557	1.381	1.194	2.445	2.623	2.813
56	1.561	1.387	1.201	2.441	2.617	2.806
57	1.564	1.392	1.207	2.437	2.612	2.799
58	1.568	1.397	1.214	2.433	2.606	2.793
59	1.572	1.402	1.220	2.429	2.601	2.786
60	1.575	1.407	1.226	2.426	2.596	2.780

Source: This table is taken from Press and Brooks (1969) with kind permission of the authors.

Table C   Significance Values for the CUSUM of Squares Test

n	α				
	0.10	0.05	0.025	0.01	0.005
1	0.40000	0.45000	0.47500	0.49000	0.49500
2	0.35044	0.44306	0.50855	0.56667	0.59596
3	0.35477	0.41811	0.46702	0.53456	0.57900
4	0.33435	0.39075	0.44641	0.50495	0.54210
5	0.31556	0.37359	0.42174	0.47692	0.51576
6	0.30244	0.35522	0.40045	0.45440	0.48988
7	0.28991	0.33905	0.38294	0.43337	0.46761
8	0.27828	0.32538	0.36697	0.41522	0.44819
9	0.26794	0.31325	0.35277	0.39922	0.43071
10	0.25884	0.30221	0.34022	0.38481	0.41517
11	0.25071	0.29227	0.32894	0.37187	0.40122
12	0.24325	0.28330	0.31869	0.36019	0.38856
13	0.23639	0.27515	0.30935	0.34954	0.37703
14	0.23010	0.26767	0.30081	0.33980	0.36649
15	0.22430	0.26077	0.29296	0.33083	0.35679
16	0.21895	0.25439	0.28570	0.32256	0.34784
17	0.21397	0.24847	0.27897	0.31489	0.33953
18	0.20933	0.24296	0.27270	0.30775	0.33181
19	0.20498	0.23781	0.26685	0.30108	0.32459
20	0.20089	0.23298	0.26137	0.29484	0.31784
21	0.19705	0.22844	0.25622	0.28898	0.31149
22	0.19343	0.22416	0.25136	0.28346	0.30552
23	0.19001	0.22012	0.24679	0.27825	0.29989
24	0.18677	0.21630	0.24245	0.27333	0.29456
25	0.18370	0.21268	0.23835	0.26866	0.28951
26	0.18077	0.20924	0.23445	0.26423	0.28472
27	0.17799	0.20596	0.23074	0.26001	0.28016
28	0.17533	0.20283	0.22721	0.25600	0.27582
29	0.17280	0.19985	0.22383	0.25217	0.27168
30	0.17037	0.19700	0.22061	0.24851	0.26772
31	0.16805	0.19427	0.21752	0.24501	0.26393
32	0.16582	0.19166	0.21457	0.24165	0.26030
33	0.16368	0.18915	0.21173	0.23843	0.25683
34	0.16162	0.18674	0.20901	0.23534	0.25348
35	0.15964	0.18442	0.20639	0.23237	0.25027
36	0.15774	0.18218	0.20387	0.22951	0.24718
37	0.15590	0.18003	0.20144	0.22676	0.24421
38	0.15413	0.17796	0.19910	0.22410	0.24134
39	0.15242	0.17595	0.19684	0.22154	0.23857
40	0.15076	0.17402	0.19465	0.21906	0.23589

Note: Enter the table at $n = \frac{1}{2}(T - k) - 1$. The level of significance for a one-sided test is α; for a two-sided test it is 2α.

		α			
n	0.10	0.05	0.025	0.01	0.005
41	0.14916	0.17215	0.19254	0.21667	0.23310
42	0.14761	0.17034	0.19050	0.21436	0.23081
43	0.14611	0.16858	0.18852	0.21212	0.22839
44	0.14466	0.16688	0.18661	0.20995	0.22605
45	0.14325	0.16524	0.18475	0.20785	0.22377
46	0.14188	0.16364	0.18295	0.20581	0.22157
47	0.14055	0.16208	0.18120	0.20383	0.21943
48	0.13926	0.16058	0.17950	0.20190	0.21735
49	0.13800	0.15911	0.17785	0.20003	0.21534
50	0.13678	0.15769	0.17624	0.19822	0.21337
51	0.13559	0.15630	0.17468	0.19645	0.21146
52	0.13443	0.15495	0.17316	0.19473	0.20961
53	0.13330	0.15363	0.17168	0.19305	0.20780
54	0.13221	0.15235	0.17024	0.19142	0.20604
55	0.13113	0.15110	0.16884	0.18983	0.20432
56	0.13009	0.14989	0.16746	0.18828	0.20265
57	0.12907	0.14870	0.16613	0.18677	0.20101
58	0.12807	0.14754	0.16482	0.18529	0.19942
59	0.12710	0.14641	0.16355	0.18385	0.19786
60	0.12615	0.14530	0.16230	0.18245	0.19635
62	0.12431	0.14316	0.15990	0.17973	0.19341
64	0.12255	0.14112	0.15760	0.17713	0.19061
66	0.12087	0.13916	0.15540	0.17464	0.18792
68	0.11926	0.13728	0.15329	0.17226	0.18535
70	0.11771	0.13548	0.15127	0.16997	0.18288
72	0.11622	0.13375	0.14932	0.16777	0.18051
74	0.11479	0.13208	0.14745	0.16566	0.17823
76	0.11341	0.13048	0.14565	0.16363	0.17604
78	0.11208	0.12894	0.14392	0.16167	0.17392
80	0.11079	0.12745	0.14224	0.15978	0.17188
82	0.10955	0.12601	0.14063	0.15795	0.16992
84	0.10835	0.12462	0.13907	0.15619	0.16802
86	0.10719	0.12327	0.13756	0.15449	0.16618
88	0.10607	0.12197	0.13610	0.15284	0.16440
90	0.10499	0.12071	0.13468	0.15124	0.16268
92	0.10393	0.11949	0.13331	0.14970	0.16101
94	0.10291	0.11831	0.13198	0.14820	0.15940
96	0.10192	0.11716	0.13070	0.14674	0.15783
98	0.10096	0.11604	0.12944	0.14533	0.15631
100	0.10002	0.11496	0.12823	0.14396	0.15483

Source: This table is taken from Durbin (1969) with the kind permission of
Biometrika and the author.

Answers to Selected Exercises

Chapter 1. — 4. Δy_t follows a non-invertible MA(1) process.
    5. $\alpha/(1 - \phi)$.
    6. Yes; roots of (5.4) are -0.8 and -0.9.

Chapter 2. — 1. $\Sigma x_t^2 = x_1^2/(1 - \kappa^2)$. Therefore Var($b$) is of $0(1)$, (3.9) does not hold and b is not consistent.
    2. Regress $\log y_t$ on $1/x_t$.

Chapter 3. — 2. $\bar{y}, \tilde{y} \sim AN(\theta, \theta^2/T)$.
    3. Avar($\tilde{\sigma}$) = $\sigma^2/2T$.
    5. The exponential distribution is a special case of the Gamma distribution; see Amemiya (1973).
    8. (i) Yes; (ii) No; (iii) Yes; (iv) No.
    9. Zeckhauser and Thompson (1970).
   10. CSS with $\epsilon_0 = \epsilon_{-1} = 0$.

Chapter 5. — 5. LR is χ^2 with $N - 1$ degrees of freedom.
    7. Regress y on x, z and $z \log z$ and carry out a t-test on $z \log z$.

Chapter 6. — 3. (a) See Kadiyala (1968). (b) C—O now better.
    7. (ii) $\tilde{y}_{T+1/T} = 11.5$; $\tilde{y}_{T+2/T} = 11.75$.

Chapter 7. — 2. IV estimation with x_{t-1} and x_{t-2} as instruments for y_{t-1} and y_{t-2} respectively, followed by one iteration of Gauss—Newton.
    3. Estimate α and β by instrumental variables and estimate θ from the residuals; cf. exercise (4.6).
    5. $\hat{\alpha} = -0.5$, $\hat{\beta}_0 = 2.0$, $\hat{\beta}_1 = 1.8$.
    8. IV estimators, $\tilde{b} = 0.096$, $\tilde{\alpha} = 0.540$.

Chapter 8. — 1. (a) 5, 1.50, 0.36. (b) $\delta_j = \delta^{j-1}(\alpha\beta_0 + \beta_1)$ for $j \geqslant 1$, $\delta_0 = \beta_0$. (d) 10, 1.6.

4. LM test of $H_0 : \alpha = \phi$ in (4.2). Regress (4.3) on its derivatives evaluated under H_0 and test $T \cdot R^2$ as χ_1^2.

7. $h_4 = (1 - \frac{1}{2}d_4) \{ T/(1 - T) \cdot a^6 \cdot \text{avar}(a) \}$.

10. See Davidson *et al.* (1979, p. 681).

11. (i) $(1 - A)^{-1}(B_0 + B_1)$. (ii) $A^2 B_0 + AB_1$.

Chapter 9. – 1. Same.

2. $\gamma_{11} = \pi_{11}/\pi_{21} = \pi_{12}/\pi_{22}$.

3. The estimator is known as indirect least squares; see Johnston (1972, pp. 344–346).

References

Akaike, H. (1973), Information theory and an extension of the maximum likelihood principle, in B. N. Petrov and F. Csaki (eds.), *2nd International Symposium on Information Theory*, Budapest, Akademini Kiado, pp. 267–281.

Almon, S. (1965), The distributed lag between capital appropriations and expenditures, *Econometrica*, 33, pp. 178–196.

Amemiya, T. (1973), Generalized least squares with an estimated autocovariance matrix, *Econometrica*, 41, pp. 723–732.

Amemiya, T. (1977), A note on a heteroscedastic model, *Journal of Econometrics*, 6, pp. 365–370.

Anderson, T. W. (1971), *The Statistical Analysis of Time Series*, Wiley, New York.

Andrews, D. F. (1974), A robust method for multiple linear regression, *Technometrics*, 16, pp. 523–531.

Andrews, D. F., Bickel, P. J., Hampel, F. R., Huber, P. J., Rogers, W. H. and Tukey, J. W. (1972), *Robust Estimates of Location*, Princeton University Press.

Armstrong, R. D. and Frome, E. L. (1976), A comparison of two algorithms for absolute deviation curve fitting, *Journal of the American Statistical Association*, 71, pp. 328–330.

Atkinson, A. C. (1970), A method for discriminating between models, *Journal of the Royal Statistical Society*, B, 32, pp. 323–344.

Baillie, R. T. (1979), The asymptotic mean squared error of multistep prediction from the regression model with autoregressive errors, *Journal of the American Statistical Association*, 74, pp. 175–184.

Basmann, R. L. (1960), On finite sample distributions of generalized classical linear identifiability test statistics, *Journal of the American Statistical Association*, 55, pp. 650–659.

Bassett, G. and Koenker, R. (1978), Asymptotic theory of least absolute error regression, *Journal of the American Statistical Association*, 73, pp. 618–622.

Beach, C. M. and MacKinnon, J. G. (1978), A maximum likelihood procedure for regression with auto-correlated errors, *Econometrica*, 46, pp. 51–58.

Berndt, E. K., Hall, B., Hall, R. E. and Hausman, J. A. (1974), Estimation and inference in non-linear structural models, *Annals of Economic and Social Measurement*, 3, pp. 653–665.

Berndt, E. R. and Savin, N. E. (1977), Conflict among criteria for testing hypotheses in the multivariate regression model, *Econometrica*, 45, pp. 1263–1278.

Box, G. E. P. and Cox, D. R. (1964), An analysis of transformations, *Journal of the Royal Statistical Society*, Series B, 26, pp. 211–252.

Box, M. J., Davies, D. and Swann, W. H. (1969), *Non-Linear Optimization Techniques*, Oliver and Boyd.

Box, G. E. P. and Jenkins, G. M. (1976), *Time Series Analysis: Forecasting and Control*, revised edition, Holden-Day, San Francisco.

Breusch, T. S. (1979), *Indirect Calculation of Lagrange Multiplier Test Statistics with Application to Testing for Autocorrelation in Dynamic Simultaneous Equations Systems*, University of Southampton Discussion Paper.

Breusch, T. S. and Pagan, A. R. (1979), A simple test for heteroscedasticity, and random coefficient variation, *Econometrica*, 47, pp. 1287–1294.

Breusch, T. S. and Pagan, A. R. (1980), The Lagrange multiplier test and its applications to model specification in econometrics, *Review of Economic Studies*, XLVII, pp. 239–253.

Brown, R. L., Durbin, J. and Evans, J. M. (1975), Techniques for testing the constancy of regression relationships over time (with discussion), *Journal of the Royal Statistical Society*, B, 37, pp. 149–192.

Brundy, J. M. and Jorgenson, D. W. (1971), Efficient estimation of simultaneous equations by instrumental variables, *Review of Economics and Statistics*, 53, pp. 207–224.

Chipman, J. S. (1979), Efficiency of least squares estimation of linear trend when residuals are autocorrelated, *Econometrica*, 47, pp. 115–128.

Chow, G. C. (1960), Tests for equality between sets of coefficients in two linear regressions, *Econometrica*, 28, pp. 591–605.

Chow, G. C. and Fair, R. C. (1973), Maximum likelihood estimation of linear equation systems with auto-regressive residuals, *Annals of Economic and Social Measurement*, 2, pp. 17–28.

Chow, G. C. (1975), *Analysis and Control of Dynamic Economic Systems*, John Wiley, New York.

Cox, D. R. (1961), Tests of separate families of hypotheses, *Proceedings of the Fourth Berkeley Symposium*, I.

Cox, D. R. (1962), Some further results on separate families of hypotheses, *Journal of the Royal Statistical Society*, B, 38, pp. 45–53.

Cox, D. R. and Hinkley, D. V. (1974), *Theoretical Statistics*, Chapman and Hall.

Cramér, H. (1946), *Mathematical Methods of Statistics*, Princeton University Press.

Davidson, J., Hendry, D. F., Srba, F. and Yeo, S., Econometric modelling of the aggregate time-series relationship between consumers' expenditure and income in the United Kingdom, *Economic Journal*, 88, pp. 661–692.

Deistler, M. (1976), The identifiability of linear econometric models with autocorrelated errors, *International Economic Review*, 17, pp. 26–45.

Dhrymes, P. J. (1970), *Distributed Lags: Problems of Formulation and Estimation*, Holden-Day, San Francisco.

Dixon, L. C. W. (1972), *Nonlinear Optimisation*, English Universities Press, London.

Domencich, T. and McFadden, D. (1975), *Urban Travel Demand: A Behavioural Analysis*, North-Holland Publishing Co.

Durbin, J. (1960), Estimation of parameters in time-series regression models, *Journal of the Royal Statistical Society*, B, 22, pp. 139–153.

Durbin, J. (1969), Tests for serial correlation in regression analysis based on the periodogram of least-square residuals, *Biometrika*, 56, pp. 1–15.

Durbin, J. and Watson, G. S. (1950), Testing for serial correlation in least squares regression, I, *Biometrika*, 37, pp. 409–428.

Durbin, J. and Watson, G. S. (1951), Testing for serial correlation in least squares regression, II, *Biometrika*, 38, pp. 159–178.

Durbin, J. and Watson, G. S. (1971), Testing for serial correlation in least squares regression, III, *Biometrika*, 58, pp. 1–19.

Engle, R. F., Hendry, D. F. and Richard, J. F. (1979), *Exogeneity*, Unpublished paper.

Friedman, M. (1953), The methodology of positive economics, in *Essays in Positive Economics*, University of Chicago Press, pp. 3–43.

Fuller, W. A. (1976), *Introduction to Statistical Time Series*, John Wiley, New York.

Gadd, A. and Wold, H. (1964), The Janus quotient: a measure for the accuracy of prediction, in H. Wold (ed.) *Economic Model Building*, North-Holland Publishing Co.

Geary, R. C. (1966), A note on residual heterovariance and estimation efficiency in regression, *American Statistician*, October, pp. 30–31.

Gentleman, W. M. (1973), Least squares computation by Givens transformations without square roots, *Journal of the Institute of Mathematics and Its Applications*, 12, pp. 329–336.

Gill, P. E. and Murray, W. (1972), Quasi-Newton methods for unconstrained optimization, *Journal of the Institute of Mathematics and Its Applications*, 9, pp. 91–108.

Gill, P. E., Murray, W. and Pitfield, R. A. (1972), The implementation of two revised quasi-Newton algorithms for unconstrained optimization, *National Physical Laboratory Report, NAC 11*.

Godfrey, L. G. (1973), A note on the treatment of serial correlation, *Canadian Journal of Economics*, 6, pp. 567–573.

Godfrey, L. (1978), A note on the use of Durbin's h test when the equation is estimated by instrumental variables, *Econometrica*, 46, pp. 225–228.

Godfrey, L. G. (1978), Testing against general autoregressive and moving average error models when the regressors include lagged dependent variables, *Econometrica*, 6, pp. 1293–1302.

Godfrey, L. G. and Poskitt, D. S. (1975), Testing the restrictions of the Almon lag technique, *Journal of the American Statistical Association*, 70, pp. 105–108.

Goldberger, A. S. (1962), Best linear unbiased prediction in the generalised linear regression model, *Journal of the American Statistical Association*, 57, pp. 369–375.

Goldberger, A. S., Nagar, A. L. and Odeh, H. S. (1961), The covariance matrices of reduced-form coefficients and of forecasts for a structural econometric model, *Econometrica*, 29, pp. 556–573.

Goldfeld, S. M. and Quandt, R. E. (1972), *Nonlinear Methods in Econometrics*, North-Holland Publishing Co.

Goldfeld, S. M., Quandt, R. E. and Trotter, H. F. (1966), Maximization by quadratic hill-climbing, *Econometrica*, 34, pp. 541–551.

Granger, C. W. J. (1969), Investigating causal relations by econometric models and cross-spectral methods, *Econometrica*, 37, pp. 424–438.

Granger, C. W. J. and Newbold, P. (1977), *Forecasting Economic Time Series*, Academic Press, New York.

Griliches, Z. and Rao, P. (1969), Small sample properties of several two-stage regression methods in the context of autocorrelated errors, *Journal of the American Statistical Association*, 64, pp. 253—272.

Griliches, Z. and Ringstad, V. (1971), *Economies of Scale and the Form of the Production Function*, North-Holland Publishing Co.

Grolub, G. H. and Styan, G. P. H. (1973), Numerical computations for univariate linear models, *Journal of Statistical Computation and Simulation*, 2, pp. 253—274.

Hannan, E. J. (1970), *Multiple Time Series*, John Wiley, New York.

Harrison, M. J. and McCabe, B. P. M. (1979), A test for heteroscedasticity based on ordinary least squares residuals, *Journal of the American Statistical Association*, 74, pp. 474—499.

Hart, R. A., Hutton, J. and Sharot, T. (1975), A statistical analysis of association football attendances, *Applied Statistics*, 24, pp. 17—27.

Harvey, A. C. (1976), Estimating regression models with multiplicative heteroscedasticity, *Econometrica*, 44, 461—465.

Harvey, A. C. (1980), On comparing regression models in levels and first differences, *International Economic Review*, (forthcoming).

Harvey, A. C. and Collier, P. (1977), Testing for functional misspecification in regression analysis, *Journal of Econometrics*, 6, pp. 103—119.

Harvey, A. C. and McAvinchey, I. D. (1978), *The Small Sample Efficiency of Two-Step Estimators in Regression Models with Autoregressive Disturbances*, Discussion Paper No. 78—10, University of British Colombia, April.

Harvey, A. C. and McAvinchey, I. D. (1980), On the relative efficiency of various estimators of regression models with moving average disturbances, in G. Charatsis (ed.), *Proceedings of European Meeting of the Econometric Society*, Athens, 1979.

Harvey, A. C. and Phillips, G. D. A. (1974b), A comparison of the power of some tests for heteroscedasticity in the general linear model, *Journal of Econometrics*, 2, pp. 307—316.

Harvey, A. C. and Phillips, G. D. A. (1976), *Testing for Stochastic Parameters in Regression Models*, Paper presented at the European Meeting of the Econometric Society, Helsinki.

Harvey, A. C. and Phillips, G. D. A. (1980a), Testing for serial correlation in simultaneous equation models, *Econometrica*, 48, pp. 747—759.

Harvey, A. C. and Phillips, G. D. A. (1980b), Testing for heteroscedasticity in simultaneous equation models, *Journal of Econometrics*, (forthcoming).

Hatanaka, M. (1974), An efficient two-step estimator for the dynamic adjustment model with autoregressive errors, *Journal of Econometrics*, 2, pp. 199—220.

Hatanaka, M. (1975), On the global identification of the dynamic simultaneous equations model with stationary disturbances, *International Economic Review*, 16, pp. 545—554.

Hatanaka, M. (1976), Several efficient two-step estimators for the dynamic simultaneous equation model with autoregressive disturbances, *Journal of Econometrics*, 4, pp. 189—204.

Haugh, L. D. and Box, G. E. P. (1977), Identification of dynamic regression (distributed lag) models connecting two time series, *Journal of the American Statistical Association*, 72, pp. 121—130.

Hausman, J. A. (1975), An instrumental variable approach to full information estimators for linear and certain nonlinear econometric models, *Econometrica*, 43, pp. 727—738.

Hendry, D. F. (1971), Maximum likelihood estimation of systems of simultaneous regression equations with errors generated by a vector autoregressive process, *International Economic Review*, 12, pp. 257—272.

Hendry, D. F. (1976), The structure of simultaneous equations estimators, *Journal of Econometrics*, 4, pp. 51—88.

Hendry, D. F. (1980), Predictive failure and econometric modelling in macroeconomics: the transactions demand for money, in P. Ormerod (ed.), *Modelling the Economy*, Heinemann Educational Books.

Hendry, D. F. and Mizon, G. E. (1978), Serial correlation as a convenient simplification not a nuisance: a comment on a study of the demand for money by the Bank of England, *Economic Journal*, 88, pp. 549—563.

Hendry, D. F. and Tremayne, A. (1976), Estimating systems of dynamic reduced form equations with vector autoregressive errors, *International Economic Review*, 17, pp. 463—471.

Hendry, D. F. and T. von Ungern-Sternberg (1980), Liquidity and inflation effects on consumers' expenditure, in A. S. Deaton (ed.), *Essays in the Theory and Measurement of Consumers' Behaviour*, Cambridge University Press.

Huber, P. J. (1973), Robust regression: asymptotics, conjectures and Monte Carlo, *Annals of Statistics*, 1, pp. 799—821.

Johnston, J. (1972), *Econometric Methods*, 2nd edition, McGraw-Hill, New York.

Jorgenson, D. W. (1966), Rational distributed lag functions, *Econometrica*, 34, pp. 135—149.

Kadiyala, K. R. (1968), A transformation used to circumvent the problem of autocorrelation, *Econometrica*, 36, pp. 93—96.

Kendall, M. G. and Stuart, A. (1973), *The Advanced Theory of Statistics*, Vol. II, Charles Griffin, London.

Kenward, L. R. (1976), A new test for residual randomness in a class of dynamic autocorrelated econometric models, *Canadian Journal of Statistics*, C, 4, pp. 51—64.

Kmenta, J. and Gilbert, R. F. (1968), Small sample properties of alternative estimators of seemingly unrelated regressions, *Journal of the American Statistical Association*, 63, pp. 1180—1200.

Kmenta, J. and Gilbert, R. F. (1970), Estimation of seemingly unrelated regressions with autoregressive disturbances, *Journal of the American Statistical Association*, 65, pp. 186—197.

Koerts, J. and Abrahamse, A. P. J. (1969), *On the Theory and Application of the General Linear Model*, Rotterdam University Press.

Konstas, P. and Khouja, M. (1969), The Keynesian demand for money function, *Journal of Money, Credit and Banking*, 1, pp. 765—777.

Leamer, E. (1978), *Specification Searches*, John Wiley, New York.

Ljung, G. M. and Box, G. E. P. (1978), On a measure of lack of fit in time series models, *Biometrika*, 66, pp. 67—72.

Maddala, G. S. and Rao, A. S. (1971), Maximum likelihood estimation of Solow's and Jorgenson's distributed lag models, *Review of Economics and Statistics*, 53, pp. 80—88.

Maddala, G. S. and Rao, A. S. (1973), Tests for serial correlation in regression models with lagged dependent variables and serially correlated errors, *Econometrica*, 41, pp. 761—774.

Maeshiro, A. (1976), Autoregressive transformation, trended independent variables and autocorrelated disturbance terms, *Review of Economics and Statistics*, 58, pp. 497—500.

Malinvaud, E. (1970), *Statistical Methods of Econometrics*, 2nd edition, North-Holland Publishing Co.

Mann, H. B. and Wald, A. (1943), On the statistical treatment of linear stochastic difference equations, *Econometrica*, 11, pp. 173–220.

Marquardt, D. W. (1963), An algorithm for least squares estimation of nonlinear parameters, *Journal of Society of Industrial Applied Mathematics*, 11, pp. 431–441.

Mizon, G. E. (1977a), Inferential procedures in nonlinear models: an application in a UK industrial cross section study of factor substitution and returns to scale, *Econometrica*, 45, pp. 1221–1242.

Mizon, G. E. (1977b), Model selection procedures, in *Studies in Modern Economic Analysis*, M. J. Artis and A. R. Nobay (eds), Basil Blackwell.

Mizon, G. E. and Hendry, D. F. (1980), An empirical and Monte Carlo analysis of tests of dynamic specification, *Review of Economic Studies*, XLVII, pp. 21–45.

McNees, S. (1980), The accuracy of macroeconometric models and forecasts of the US economy, in P. A. Ormerod, *Economic Modelling*, Heinemann.

Nelder, J. A. and Mead, R. (1965), A simplex method for function minimization, *Computer Journal*, 5, pp. 308--313.

Nelson, C. R. (1972), The prediction performance of the FRB–MIT–PENN model of the US economy, *American Economic Review*, LXII, pp. 902–917.

Nelson, C. R. (1975), Rational expectations and the predictive efficiency of economic models, *The Journal of Business*, (University of Chicago), July, pp. 331–343.

Oberhofer, W. and Kmenta, J. (1974), A general procedure for obtaining maximum likelihood estimates in generalized regression models, *Econometrica*, 42, pp. 579–590.

Pagan, A. R. and Byron, R. P. (1977), A synthetic approach to the estimation of models with autocorrelated disturbance terms, in R. Bergstrom, (ed.), *Stability and Inflation*, John Wiley, New York.

Pesaran, M. H. (1973), The small sample problem of truncation remainders in the estimation of distributed lag models with autocorrelated errors, *International Economic Review*, 14, pp. 120–131.

Pesaran, H. (1974), On the general problem of model selection, *Review of Economic Studies*, 41, pp. 153–171.

Phillips, G. D. A. and Harvey, A. C. (1974), A simple test for serial correlation in regression analysis, *Journal of the American Statistical Association*, 69, pp. 935–939.

Phillips, P. C. B. and Wickens, M. R. (1978), *Exercises in Econometrics*, Vols. 1 and 2, Philip Allan.

Pierce, D. A. (1971a), Least squares estimation in the regression model with autoregressive-moving average errors, *Biometrika*, 58, pp. 299–312.

Pierce, D. A. (1971b), Distribution of the residual autocorrelations in the regression model with autoregressive-moving average errors, *Journal of the Royal Statistical Society*, B, 33, pp. 140–146.

Pierce, D. A. (1975), Forecasting in dynamic models with stochastic regressors, *Journal of Econometrics*, 3, pp. 349–374.

Pierce, D. A. (1977), Relationships — and the lack thereof — between economic time series, with special reference to money and interest rates, *Journal of the American Statistical Association*, 72, pp. 11–22.

Plackett, R. L. (1950), Some theorems in least squares, *Biometrika*, 37, pp. 149–157.

Plosser, C. I. and Schwert, G. W. (1977), Estimation of a non-invertible moving average process: the case of overdifferencing, *Journal of Econometrics*, 6, pp. 199—224.

Popper, K. F. (1959), *The Logic of Scientific Discovery*, Hutchinson.

Powell, M. J. D. (1964), An efficient method for finding the minimum of a function of several variables without calculating derivatives, *Computer Journal*, 7, pp. 155—162.

Prais, S. J. and Houthakker, H. S. (1955), *The Analysis of Family Budgets*, Cambridge University Press.

Press, S. J. and Brooks, R. B. (1969), *Testing for Serial Correlation in Regression*, Report No. 6911, Center for Mathematical Studies in Business and Economics, University of Chicago.

Prothero, D. L. and Wallis, K. F. (1976), Modelling macroeconomic time series (with discussion), *Journal of the Royal Statistical Society*, Series A, 139, pp. 468—500

Ramsey, J. B. (1969), Tests of specification error in the general linear model, *Journal of the Royal Statistical Society*, B, 31, pp. 250—271.

Ramsey, J. B. and Schmidt, P. (1976), Some further results on the use of OLS and BLUS residuals in specification error tests, *Journal of the American Statistical Association*, 71, pp. 389—90.

Reinsel, G. (1979), FIML estimation of the dynamic simultaneous equations model with ARMA disturbances, *Journal of Econometrics*, 9, pp. 263—282.

Revankar, N. S. (1971), A class of variable elasticity of substitution production functions, *Econometrica*, 39, pp. 61—71.

Rothenberg, T. J. and Leenders, C. T. (1964), Efficient estimation of simultaneous equation systems, *Econometrica*, 32, pp. 57—76.

Rothenberg, T. J. (1971), Identification in parametric models, *Econometrica*, 39, pp. 577—591.

Rothenberg, T. J. (1979), *Second Order Efficiency of Estimators and Tests in Simultaneous Equations Models*, Paper presented at European Meeting of the Econometric Society, Athens, August.

Sargan, J. D. (1958), The estimation of economic relationships using instrumental variables, *Econometrica*, 26, pp. 393—415.

Sargan, J. D. (1961), The maximum likelihood estimation of economic relationships with autoregressive residuals, *Econometrica*, 29, pp. 414—426.

Sargan, J. D. (1964a), Wages and prices in the United Kingdom: a study in econometric methodology, in *Econometric Analysis for National Economic Planning*, P. E. Hart, G. Mills and J. K. Whitaker (eds), pp. 25—54, Butterworth.

Sargan, J. D. (1972), *The Identification and Estimation of Sets of Simultaneous Stochastic Equations*, LSE Discussion Paper No. A1, LSE Econometrics Programme.

Sargan, J. D. (1980a), Some tests of dynamic specification for a single equation, *Econometrica*, 48, pp. 879—897.

Sargan, J. D. (1980b), The consumer price equation in the post-war British economy: an exercise in equation specification testing, *Review of Economic Studies*, XLVII, pp. 113—135.

Schlossmacher, E. J. (1973), An iterative technique for absolute deviations curve fitting, *Journal of the American Statistical Association*, 68, pp. 857—859.

Schmidt, P. (1972), A generalization of the Durbin—Watson test, *Australian Economic Papers*, 11, pp. 203—209.

Schmidt, P. (1974), The asymptotic distribution of forecasts in an econometric model, *Econometrica*, 42, pp. 303—309.

Schmidt, P. (1975), The small sample effects of various treatments of truncation remainders in the estimation of distributed lag models, *Review of Economics and Statistics*, 57, pp. 387—389.

Schmidt, P. (1977), Some small sample evidence on the distribution of dynamic simulation forecasts, *Econometrica*, 45, pp. 997—1006.

Schweppe, F. (1965), Evaluation of likelihood functions for Gaussian signals, *IEEE Transactions on Information Theory*, 11, pp. 61—70.

Seber, G. A. F. (1966), *The Linear Hypothesis*, Charles Griffin, London.

Seber, G. A. F. (1977), *Linear Regression Analysis*, John Wiley, New York.

Shiller, R. J. (1978), Rational expectations and the dynamic structure of macroeconomic models, *Journal of Monetary Economics*, 4, pp. 1—44.

Silvey, S. D. (1970), *Statistical Inference*, Chapman and Hall, London.

Sims, C. A. (1972), Money, income and causality, *American Economic Review*, 62, pp. 540—552.

Spencer, B. G. (1975), The small sample bias of Durbin's tests for serial correlation, *Journal of Econometrics*, 3, pp. 249—254.

Spencer, D. E. (1979), Estimation of a dynamic system of seemingly unrelated regressions with autoregressive disturbances, *Journal of Econometrics*, 10, pp. 227—241.

Spitzer, J. J. (1976), The Keynesian demand for money function revisited, *Journal of Monetary Economics*, 2, pp. 381—387.

Theil, H. (1971), *Principles of Econometrics*, John Wiley, New York.

Wallis, K. F. (1972), Testing for fourth order autocorrelation in quarterly regression models, *Econometrica*, 40, pp. 617—636.

Wallis, K. F. (1974), Seasonal adjustment and relations between variables, *Journal of the American Statistical Association*, 69, pp. 18—31.

Wallis, K. F. (1977), Multiple time series analysis and the final form of econometric models, *Econometrica*, 45, pp. 1481—1497.

Wallis, K. F. (1980), Econometric implications of the rational expectations hypothesis, *Econometrica*, 48, pp. 49—73.

Wolfe, M. A. (1978), *Numerical Methods for Unconstrained Optimization*, Van Nostrand Reinhold.

Zeckhauser, R. and Thompson, M. (1970), Linear regression with non-normal error terms, *Review of Economics and Statistics*, 52, pp. 280—287.

Zellner, A. (1963), Estimators for seemingly unrelated regression equations: some exact finite sample results, *Journal of the American Statistical Association*, 58, pp. 977—992.

Zellner, A. and Palm, F. (1974), Time series analysis and simultaneous equation econometric models, *Journal of Econometrics*, 2, pp. 17—54.

Author Index

Subject Index

379